Al

Elena N Ieno

Anatoly A Saveliev

Beginner's Guide to

Spatial, Temporal, and Spatial-Temporal Ecological Data Analysis with R-INLA

Volume I: Using GLM and GLMM

Published by Highland Statistics Ltd.
Highland Statistics Ltd.
Newburgh
United Kingdom
highstat@highstat.com

ISBN: 978-0-9571741-9-1
First published in February 2017

www.highstat.com

I would like to thank my wife Nandani for giving me the most beautiful present a man can wish for. They say that storks bring babies. It is true!

– Alain F Zuur –

To Norma, Juan Carlos, and Walter for their constant support

– Elena N Ieno –

Special thanks to my wife, friends, and colleagues who make my life more interesting

– Anatoly A Saveliev –

Preface

The authors of this book have been giving statistics courses to ecologists for 15 years. We have taught more than 8,000 scientists. During our courses we cover topics such as R, data exploration, data visualizsation, multiple linear regression, generalised linear models, linear mixed-effects models, generalised linear mixed-effects modelling (GLMM), generalised additive models (GAM), generalised additive mixed-effects models (GAMM), Bayesian analysis and MCMC, and multivariate analysis, among many other topics. Over the years a large number of participants have asked us to teach a module that covers the analysis of spatial, temporal, and spatial-temporal data. Although random effects in GLMM and GAMM can be used to deal with dependency, such an approach is not optimal for spatial, temporal or spatial-temporal data. Although there were various tools available in R, they either required expertise knowledge or required extensive computing time (e.g. MCMC in WinBUGS or OpenBUGS). We therefore elected to stay away from teaching and writing about spatial, temporal, and spatial-temporal data analysis.

It was only after we became aware of material described in Lindgren et al. (2011) that we realised that GLMs and GLMMs, and all their zero-inflated cousins and smoothing cousins, can be extended to spatial, temporal, and spatial-temporal data.

Unfortunately, the literature describing the approach (Integrated Nested Laplace Approximation, abbreviated as INLA) is rather technical. A book published in 2015 by Blangiardo and Cameletti helped us understand the INLA world better. Although we find it an excellent book, it still requires a fair amount of statistical knowledge in order to fully comprehend the material.

Availability of the software package R-INLA has put the application of GLMs and GLMMs on spatial, temporal, and spatial-temporal data within the reach of every scientist. We therefore decided to extend our *Beginner's Guide* book series with a book on the use of R-INLA to analyse spatial, temporal, and spatial-temporal data.

Acknowledgements

We are greatly indebted to all scientists who supplied data for this book. Alexandre Roulin supplied the owl data. Bob Steidl provided the osprey data. Robert Cruikshanks allowed us to use the Irish pH dataset. Christophe Barbraud gave us the Adelie penguins data. Boudjéma Samraoui provided the White Storks data. Juan Timi gave us the Brazilian sand perch parasite data. Yusuke Fukuda provided the crocodile data. Matias Maggi supplied the honey bee mites data. Allesandro Ligas provided the crayfish data. Mette Mauritzen gave us the polar bears data. Chris Smeenk supplied the sperm whale data. Michael Reed gave us the Hawaiian bird data, and Helen Sofaer supplied the orange-crowned warbler data. We also thank the following authors for making their data publically available: Petty et al. (2015) for the subnivium temperature data, Hopkins et al. (2013) for the chimpanzees data, Sturrock et al. (2015) for the otolith plaice data, Irl et al. (2015) for the plant richness data on La Palma (and who kindly emailed a modified data set), Crozier et al. (2011) for the sockeye salmon, Etheridge et al. (1998) for the historical CO_2 records, and Muller and van Woesik (2014) for the white-pox disease data.

We greatly appreciate the efforts of those who wrote R (R Development Core Team 2016) and its many packages. This book would not have been possible without the efforts of the R-INLA programmers (Rue et al. 2009; www.r-inla.org; Lindgren et al. 2011). We hope that they will keep up the excellent work.

We thank Joseph Hilbe and Thierry Onkelinx for helpful comments on an earlier draft. Special thanks to Christine Andreasen for editing this book.

Data sets and R code used in this book

All data sets used in this book may be downloaded from www.highstat.com/books.htm. All R code also may be downloaded from the website for this book. To open the ZIP file with R code, use the password **Whales123454321**

Cover art

The cover drawing is by Jon Thompson (www.yellowbirdgallery.org). Mr Thompson was born in 1939 to Irish parents and has lived most of his life in Scotland. In the 1980s, he was drawn to the Orkney Islands. He is continually inspired by the landscape and bird life of Orkney. He has been creating bird art for 30 years in a variety of media, including drawing, painting, sculpture, and jewellery, never attempting to reproduce nature, but to draw parallels with it. A close-up view of a bird feather is all the inspiration he needs.

Alain F Zuur,
Newburgh, Scotland

Elena N Ieno,
Alicante, Spain

Anatoly A Saveliev,
Kazan, Russia

February 2017

Contents

1 Overview of This Book

1.1 Volumes I and II

This book, *Beginner's Guide to Spatial, Temporal, and Spatial-Temporal Ecological Data Analysis with R-INLA*, consists of two volumes. You are reading Volume I, *Using GLM and GLMM*. Volume II is entitled *Using GAM and Zero-Inflated Models*.

1.1.1 Volume I

In Volume I we explain how to apply linear regression models, generalised linear models (GLM), and generalised linear mixed-effects models (GLMM) to spatial, temporal, and spatial-temporal data. The models that will be employed use the Gaussian and gamma distributions for continuous data, the Poisson and negative binomial distributions for count data, the Bernoulli distribution for absence–presence data, and the binomial distribution for proportional data.

1.1.2 Volume II

In Volume II we apply zero-inflated models and generalised additive (mixed-effects) models to spatial and spatial-temporal data. We also discuss models with more exotic distributions like the generalised Poisson distribution to deal with underdispersion and the beta distribution to analyse proportional data.

1.2 What type of spatial data do we analyse in this book?

The short answer to this question is 'geostatistical' data. The long answer explains the three types of spatial data and then comes to the same answer. The spatial statistical literature distinguishes three types of spatial data, namely (i) lattice and areal data, (ii) geostatistical data, and (iii) spatial point pattern data (Cressie 1991; Schabenberger and Pierce 2002; Haining 2003). We briefly discuss each of these types of data.

1.2.1 Areal and lattice data

Suppose that we sample the number of tuberculosis cases in cattle in each county in England, or the total air pollution for each European country, or the total number of fish catches per European country, or the number of deaths per health clinic in a large city, or the number of babies born per country, or the crime rates per city block.

In a general notation we can state that we sample the realisations $y(\boldsymbol{s})$ of a stochastic process $Y(\boldsymbol{s})$, where $\boldsymbol{s}$ is part of a study area D. The study area in spatial studies is typically two-dimensional, which means that we can write $D \in \mathbb{R}^2$. If we have N samples in our study area D, then we can write the observed values as $y(\boldsymbol{s}_1)$, ..., $y(\boldsymbol{s}_N)$. This set of observations is one realisation of $Y(\boldsymbol{s}_1)$, ..., $Y(\boldsymbol{s}_N)$. The latter is also called a random field.

In all examples that we gave in the first paragraph of this subsection the $\boldsymbol{s}$ represents an areal unit (country, county, or health clinic) and $y(\boldsymbol{s})$ is the aggregated or averaged value for that areal unit. In all these examples D is a countable collection of spatial units. If the areas are regular placed then we call it lattice data, otherwise it is areal data.

Areal data are common in medical science because quite often only aggregated data (per hospital or county) are available due to patient confidentiality. In ecology, one has to search much harder to find areal data. Two examples are the aggregation of numbers of species per area (e.g. fish landings per area of the sea; see www.ices.dk for examples) and average air pollution levels per country (see for examples www.eea.europa.eu/). In this book we will not focus on areal or lattice data.

1.2.2 Geostatistical data

In geostatistical data analysis we do not work with a value $y(\boldsymbol{s})$ for a specific areal unit $\boldsymbol{s}$. Instead $\boldsymbol{s}$ represents a spatial index like latitude and longitude. In Chapters 2, 4, and 12 we will use a field study in which pH is sampled at 253 locations in Ireland (see Figure 1.1); hence we sample $pH(\boldsymbol{s}_1), pH(\boldsymbol{s}_2)$ to $pH(\boldsymbol{s}_{253})$ in the study area D, and the $\boldsymbol{s}_i$ contain the spatial coordinates.

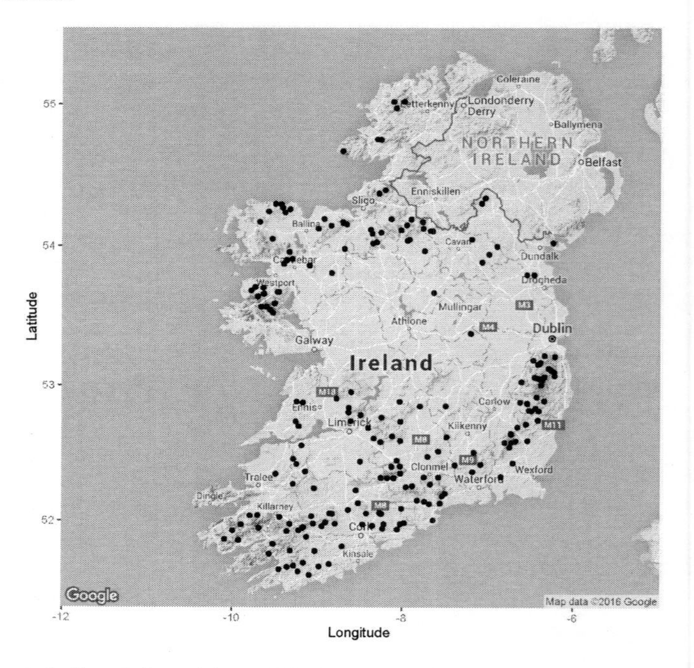

Figure 1.1. Spatial positions (black dots) of the 257 sampling locations in Ireland.

Other examples of geostatistical data that are analysed in Volume I are a plant species diversity index sampled at 890 sites on La Palma, Spain (Chapter 13), numbers of fledged bird chicks sampled at 181 nests on Santa Catalina Island, California (Chapter 15), and absence–presence of coral disease in 68 reef colonies in Haulover Bay on the northeast side of St. John, US Virgin Islands (Chapter 16).

1.2.3 Spatial point pattern data

Suppose that in a bird study we sample N areas. If a particular species is present in area i, then we set $y(\boldsymbol{s}_i) = 1$; otherwise it is 0. Or suppose we sample N areas for the presence of a disease in humans. Again we end up with $y(\boldsymbol{s}_1)$ to $y(\boldsymbol{s}_N)$ values that are either 0 (absent) or 1 (present).

The null hypothesis in these studies is complete spatial randomness of the areas where we have a 1. Interest is then whether there is any spatial clustering of areas where the disease occurs or where the bird is present. It also possible to include covariates in this type of analysis. For example, we can investigate whether the patterns where a certain disease appears differ in terms of the covariate effects. Or a pollution effect may cause birds to appear in only certain parts of the study area.

In spatial point pattern data analysis the sampling locations $\boldsymbol{s}$ are random, and we investigate the patterns in the positions of the points.

In this book we will not focus on spatial point data.

1.3 Outline of this book

In Chapter 2 we discuss an important topic: dependency. Ignoring this means that we have pseudoreplication. We present a series of examples and discuss how dependency can manifest itself.

We briefly discuss frequentist tools that are available for the analysis of temporal and spatial data in Chapters 3 and 4, and we will conclude that their application is rather limited, especially if non-Gaussian distributions are required. We will therefore consider alternative models, but these require Bayesian techniques.

In Chapter 5 we discuss linear mixed-effects models to analyse hierarchical (i.e. clustered or nested) data, and in Chapter 6 we outline how we add spatial and spatial-temporal dependency to regression models via spatial (and/or temporal) correlated random effects.

In Chapter 7 we introduce Bayesian analysis, Markov chain Monte Carlo techniques (MCMC), and Integrated Nested Laplace Approximation (INLA). INLA allows us to apply models to spatial, temporal, or spatial-temporal data.

In Chapters 8 through 16 we present a series of INLA examples. We start by applying linear regression and mixed-effects models in INLA (Chapters 8 and 9), followed by GLM examples in Chapter 10. In Chapters 11 through 13 we show how to apply GLM models on spatial data. In Chapter 14 we discuss time-series techniques and how to implement them in INLA. Finally, in Chapters 15 and 16 we analyse spatial-temporal models in INLA.

1.4 Prerequisites

We assume that readers are familiar with R (see for example Zuur et al. 2009b) and multiple linear regression. Working knowledge of Poisson, negative binomial, and Bernoulli GLM is also recommended though we do provide a short revision of these topics in Chapter 10.

1.5 Availability of the R code and data

All R code and data sets are available on the website for the book. In the preface we explain how to open the R files.

2 Recognising statistical dependency

We start this chapter with an explanation of pseudoreplication. Ignoring pseudoreplication can have serious consequences for the conclusions based on the statistical analyses. In our experience, nearly every ecological study has some form of pseudoreplication. We will present three examples in this chapter that will help you recognise the presence of pseudoreplication. In later chapters we will provide solutions.

Prerequisites for this chapter: Knowledge of R, multiple linear regression, and Poisson GLM is required. Familiarity with GAM and mixed-effects models is recommended. In later chapters we will explain GAM and linear mixed-effects models in more detail.

2.1 Pseudoreplication

The term pseudoreplication has a somewhat scary aura surrounding it. It goes back to a publication by Hurlbert (1984) and a series of follow-up papers, sometimes shaming authors who ignored pseudoreplication. The debate whether pseudoreplication is still a problem, 30 years after being introduced, can sometimes still be heated (Oksanen 2001; Freeberg and Lucas 2009; Hurlbert 2004).

So what is pseudoreplication? Hurlbert (1984) defined it as: '… the use of inferential statistics to test for treatment effects with data from experiments where either treatments are not replicated (though samples may be) or replicates are not statistically independent'. Hurlbert (1984) then went on to define three forms of pseudoreplication in the context of ecological field experiments (where experiments can be manipulated). Millar and Anderson (2004) provide a more modern discussion on pseudoreplication, and use examples from fisheries science. There is no need to know the names of all the different forms of pseudoreplication; at the end of the day it means that the observations on the response variable are not independent and this aspect is ignored during the statistical analysis. Let's present some examples.

A classical example of pseudoreplication is as follows. Suppose you investigate whether a certain diet for weight loss is effective. A possible approach is to sample the weight of a person a couple of time before he starts a diet, while he is doing the diet, and after the diet. Let's say seven measurements per person in total. And suppose this is done for 100 people. So we have 700 weight observations. Because the same person is measured seven times it would be wrong to assume that we have 700 independent observations to test for a diet effect. Pretending that we have 700 independent observations results in standard errors that are too small.

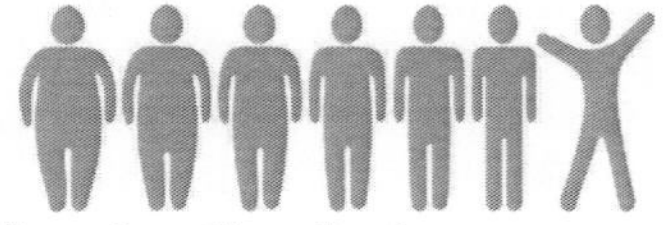

The correct analysis approach is a mixed-effects model; see, for example, Pinheiro and Bates (2000).

Zuur et al. (2013) used a turnstone (*Arenaria interpres*) data set (originally published in Fuller et al. (2013)). A flock of birds on the beach was identified. From this flock one focal bird was selected and observed for 30 seconds. The number of times it raised its head (to look for danger) was recorded. Then a second bird from the *same* flock was selected and the ecologists counted again the number of 'heads-up'.

This process was repeated until the flock flew away. A large number of flocks were sampled in the same way. It would be wrong to assume that the observations from the same flock are independent. If a fox is watching the birds from a nearby dune then most likely every observation on the number of heads-up from that flock will be very high! So we have pseudoreplication. The correct analysis approach is a generalised linear mixed-effects model using flock as random effect. See Zuur et al. (2014) for a fully worked-out solution.

Sick et al. (2014) investigated social strategies throughout the course of the day in chacma baboons (*Papio ursinus*). The data were also used in Ieno and Zuur (2015) and Zuur et al. (2016a). Data were collected from 60 baboons from two troops in the Tsaobis Leopard Park in Namibia. An individual baboon was followed for 1 h (focal hour), during which its grooming and dominance interactions were recorded, including the identity of the baboon being groomed (receiver). Multiple observations of the same baboon in a focal hour was the first source of dependency. During the 6-month sampling period, each baboon was repeatedly sampled, which is another source of dependency. To increase complexity, the receiver represents another level of dependency. Dealing with the pseudoreplication requires a mixed effects model with a two-way nested and crossed random effect.

Reed et al. (2011) analysed long-term survey data for three endangered waterbirds endemic to the Hawaiian Islands, the Hawaiian moorhen (*Gallinula chloropus sandvicensis*), Hawaiian coot (*Fulica alai*), and Hawaiian stilt (*Himantopus mexicanus knudseni*). The data set consists of annual (winter) counts covering a time span of 1956–2007. Each time series represented one species on one island: coot and stilt numbers on Oahu, Maui, and Hawaii, and moorhen numbers on Oahu and Kauai. Bird counts of a species in a particular year on a particular island are likely to depend on the counts in the previous year for the same island, but potentially also on the counts for other islands, and counts of other bird species. So we have temporal (and potentially spatial) pseudoreplication.

To deal with the temporal pseudoreplication we need to include a temporal dependency structure in the GLM. Correlation between species can be dealt with using a multivariate GLM (Berridge and Crouchley 2011; Zuur et al. 2016a).

Zuur et al. (2014) contains a chapter on the analysis of zero-inflated and spatial correlated Common Scoter data. The data set consists of counts of nonflying sea ducks, collected at a large number of spatial locations in front of a specific part of the Dutch coast. If you count large number of sea ducks at a certain location then most likely you will also count them in nearby positions. Or perhaps it is just the other way around: large numbers of ducks in one place (clustering) and zero ducks in nearby positions. Hence, we have spatial pseudoreplication. To deal with this we need a GLM that contains a residual spatial component.

We could fill another 10 pages with examples of dependency. Nearly every data set that is used in our mixed modelling book (Zuur et al. 2009a) or in our GAMM book (Zuur et al. 2014) contains dependency problems.

In our experience nearly every ecological data set has a dependency element somewhere. The solution is to use statistical techniques that can cope with dependency. Our mixed modelling books (Zuur et al. 2009a, 2013) show how to do this for certain types of dependency, namely for nested data. In this book we focus on dealing with spatial, temporal, and spatial-temporal dependency (and if necessary also with random effects).

Next we present three examples and discuss in more detail how to recognise dependency.

2.2 Linear regression applied to spatial data

2.2.1 Irish pH data

The data used in this section are a subset of the data analysed in Cruikshanks et al. (2007), a technical report by the Environmental Protection Agency in Ireland. These data were also used in Zuur et al. (2009a) to explain regression models with residual spatial dependency structures. The original research used sampled data from 257 rivers in Ireland during 2002 and 2003. We use only the 2003 data.

One of the aims underlying the study was to find a different tool for identifying acid-sensitive waters, which currently uses measures of pH. The problem with pH is that it is extremely variable within a catchment, and depends on both flow conditions and underlying geology. As an alternative measure, the Sodium Dominance Index (SDI) is proposed as an indicator of the acid sensitivity of rivers. SDI is defined as the contribution of sodium (Na^+) to the sum of the major cations (positively charged atoms or molecules).

The motivation for this research is the increase in plantation forestry cover in Irish landscapes and its potential impacts on aquatic resources. The underlying biological question is whether we can find a relationship between pH and SDI, while taking into account altitude and whether a site is forested.

2.2.2 Protocol from Zuur et al. (2016b)

Zuur and Ieno (2016b) formulated a 10-step protocol for conducting and presenting results of regression-type analyses. The 10 steps discussed in the paper are as follows.

1. State appropriate questions.
2. Visualise the experimental design.
3. Conduct data exploration.
4. Identify the dependency structure in the data.
5. Present the statistical model.
6. Fit the model.
7. Validate the model.
8. Interpret and present the numerical output of the model.
9. Create a visual representation of the model.
10. Simulate from the model.

Throughout this book we will aim to follow this protocol. It streamlines analysis of data that will enhance understanding of the data, the statistical models, and the results, and optimise communication with the reader with respect to both the procedure and the outcomes.

With respect to the content of this book and the previous section, note step 4. It says 'Identify the dependency structure'. By this we mean that we need to figure out which observations of the response variable (pH in this case) are potentially dependent. Note that you should think about this before fitting the model because the answer decides which statistical technique should be fitted (e.g. a linear regression model or a linear mixed-effects model). Let's discuss a couple of scenarios.

Before starting the analysis carefully consider which observations for your response variable can be dependent. If dependency is present, apply a statistical technique that is able to cope with it.

Suppose that the 257 water samples are collected by 10 different laboratories. Due to different sampling techniques or skills of the scientists involved, the pH values from the same laboratory may be more similar than those from different laboratories. Hence we have dependency.

Suppose that some of the pH samples are taken along the same river. If the river has rather high acidity values, then most likely all pH values taken from this river are high. Hence we have dependency.

Suppose that the pH values are all taken from different rivers, but some sampling locations are rather close to one another. We may still have (spatial) dependency due to a catchment effect.

It is unlikely that all samples were taken at the same moment in time. Suppose that 10 sites were sampled per working day. A heavy rainfall (it is Ireland after all!) may cause all pH values sampled on a certain day to be more similar than pH values sampled on different days. Hence we have temporal dependency.

Let's now discuss a situation that looks like dependency, but is in fact independence. We have two sites with a very low pH value. The sites are not geographically close to each other, they are not sampled by the same laboratory, or on the same day. However, both sites have a very high altitude value, which is the cause of the low pH. Hence a covariate explains the similar pH values.

Deciding whether we have dependency for the Irish pH data requires that we know the exact details of the experimental design. It is a matter of common sense and no Harry Potter magic is involved. It helps visualise the experimental design. See also step 2 in the protocol and the next subsection.

2.2.3 Visualisation of the experimental design

Figure 1.1 shows the spatial position of the 257 sites in Ireland that were sampled in 2003. This graph should immediately set off all the pseudoreplication alarm bells.

As discussed in the previous subsection we need to ask ourselves whether there is pseudoreplication. We already listed a series of causes in the previous subsection. To answer them we need to speak to the owners of the data and ask them about the ins and outs of the sampling process. The problem is that we obtained this specific data set more than 10 years ago. The owners of the data have new jobs at new places with new email addresses with new job descriptions and new deadlines.

If you have historical data going back decades or centuries then it may be difficult to figure out how the data were collected. So make sure to carefully document your own data for future use. Don't forget to include yourself as a future potential user.

Figure 1.1 has one big (invisible) mark written on it, that says 'spatial dependency'; pH measures acidity of water, and water comes from rain, and the rain comes from the Atlantic Ocean hitting Ireland from the west. So, there may be spatial dependency! This means that we need to consider applying regression models with spatial dependency.

The next question is whether we should start with a model that contains such a dependency structure or should we start simple. 'Simple' in this case means applying a multiple linear regression model without spatial correlation and seeing whether it violates any underlying assumptions. Obviously, the first approach would be better but it also requires a much greater degree of statistical knowledge and access to sophisticated software.

2.2.4 Data exploration

Zuur et al. (2010) provided another protocol, this time for data exploration. The eight points discussed in the paper were as follows.

1. Are there outliers?
2. Do we have homogeneity of variance?
3. Are the data normally distributed?
4. Are there lots of zeros in the data?
5. Is there collinearity among the covariates?
6. What are the relationships between Y and X variables?
7. Should we consider interactions?
8. Are observations of the response variable independent?

Data exploration avoids type I and type II errors, among other problems, thereby reducing the chance of making erroneous ecological conclusions and poor recommendations. It is therefore essential for good quality management and policy based on statistical analyses.

Because this section is not about the analysis of the Irish pH data, but merely to discuss dependency problems, we only present one data exploration graph. Figure 2.1 shows scatterplots of pH versus altitude and SDI. Regression lines for forested and non-forested data have been added. Note that it is an option to transform altitude because it has some large values. In general we are not in favour of transforming variables. The only two exceptions we are willing to make are (i) transforming a covariate in the event it has a few large values, and (ii) transforming the response variable in the event it considerably simplifies the statistical analysis. An example of the first scenario is given in the left panel of Figure 2.1. As to simplifying the statistical analysis, in Volume II of this book we will analyse species richness of moss species. We concluded that an underdispersed Poisson GLM with spatial correlation has to be applied. Finding software to fit such a model is a major challenge. The square root (or log) transformed data can easily be analysed with a Gaussian model containing a spatial correlation component.

Let us catch up with the R code. We import the data from a text file using the `read.table` function and we load all required packages and our own support file `HighstatLibV10.R` (which is available on the website for this book).

```
> iph <- read.table(file = "IrishPh.txt",
                    header = TRUE,
                    dec = ".")
> library(ggplot2)
> library(lattice)
> library(sp)
> library(gstat)
> library(ggmap)
> source("HighstatLibV10.R")
```

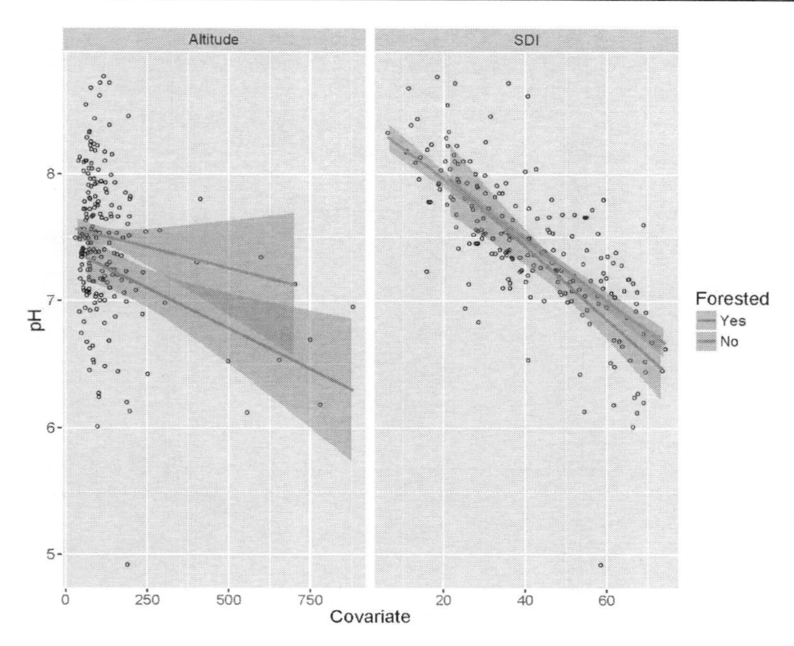

Figure 2.1. Left: Scatterplot of altitude versus pH. Right: Scatterplot of SDI versus pH. Regression lines for forested and non-forested data have been added.

The categorical covariate Forested is coded numerically. We recode it using clearer labels.

```
> iph$fForested <- factor(iph$Forested,
                          levels = c(1, 2),
                          labels = c("Yes", "No"))
```

We used the `ggplot2` package (Wickham 2009) to create Figure 2.1. Because we want to create a graph in which both SDI and altitude are along the x-axis, we need to create a new data frame (`MyData`). We repeat the variable pH twice with the `rep` function, concatenate SDI and altitude, repeat the Forested variable twice, and create a variable identifying the two blocks of data.

```
> MyData <- data.frame(
            Y         = rep(iph$pH, 2),
            Cov       = c(iph$SDI, iph$Altitude),
            Forested  = rep(iph$fForested, 2),
            ID        = rep(c("SDI", "Altitude"),
                            each = nrow(iph)))
```

The rest is a matter of straightforward `ggplot2` coding. See Wickham (2009) for a detailed explanation of `ggplot2` and Ieno and Zuur (2015) for various case studies.

```
> p <- ggplot()
> p <- p + geom_point(data = MyData,
                      aes(y = Y,
                          x = Cov),
                      shape = 1, size = 1)
> p <- p + xlab("Covariate") + ylab("pH")
> p <- p + theme(text = element_text(size = 15))
> p <- p + geom_smooth(data = MyData,
                       aes(x = Cov,
                           y = Y,
                           group = Forested),
                       method = "lm")
> p <- p + facet_grid(.~ID, scales = "free_x")
> p
```

It is our personal preference to code a ggplot2 graph line by line (as above). This is not the most efficient approach as can be seen from the following, more compact code.

```
> theme_set(
    theme_gray() +
    theme(text = element_text(size = 15)))
> update_geom_defaults(
    "point", list(shape = 16, size = 2))
> update_geom_defaults(
    "line", list(lwd = 3))
```

We have now specified some default settings, which we can use in all the remaining ggplot2 code of this book. Figure 2.1 can then be created with a much more compact code.

```
> ggplot(MyData, aes(x = Cov,
                     y = Y,
                     group = Forested,
                     col = Forested)) +
  geom_point() +
  geom_smooth(method = "lm") +
  facet_grid(. ~ ID, scales = "free_x") +
  xlab("Covariate") + ylab("pH")
```

Although this code is shorter we prefer the first and longer coding approach.

2.2.5 Dependency

The alarm bell for pseudoreplication already sounded two subsections ago. We argued that sites close to one another are bound to have similar pH values. Hence we have dependency. So far we have used the word 'dependency' in a very loose way. We relied on common sense. Within the context of a linear regression model, dependency is more formally defined as follows. Suppose we fit the following linear regression model (for ease of visualisation and explanation we only use SDI as the covariate).

$$pH_s = \beta_1 + SDI_s \times \beta_2 + \varepsilon_s$$
$$\varepsilon_s \sim N(0, \sigma^2)$$

pH_s is the observed pH value at site s, and SDI_s is the corresponding SDI value. The term ε_s is a residual and represents unexplained information. Instead of this model notation we use a slightly different mathematical notation, but the model is exactly the same!

$$\begin{aligned} & pH_s \sim N(\mu_s, \sigma^2) \\ & E(pH_s) = \mu_s \quad \text{and} \quad \text{var}(pH_s) = \sigma^2 \\ & \mu_s = \beta_1 + SDI_s \times \beta_2 \end{aligned} \qquad (2.1)$$

This reads as follows. We assume that each pH_s value is normal distributed with mean μ_s and variance σ^2. The mean μ_s is modelled as a function of SDI and an intercept. Figure 2.2A shows the model fit, which represents the μ_s component of the model. We will discuss model validation and interpretation of the full model later.

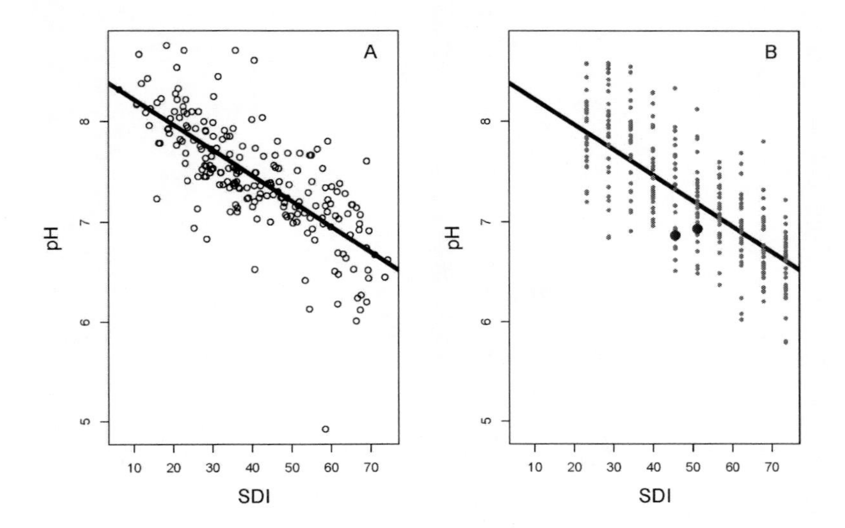

Figure 2.2. A: Observed pH and SDI data. The fit of a linear regression model has been added. B: Fit of the linear regression model. At 10 covariate values we have added 30 pH values that are simulated from a normal distribution with the mean equal to the fitted line and the standard deviation is taken from the estimated σ from the linear regression model. See the text for an explanation of the two large dots.

While we are presenting the results of a linear regression model, we might as well take the opportunity and discuss its underlying assumptions and model validation steps in more detail. This allows us to ensure that all readers are fully familiar with these assumptions.

Linear regression is based on a series of assumptions, namely independence, homogeneity of variance, and normality (and no noise on the covariates). Let us start with the least important assumption: normality.

Normality

Quite often ecologists make a histogram of the response variable and if the shape of the histogram diverges from a symmetric bell-shaped curve then they apply all sorts of transformations on the response variable. This is a major misconception in ecological science. Figure 2.3 shows what exactly the normal assumption means in a linear regression model. At each covariate value we assume that the realisations (other potential pH values that could have been measured) are normal distributed. Unless one has a large number of pH observations at the same covariate value, one cannot verify this assumption (Quinn and Keough 2002)! The best we can do is to aggregate residuals from different covariate values and hope that these are normal distributed.

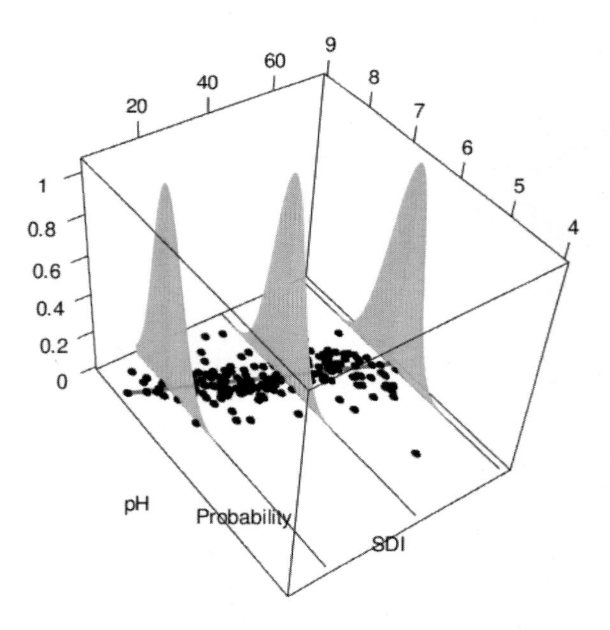

Figure 2.3. Graphical presentation of the normality assumption. At three (arbitrary) covariate values we have visualised the normal distribution. It defines the probability of finding other pH realisations.

Independence

Let us now discuss the independence assumption. Figure 2.2B shows again the model fit of the linear regression model. We have simulated 30 possible pH values (realisations) at each covariate value. Independence means that the pH value at one covariate value (SDI_i) should not influence the pH value at another covariate value (SDI_j). Expressed differently, the position of a pH realisation at a particular covariate value should be independent of pH realisations at other covariate values. We have visualised this in Figure 2.2B; note the two large black dots. These represent the pH values at two different sites. At both sites the pH value is below the fitted line. If these two sites are far away from each other then it is unlikely that the pH value at one site gives us any information about the pH at the other site. Hence independence is a plausible assumption. But if these two sites are from the same river or are close to each other, then it is likely that the pH values are dependent. And because the model does not take this into account we have pseudoreplication. The consequence of this is that the standard errors of the regression parameters are too small. You may end up stating that a covariate is significant, whereas in reality it is not. The correct approach is to apply a model that contains a spatial correlation structure.

Assessing whether we have violation of independence can be difficult. First and foremost, one should use knowledge of the experimental design to assess whether there is potentially pseudoreplication. In this specific example we need to ask the question whether we have multiple pH values from the same river or from the same catchment. These are likely to be dependent as they represent the same water mass. See Subsection 2.2.2 for other sources of dependency.

If the status of dependency for the response variable is unclear, then we can fit a model without a dependency structure and assess the residuals for spatial (or, if relevant, temporal or spatial-temporal) residual dependency. If sites close to one another tend to have similar values for the residuals, and the further sites are separated in space, the smaller the dependency, then we may be able to detect this with a so-called variogram function (or auto-correlation functions for time series). We will explain variograms later in this section. We can then modify the model and take this dependency into account.

2.2.6 Statistical model

To answer the underlying biological question formulated in Subsection 2.2.1, we model pH as a function of SDI, altitude, and whether a site is forested or not. This means that we have the following statistical model in mind:

$$
\begin{aligned}
&pH_s \sim N(\mu_s, \sigma^2) \\
&E(pH_s) = \mu_s \\
&\text{var}(pH_s) = \sigma^2 \\
&\mu_s = Intercept + SDI_s + Altitude_s + Forested_s + \\
&\quad SDI_s \times Altitude_s + SDI_s \times Forested_s + Altitude_s \times Forested_s + \\
&\quad SDI_s \times Altitude_s \times Forested_s
\end{aligned}
\tag{2.2}
$$

pH_s is the observed pH value at site s, and SDI_s, $Altitude_s$, and $Forested_s$ (categorical variable with values yes and no) are the three covariates. The model contains all main terms, two-way interactions, and a three-way interaction term. To simplify notation we omitted the regression parameters. Obviously, they are present in the model. Note that the model does not yet contain a spatial dependency component. The reason for this is that we don't have the knowledge yet to implement this. We will show how to do this in Chapter 12.

2.2.7 Fit the model

The following R code fits a multiple linear regression model with all main terms, two-way interactions, and the three-way interaction.

```
> iph$LOGAltitude <- log10(iph$Altitude)
> M2 <- lm(pH ~ LOGAltitude + SDI + fForested +
                LOGAltitude : SDI  +
                LOGAltitude : fForested +
                SDI : fForested +
                LOGAltitude : SDI : fForested,
           data = iph )
```

The three-way interaction is not significant (results are not shown here), leaving us with the problem of what to do with respect to model selection. This is a highly controversial field. Some scientists routinely apply model selection using tools like the AIC (e.g. with backwards or forwards selection routines), others use information-theoretic approaches (Burnham and Anderson 2002), and some don't apply model selection at all, or only on interactions.

Using a classical backwards model selection with the AIC as implemented with the `step` function in R, we end up with the following model:

```
> M3 <- lm(pH ~ LOGAltitude + SDI + fForested +
                LOGAltitude:fForested,
           data = iph)
```

This model contains all three main terms and one two-way interaction.

2.2.8 Model validation

To verify whether the model complies with all underlying assumptions we need to apply model validation. In this process we plot the residuals versus the fitted values to assess homogeneity, the residuals versus each covariate in the model and each covariate not in the model to assess model misfit, and we inspect the residuals for spatial, temporal, or spatial-temporal dependency (if relevant for the experimental design). And as a final step we (should) assess normality of residuals with a histogram of QQ-plot.

In this case, the graphs of residuals versus fitted values and residuals versus the covariates do not indicate any problems, and they are not presented here. We focus on the independence assumption. Before diving into variograms we plot the residuals versus their spatial positions; see Figure 2.4. Black points are positive residuals and open circles are negative residuals. The larger a point, the large the (absolute) value of the residual. We should not be able to see any pattern (i.e. clustering of large points or small points, or open or closed circles) in this graph. It seems that the majority of the negative residuals are along the southeast coast. This may indicate violation of spatial independence.

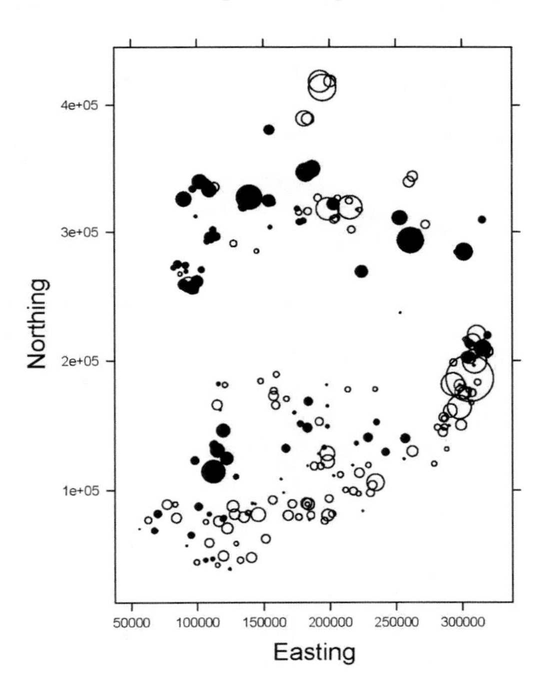

Figure 2.4. Residuals plotted versus spatial position. The width of a point is proportional to the (absolute) value of a residual. Filled circles are positive residuals and open circles are negative residuals. It would be useful to add the contour lines of the Irish borders.

The R code to create Figure 2.4 is as follows. First we extract the standardised residuals. Then we define vectors `MyCex` and `MyPch` for the point sizes and point characters respectively. We use the `xyplot` function from the `lattice` package (Sarkar 2008). The `aspect = "iso"` argument ensures that the spatial positions are not distorted by the plotting function.

```
> E3     <- rstandard(M3)
> MyCex <- 3 * abs(E3) / max(E3) + 0.5
> Sign   <- as.numeric(E3 >=0) + 1
> MyPch <- c(1, 16)[Sign]
> xyplot(Northing ~ Easting,
       data = iph, aspect = "iso",
       cex = MyCex, pch = MyPch, col = 1,
       xlab = list(label = "Easting", cex = 1.5),
       ylab = list(label = "Northing", cex = 1.5))
```

Variogram

Judging whether there is spatial dependency from Figure 2.4 is a rather arbitrary process, and most of the time it is not as clear as in Figure 2.4. The variogram is a more objective tool to assess the presence of spatial (or temporal) dependency in the residuals. When we calculate the variogram for data it is actually called an 'experimental' variogram or sample variogram (Cressie 1993). We will discuss tests for spatial correlation later in this book.

In layman's terms, the variogram works as follows. Figure 2.5 shows four graphs. Each graph shows combinations of two sites (for the Irish pH data) for which the distance between the sites is within a certain interval. For example, the upper-left panel connects any two sites with a line that are separated by less than 10 km. In the upper-right panel all sites separated by a distance between 10 and 20 km are connected with a line. Similar graphs are presented in the lower two panels.

Let us start with the upper-left panel. The code for the variogram function takes the residuals at both ends of a line, subtracts the residual values from each other, squares the difference, and does this for any two sites separated by less than 10 km. Finally, it adds up all the squared differences and divides this by the number of connections. This is then repeated for other distances.

The idea behind the variogram is that if there is spatial dependency in the residuals, then sites close to one another are likely to have similar residual values, and therefore the value of the experimental variogram for such distances will be small. But the larger the distances between sites, the more dissimilar the residuals, and the larger the experimental variogram values. But at a certain distance there will be no dependency anymore, and the experimental variogram will plateau.

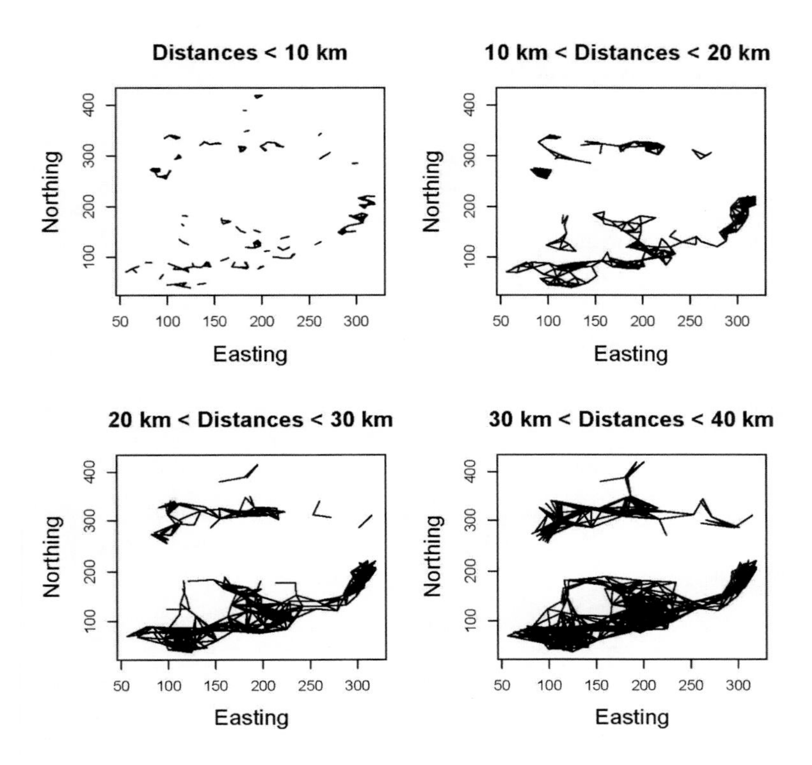

Figure 2.5. Each panel shows combinations of any two sampling locations with distances of certain threshold values.

So when inspecting an experimental variogram, the key question is whether we have a horizontal band of points (indicating residual spatial independency) or a line that first increases and then plateaus (indication of spatial dependency). Figure 2.6 shows the two scenarios for simulated data.

There is one small detail: Do we assume that the spatial dependency in the residuals is the same in each direction (isotropic) or does it differ per direction (anisotropic)? The Atlantic Ocean hits Ireland on the west coast, and it is therefore likely that residual patterns in north-south directions are different from residual patterns in east-west directions. The experimental variogram in Figure 2.7 is divided into different directions (0 = north, 90 = east). Note that there is clear dependency in the east-west direction, and this means that we can stop with the current analysis at this point. We have spatial dependency and the model does not take this into account. And that is exactly pseudoreplication. Later in this book we will re-analyse the Irish pH data with a multiple linear regression model that contains a spatially correlated residual term.

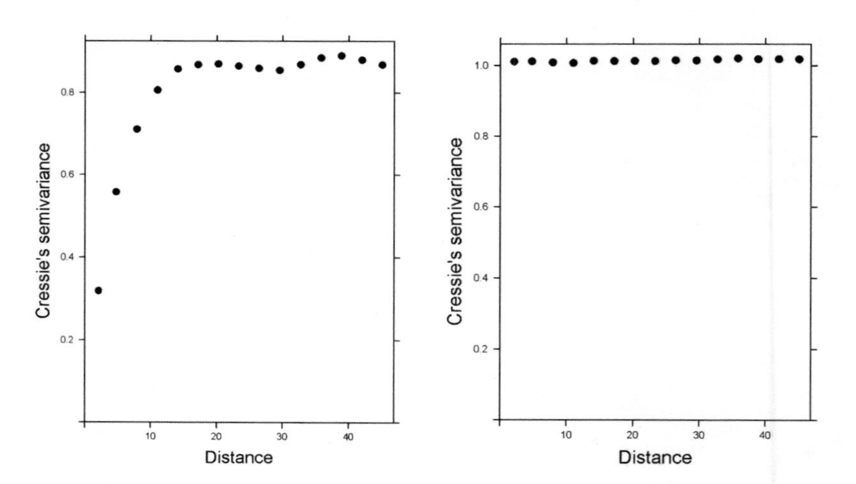

Figure 2.6. Left: Experimental variogram showing spatial dependency. Right: Experimental variogram showing no spatial dependency.

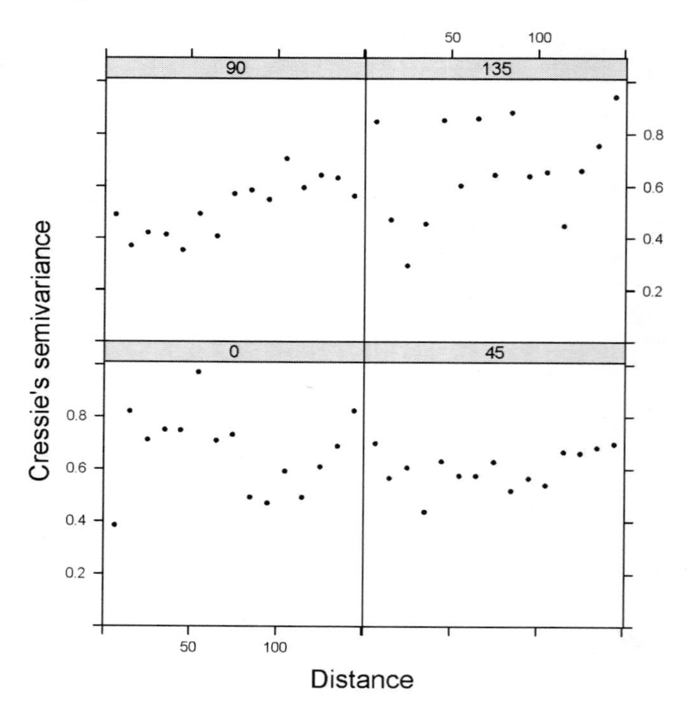

Figure 2.7. Variogram of the residuals. The panel with the label '0' uses the combinations of sites in north-south directions. '90' is for east-west directions. The directions 0 and 180 are the same.

When making a sample variogram it is important to have a decent number of pairs for each distance bin, say at least 100. If this is not the case then increase the bins. In Figure 2.7 the smallest distance in each panel has less than 100 pairs.

2.3 GAM applied to temporal data

2.3.1 Subnivium temperature data

In this section we use data from Petty et al. (2015). The authors looked at the temperature in the subnivium, the area between the soil and snow. It is a habitat used by many organisms during the winter. It is important to understand how changes in snow thickness (e.g. due to climate change) affect the temperature in the subnivium.

Three sites in south-central Wisconsin (US) with three different habitats (tall grass prairie, deciduous (mainly oaks), coniferous) were selected during the winter of 2013–2014. At each site four data loggers were installed in the subnivium. The paper also uses ambient data loggers (so that ambient and subnivium temperature can be compared), but we will not use these here. At each site the loggers are distributed at a distance of 5 m in each cardinal direction from a predetermined center point.

Each logger recorded daily average temperature. Hence we have 12 time series. The main underlying question is whether we have different trends per habitat. Figure 2.8 shows a scatterplot of the average daily temperature versus day for each habitat type. It seems that the loggers behave similarly. Unfortunately, not all loggers were active during the entire sampling period.

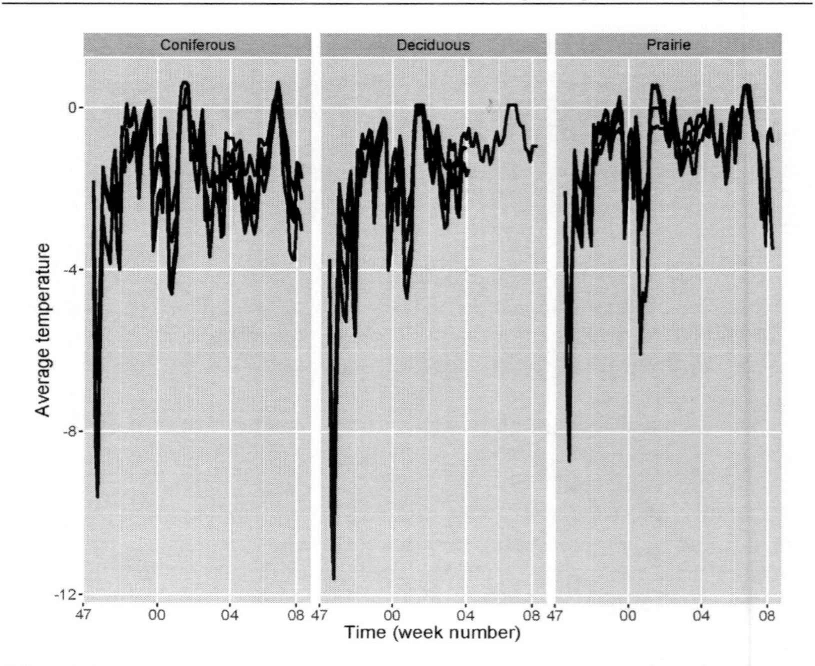

Figure 2.8. Average daily temperature plotted versus time (day). Each panel corresponds to a habitat. There are four time series per panel.

2.3.2 Sources of dependency

We have daily average temperature per logger (for 3 months), four loggers per site, and three sites in total. That gives us 12 time series of daily average temperature. Following our 2016 protocol, we first ask the following question: Do we have dependency in the response variable?

We may have dependency for the temperature data from the same logger, for the data from the same site, and even for data from the three different sites.

Dependency between sites

The UTM coordinates of the three sites are (301334E, 4767977N) for coniferous, (301991E, 4768093N) for prairie, and (301114E, 4767466N) for deciduous. This means that the sites are within 3 km of one another.

We are by no means specialists in the science of subnivium temperature. Based on the information in the Petty et al. (2015) paper, the study area has an average snow depth of 125 cm. Can we assume that the average daily temperature measured by a particular logger on day i under a snow pack of more than 1 m is independent of the temperature sampled by a different logger, 3 km away, on the same day? Maybe it is, in which case we can eliminate the pseudoreplication due to dependency between sites.

Dependency within sites

The four loggers per site are all within 10 m of one another. Assuming that the average daily temperature sampled by a logger on day i is independent of the temperature sampled by a different logger positioned less than 10 m away, on the same day, seems a strong assumption to us. But let's assume for the moment that they are independent. Later in this book we will address spatial dependency structures.

Dependency within a logger

The last source of independence we need to consider is temporal. Does the subnivium daily temperature at logger i on day s depend on the temperature at day $s - 1$ at the same logger? It is unlikely that we can ignore this dependency. This means that the starting point of the analysis should be a regression model containing a temporal dependency structure to avoid pseudoreplication. Such a model will be discussed in later chapters.

Summarising, on a gradient from 'very serious' to 'not so serious' we need to consider the following sources of dependency:

1. Temporal dependency within a logger
2. Spatial dependency within a site
3. Spatial dependency between the sites

In the remaining part of this section we show how we can detect these three sources of dependency if an ordinary regression-type model is fitted. Solutions will be given later in this book.

2.3.3 The model

To analyse whether the average daily temperature differs per habitat, we can fit the following model:

$$\begin{aligned}
&T_{it} \sim N(\mu_{it}, \sigma^2) \\
&E(T_{it}) = \mu_{it} \quad \text{and} \quad \text{var}(T_{it}) = \sigma^2 \\
&\mu_{it} = Intercept + f(Day_t) + Type_i + a_i \\
&a_i \sim N(0, \sigma^2_{Logger})
\end{aligned} \tag{2.3}$$

T_{it} is the average daily temperature measured by logger i on day t. We assume that it is normal distributed with mean μ_{it} and variance σ^2. We model the mean μ_{it} as a function of an intercept plus a smoothing function $f(Day_t)$, habitat type ($Type_i$), and a random effect a_i for logger i. If you are not familiar with random effects then see Chapter 3 for a short explanation. In Chapter 3 we will explain that a linear mixed-effects model (of which the model in Equation (2.3) is a special case) with a random intercept imposes the following correlation structure between any two observations from the same logger:

$$ICC = \frac{\sigma^2_{Logger}}{\sigma^2_{Logger} + \sigma^2} \tag{2.4}$$

ICC stands for intraclass correlation. The values of σ and σ_{Logger} will be estimated by the software code, so we can easily calculate the *ICC* value once the model has been fitted.

The term $f(Day_t)$ is a smoother (technically, it is a thin plate regression spline). If you are not familiar with GAMs and splines, just consider it for the moment to be like a moving average or a line that follows the pattern of the data. We will explain GAMs in more detail in Volume II of this book.

We can also fit a model with 3 smoothers (one per habitat) and even a model with 12 smoothers (one per logger). The model with 3 smoothers (one per habitat type) is given by

$$\mu_{it} = Intercept + f_{habitat}\left(Day_t\right) + Type_i + a_i \tag{2.5}$$

To fit this model we use the following R code:

```
> M4 <- gamm(Temp ~ s(TimeNum, by = Type) + Type,
             random = list(Logger=~1),
             data = SN2)
```

We will not present the numerical output or the model fit of the GAM here. We will do that later in this book using a different data set. Let us focus on the dependency part of the analysis. The estimated values for σ and σ_{Logger} are 1.00 and 0.34 respectively. This means that the *ICC* is equal to 0.10. Hence, this model assumes that any two observations from the same logger are correlated with a value of 0.10. This seems rather small! Either the correlation between daily average temperature values from the same logger is indeed small or we are using the wrong type of dependency structure. The *ICC* assumes that the correlation between T_{it} and $T_{i,t-1}$ is 0.10, but also the correlation between T_{it} and $T_{i,t-100}$. Simply stated, we use a linear mixed-effects model with random intercepts to impose a dependency structure on the temperature data from the same logger, then we implicitly assume that the correlation between two sequential days is the same as the correlation between two days separated by a much larger time span. Let's introduce some tools to verify this.

2.3.4 Model validation

We will focus on the dependency aspect of model validation. Define e_{it} as the residual (i.e. observed minus fitted values) for logger i on day t. Hence, for the first logger we have a vector $\boldsymbol{e}_1$ with residuals for K number of days: $\boldsymbol{e}_1 = (e_{11}, \ldots, e_{1K})$. Because we have 12 loggers we have 12 such vectors: $\boldsymbol{e}_1, \ldots, \boldsymbol{e}_{12}$. The first thing we should do is to plot the residuals $\boldsymbol{e}_1$ from logger 1 versus time and inspect the residuals for any patterns; see

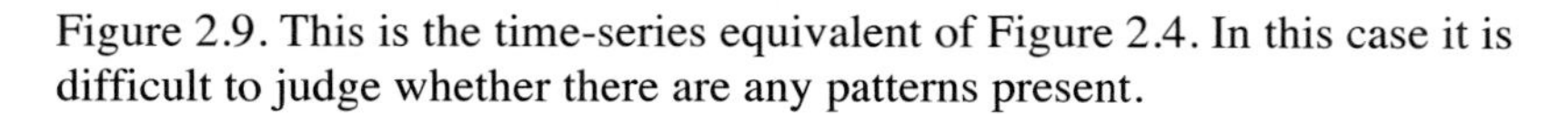
Figure 2.9. This is the time-series equivalent of Figure 2.4. In this case it is difficult to judge whether there are any patterns present.

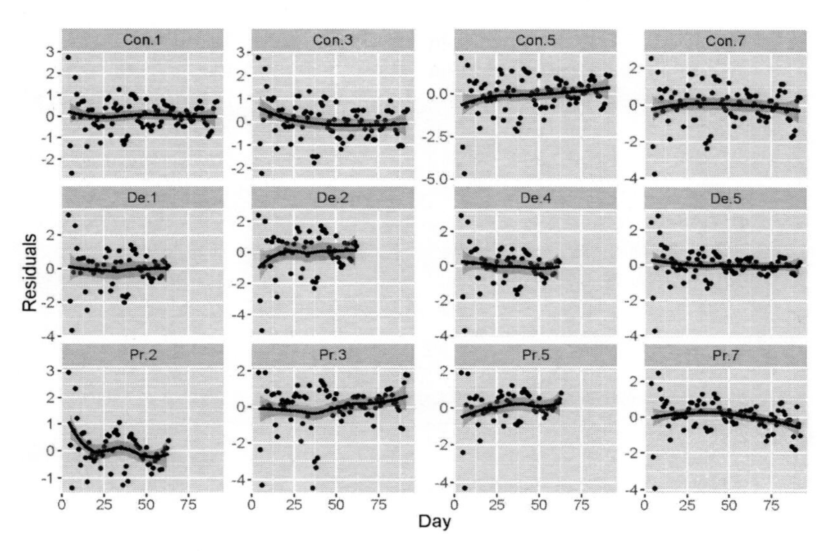

Figure 2.9. Plot of residuals versus time for each logger. A moving average smoother was added to each panel to aid visual interpretation. The top four panels are all from the same habitat. The same holds for the middle four and the lower four panels.

In the same way that we went from Figure 2.4 to the variogram function for spatial dependency we now go from Figure 2.9 to a more formal tool to assess temporal dependency, namely the auto-correlation function (acf). The auto-correlation function is the correlation of a time series with itself, after applying a shift (lag) of k days. We explain the essential steps. The (normalised) residuals from the GAMM are obtained using:

```
> E3 <- resid(M3$lme, type ="n")
```

These are the residuals from all 12 loggers. Let us extract the residuals from only logger 1.

```
> E3.Log1 <- E3[SN2$Logger == "Con.1" ]
```

The problem is that the data were not ordered by day, so let us do this now:

```
> TimeOrder <- order(SN2[SN2$Logger == "Con.1",
                         "TimeNum"])
> E3.Log1.sorted <- E3.Log1[TimeOrder]
```

The variable `E3.Log1.sorted` contains the residuals from logger 1, sorted by day number. The auto-correlation function first calculates the correlation between the variables `E3.Log1.sorted` and

`E3.Log1.sorted`; obviously this is equal to 1. Next, it calculates the correlation between the variables `E3.Log1.sorted` and `E3.Log1.sorted`, where the latter is shifted by 1 day. This is then repeated for further time lags. The upper-left panel in Figure 2.10 shows the resulting acf values as vertical lines. It also shows a pattern that we don't want to see: the residuals contain a significant time lag for 1 day. This means that the residuals (for this specific logger) on day t are correlated to the residuals on day $t - 1$, and this violates the independence assumption. And all loggers show this pattern! What we want to see is that no time lags are significant (except for time lag 0). Residuals with significant autocorrelation at certain time lags means that we can stop the analysis and fit a GAMM that contains a residual temporal correlation structure. The random intercept model was not able to capture the temporal dependency via the *ICC*.

Instead of the acf we can also use the variogram to assess whether there is temporal correlation. This is especially useful if the data are irregularly spaced in time. An example will be shown later.

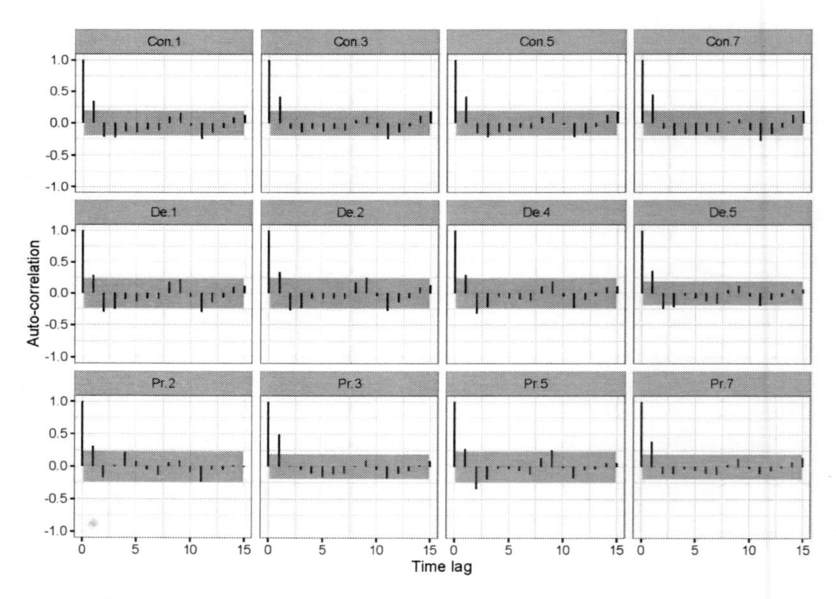

Figure 2.10. Auto-correlation function of residuals for each logger. The grey band is a 95% confidence interval.

2.4 GLMM applied on hierarchical and spatial data

Roulin and Bersier (2007) investigated vocal behaviour of barn owl siblings. One of their primary questions was whether the vocal response of chicks differs depending on the sex of the parent providing food. They designed a study in which data were collected from 27 barn owl nests. Using microphones, they sampled the number of calls (sibling negotiation) that chicks produced during a short time period. Since food availability may influence how chicks respond via sibling negotiation to the parent bird, chicks in half the nests were provided with extra prey in the morning preceding recording, and prey remains were removed from the other nests (food treatment, satiated or deprived). Sampling was carried out between 21.30 and 05.30.

The biological question is whether the relationship between sibling negotiation and sex of the parent differs with food treatment, and whether the effect of time on sibling negotiation differs with food treatment.

We have multiple observations from the same nests. By nature chicks in some nests are more vocal than chicks in other nests. Or perhaps the parents of a specific nest are excellent hunters and bring lots of food, and as a result all observations on the number of calls from this nest are low. Summarising, there is dependency and the standard approach to analyse such nested (or hierarchical) data is a mixed effects model. In this case we have count data, so a GLMM with a Poisson distribution is the starting point. Fixed covariates are sex of the parent (categorical with two levels), arrival time (continuous), and food treatment (categorical with two levels). The interaction terms are food treatment times sex of parent and food treatment times arrival time. To incorporate the dependency among observations of the same nest, we used nest as the random intercept (see Chapter 3 for a detailed explanation). The model is given in Equation (2.6).

$$
\begin{aligned}
& NCalls_{it} \sim Poisson(\mu_{it}) \\
& E(NCalls_{it}) = \mu_{it} \\
& \mu_{it} = Intercept + SexParent_{it} + FoodTreatment_{it} + \\
& \qquad ArrivalTime_{it} + SexParent_{it} \times FoodTreatment_{it} + \\
& \qquad SexParent_{it} \times ArrivalTime_{it} + \mathrm{a}_i \\
& \mathrm{a}_i \sim N(0, \sigma^2_{Nest})
\end{aligned}
\tag{2.6}
$$

where $NCalls_{it}$ is the tth observation in nest i, $i = 1, \ldots, 27$, and a_i is the random intercept, which is assumed to be normally distributed with mean 0 and variance σ^2_{Nest}.

When we fitted the model we noticed that it was overdispersed. Plotting Pearson residuals versus arrival time showed a clear non-linear arrival time effect. We extended the model and allowed for a non-linear arrival time effect via a GAMM. The model was still overdispersed. We then extended the Poisson distribution to a zero-inflated Poisson distribution (25% of the observations on *NCall* are equal to 0), but the resulting model was still overdispersed.

Figure 2.11 shows a geographical presentation on the nests. All these nests are within 10–15 km. What about spatial pseudoreplication? To answer this question we need to focus on the question: How far does an owl fly? And if the hunting territories of different owls cross, how does this affect the dependency? In Volume II of this book we will allow for spatial dependency between the random effects. It solves all problems for this data set! But we should have thought about the spatial dependency before starting the analysis.

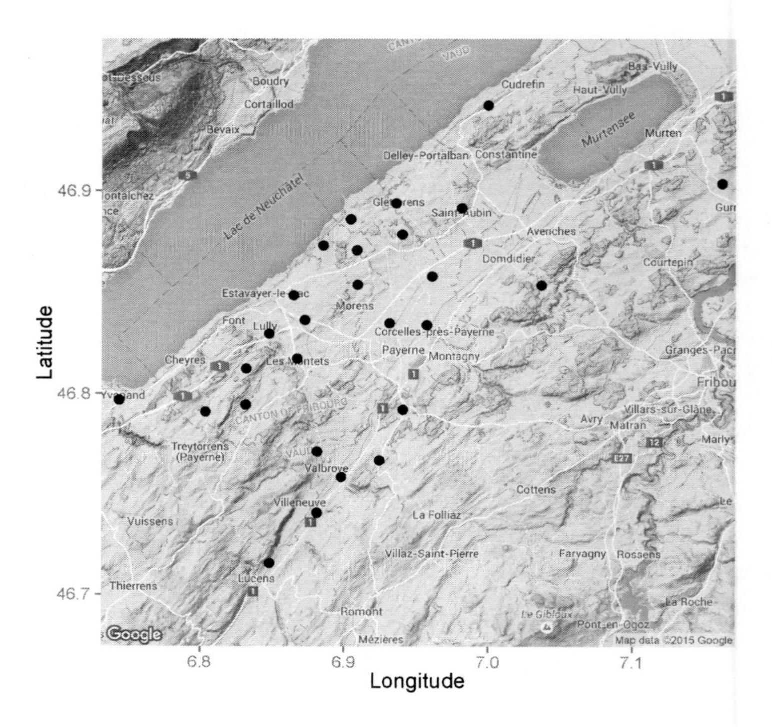

Figure 2.11. Spatial position of the nests for the owl data.

2.5 Technicalities

In the first subsection we present some elementary matrix notation that is required in later chapters. In the second subsection we show how exactly violation of independence is causing problems in a multiple linear regression model. Upon first reading of this book, you can omit this subsection (if you wish).

2.5.1 Matrix notation

Using a vector

Define the vector $\boldsymbol{\beta}$ with elements 2, 5, and 7 as

$$\boldsymbol{\beta} = \begin{pmatrix} 2 \\ 5 \\ 7 \end{pmatrix}$$

This vector has three rows and one column and therefore the dimensions of $\boldsymbol{\beta}$ are 3 by 1. We can multiply the elements of $\boldsymbol{\beta}$ with a number. For example, $2 \times \boldsymbol{\beta}$ is defined as

$$2 \times \boldsymbol{\beta} = 2 \times \begin{pmatrix} 2 \\ 5 \\ 7 \end{pmatrix} = \begin{pmatrix} 4 \\ 10 \\ 14 \end{pmatrix}$$

The dimensions of the new vector are 3 by 1.

Using a matrix

Define a matrix $\boldsymbol{X}$ with two rows and three columns, with the values 1, 2, 3 for the first row, and 4, 5, and 6 for the second row as

$$\boldsymbol{X} = \begin{pmatrix} 1 & 2 & 3 \\ 4 & 5 & 6 \end{pmatrix}$$

The dimensions of $\boldsymbol{X}$ are 2 by 3 (we have two rows and three columns). It is custom to use bold lowercase to indicate vectors and bold uppercase to indicate matrices, hence the $\boldsymbol{\beta}$ and the $\boldsymbol{X}$. A matrix with one column is a vector.

Just as for a vector we can multiply a matrix with a number. For example, $2 \times \boldsymbol{X}$ is given by

$$2 \times \boldsymbol{X} = 2 \times \begin{pmatrix} 1 & 2 & 3 \\ 4 & 5 & 6 \end{pmatrix} = \begin{pmatrix} 2 & 4 & 6 \\ 8 & 10 & 12 \end{pmatrix}$$

Multiplying a matrix with a vector

It is also possible to multiply a matrix with a vector, provided the dimensions match. With this we mean that a matrix of dimension a by b can be multiplied with a vector of dimension b by 1. For example $\boldsymbol{X} \times \boldsymbol{\beta}$ is defined as

$$\boldsymbol{X} \times \boldsymbol{\beta} = \begin{pmatrix} 1 & 2 & 3 \\ 4 & 5 & 6 \end{pmatrix} \times \begin{pmatrix} 2 \\ 5 \\ 7 \end{pmatrix} = \begin{pmatrix} 1 \times 2 + 2 \times 5 + 3 \times 7 \\ 4 \times 2 + 5 \times 5 + 6 \times 7 \end{pmatrix} = \begin{pmatrix} 33 \\ 75 \end{pmatrix}$$

The resulting matrix has the dimensions 2 by 1.

Writing a regression model as $\boldsymbol{Y} = \boldsymbol{X} \times \boldsymbol{\beta} + \boldsymbol{\varepsilon}$

Now that we have defined vectors and matrices we show how matrix notation can be used to shorten the mathematical notation for a multiple linear regression model. Let us return to the Irish pH data. Define Y_i as the pH value at site i, where i = 1, ..., 257 sites. In Equation (2.1) we modelled pH as a function of SDI. Let X_i be the SDI value at site i.

The linear regression model in Equation (2.1) for the Irish pH data is given by

$$Y_i = \beta_1 + X_i \times \beta_2 + \varepsilon_i$$

The parameters β_1 and β_2 are the intercept and slope respectively. We can write this equation for each site, resulting in

$$\begin{aligned}
Y_1 &= \beta_1 + X_1 \times \beta_2 + \varepsilon_1 \\
Y_2 &= \beta_1 + X_2 \times \beta_2 + \varepsilon_2 \\
Y_3 &= \beta_1 + X_3 \times \beta_2 + \varepsilon_3 \\
Y_4 &= \beta_1 + X_4 \times \beta_2 + \varepsilon_4 \\
&\ldots\ldots \\
Y_{257} &= \beta_1 + X_{257} \times \beta_2 + \varepsilon_{257}
\end{aligned}$$

This takes quite some space. Matrix notation can be used to simplify the notation. Define the vector $\boldsymbol{y}$, the matrix $\boldsymbol{X}$, the vector $\boldsymbol{\beta}$, and the vector $\boldsymbol{\varepsilon}$ as

$$\boldsymbol{y} = \begin{pmatrix} Y_1 \\ Y_2 \\ \vdots \\ Y_{257} \end{pmatrix} \quad \boldsymbol{X} = \begin{pmatrix} 1 & X_1 \\ 1 & X_2 \\ \vdots & \vdots \\ 1 & X_{257} \end{pmatrix} \quad \boldsymbol{\beta} = \begin{pmatrix} \beta_1 \\ \beta_2 \end{pmatrix} \quad \boldsymbol{\varepsilon} = \begin{pmatrix} \varepsilon_1 \\ \varepsilon_2 \\ \vdots \\ \varepsilon_{257} \end{pmatrix}$$

Using matrix notation we can write the linear regression model as

$$\begin{pmatrix} Y_1 \\ Y_2 \\ \vdots \\ Y_{257} \end{pmatrix} = \begin{pmatrix} 1 & X_1 \\ 1 & X_2 \\ \vdots & \vdots \\ 1 & X_{257} \end{pmatrix} \times \begin{pmatrix} \beta_1 \\ \beta_2 \end{pmatrix} + \begin{pmatrix} \varepsilon_1 \\ \varepsilon_2 \\ \vdots \\ \varepsilon_{257} \end{pmatrix}$$

And this can be written as

$$\boldsymbol{Y} = \boldsymbol{X} \times \boldsymbol{\beta} + \boldsymbol{\varepsilon}$$

The matrix $\boldsymbol{X}$ can be easily obtained in R using the `model.matrix` function:

```
> X <- model.matrix(M1)
> X

    (Intercept)         SDI
1             1 48.757764
2             1 73.529412
3             1 67.063492
4             1 53.183521
5             1 57.923497
...
```

The first column contains only ones and is used for the intercept. The second column contains the covariate SDI. We can also get the estimated regression parameters with the `coef` function.

```
> M5   <- lm(pH ~ SDI, data = iph)
> beta <- coef(M5)
```

These are the same as when you type `summary(M5)`. If you would like to read more on basic matrix operations in R, see the white paper by Prof. Joseph Hilbe: http://works.bepress.com/joseph_hilbe/.

2.5.2 How is dependency causing problems?

Once we have written a linear regression model as $\boldsymbol{y} = \boldsymbol{X} \times \boldsymbol{\beta} + \boldsymbol{\varepsilon}$, we can easily derive expressions for estimated parameters, standard errors, fitted values, and 95% confidence intervals for the fitted values. For example, the estimated regression parameters are obtained with

$$\hat{\boldsymbol{\beta}} = (\boldsymbol{X}' \times \boldsymbol{X})^{-1} \times \boldsymbol{X}' \times \boldsymbol{y} \quad (2.7)$$

The hat on $\boldsymbol{\beta}$ indicates that it is an estimator; the $'$ stands for transpose and -1 is for inverse. If you wonder where this expression is coming from, internally the software is applying ordinary least squares. This means that derivatives are taken of a quadratic function (the sum of squared residuals), and these are set to 0 and then solved. This is all high school mathematics!

Let us present the results of the model.

```
> summary(M5)

              Estimate Std. Error t value Pr(>|t|)
(Intercept)   8.475771   0.071399  118.71   <0.001
SDI          -0.025396   0.001616  -15.72   <0.001
```

The first column with estimated values comes from the expression in Equation (2.7). We now focus on the question where the standard errors come from. A little bit of mathematics gives the following expression for the covariance matrix of the estimator for $\boldsymbol{\beta}$:

$$\begin{aligned}\text{cov}\left(\hat{\boldsymbol{\beta}}\right) &= (\boldsymbol{X}'\times\boldsymbol{X})^{-1}\times\boldsymbol{X}'\times cov\left(\boldsymbol{y}\right)\times\boldsymbol{X}\times(\boldsymbol{X}'\times\boldsymbol{X})^{-1} \\ &= \sigma^2\times(\boldsymbol{X}'\times\boldsymbol{X})^{-1}\end{aligned} \tag{2.8}$$

Due to the independence assumption, the $cov(\boldsymbol{y})$ part changes into σ^2, and the whole inner part of the equation cancels out. The `lm` function for linear regression takes the diagonal elements of this matrix and then applies the square root. The resulting values are the standard errors, and these are in the second column of the `summary` output. These standard errors are used for the *t*-values and *p*-values.

In principle there is no need to know all this stuff. However, if you are familiar with matrix algebra you may have noticed that the cov($\boldsymbol{y}$) term changed into σ^2 (technically it changed into $\sigma^2 \times \boldsymbol{I}$, where $\boldsymbol{I}$ is an identity matrix). An identity matrix is a matrix that contains only 0s, except for the diagonal elements, which are all equal to 1. The cov($\boldsymbol{y}$) term stands for the covariance of the pH data. If we assume independence (and homogeneity) then this term is indeed equal to σ^2. However, if we have dependency then we need to modify the expression for the standard errors. Otherwise the standard errors are wrong. In the remainder of this book we will allow for dependency, but this tends to give larger standard errors!

2.6 Discussion

The main message of this chapter is to think about dependency before the start of the analysis. Expect dependency to be present.

3 Time series and GLS

In this chapter we show how a regression model can be extended to deal with temporal dependency. We will use R functions like `lm` and `gls` from the `nlme` package (Pinheiro et al. 2016). These functions apply statistical methods that are called 'frequentist' techniques. In Chapter 7 we will explain where this name comes from.

Although the techniques explained in this chapter are not the core part of this book, and in our experience they don't always work, these methods do allow us to explain the basics of essential tools that we need later.

Prerequisite for this chapter: You need to be familiar with R, multiple linear regression, the autocorrelation function, the variogram, and the matrix notation that we discussed in Chapter 2.

3.1 Ospreys

Steidl et al. (1991) examined contaminant levels of osprey eggs (*Pandion haliaetus*). In one of their analyses they modelled eggshell thickness as a function of DDD using linear regression and found a significant relationship. DDD is a breakdown product of DDT (a pesticide) used in the past to control insects in agriculture. This data set was also used in Zuur et al. (2016a) to explain Markov chain Monte Carlo (MCMC). It this is not a time series data set, but we will use it to explain the residual covariance matrix.

Let us import the data in R with the `read.csv` function.

```
> Ospreys <- read.csv(file = "Ospreys.csv",
                      dec = ".")
```

The variables of interest are THICK (eggshell thickness) and DDD.

3.2 Covariance and correlation coefficients

To investigate whether there is a relationship between THICK and DDD we can make a scatterplot, or alternatively, calculate the covariance or the correlation coefficient for these two variables. To calculate the covariance coefficient the software subtracts the mean from each variable, calculates the cross product, and averages the resulting values:

$$\text{cov}(SHELL_i, DDD_i) = \frac{1}{N-1} \times \sum_{i=1}^{N} (SHELL_i - \mu_{SHELL}) \times (DDD_i - \mu_{DDD})$$

where μ_{SHELL} and μ_{DDD} are the mean values of SHELL and DDD respectively and N is the sample size. We can easily do this in R with the `cov` function.

```
> cov(Ospreys[, c("THICK", "DDD")])

            THICK      DDD
THICK      31.060   -0.532
DDD        -0.532    0.051
```

The diagonal elements are the variances of each variable. Eggshell thickness has a much larger variance than DDD. The covariance between these two variables is –0.532. This means that if one variable is below average then most likely the other variable is also below average, and vice versa. However, the two variables are in different units and this makes the interpretation of the value of the covariance rather difficult. For example, is the –0.532 a large value or a small value? The Pearson correlation coefficient rescales both variables and then calculates the cross product. As a result we end up with a number between –1 and 1, which is much easier to interpret. The correlation coefficient is calculated as

$$\text{cor}(SHELL_i, DDD_i) = \frac{1}{N-1} \times \sum_{i=1}^{N} \frac{(SHELL_i - \mu_{SHELL})}{\sigma_{SHELL}} \times \frac{(DDD_i - \mu_{DDD})}{\sigma_{DDD}}$$

In R we can calculate the Pearson correlation with the `cor` function.

```
> cor(Ospreys[, c("THICK", "DDD")])

            THICK      DDD
THICK       1.000   -0.419
DDD        -0.419    1.000
```

The correlation between eggshell thickness and DDD is –0.419, which is relatively large. Whether the Pearson correlation coefficient is significant is another issue (and we will not address that here).

The `cov` and `cor` functions are simple tools to assess whether there is a linear relationship between two variables. If there are more than two variables, then we can produce a covariance or correlation matrix showing all the pairwise covariances or correlations at once.

The covariance and correlation coefficients tell us whether there is a linear relationship between two variables.

3.3 Linear regression model

It is important to realise that the covariance and correlation functions do not model a cause–effect relationship. To do that we apply a linear regression model in which we model eggshell thickness as a function of an intercept and a slope times DDD.

$$Thickness_i = \beta_1 + DDD_i \times \beta_2 + \varepsilon_i$$
$$\varepsilon_i \sim N(0,\sigma^2)$$

$Thickness_i$ is the eggshell thickness for observation i, and DDD_i is the value of the covariate DDD. The index i runs from 1 to 25, which means that we have a rather small sample size. The residuals ε_i are the unexplained variation, and we assume that these are independently and normal distributed with mean 0 and variance σ^2. We can fit this linear regression model in R using

```
> M1 <- lm(THICK ~ DDD, data = Ospreys)
```

As explained in Section 2.5, the linear regression model can also be written as

$$\mathbf{y} = \mathbf{X} \times \boldsymbol{\beta} + \boldsymbol{\varepsilon}$$

where the **y** vector contains all the $Thickness_1$, ..., $Thickness_{25}$ values, $\mathbf{X}$ is a matrix with the first column equal to 1 (for the intercept), and the second column contains the 25 DDD values. The vector $\boldsymbol{\varepsilon}$ contains the 25 residuals.

3.4 Focussing on the residual covariance matrix

Instead of writing that each individual residual ε_i has mean 0 and variance σ^2, we can write this in matrix notation as

$$\boldsymbol{\varepsilon} \sim N(\mathbf{0}, \sigma^2 \times \mathbf{I})$$

where $\mathbf{I}$ is an identity matrix with the dimensions 25 by 25. A more general mathematical notation for the residuals is

$$\boldsymbol{\varepsilon} \sim N(\mathbf{0}, \boldsymbol{\Sigma})$$

where $\boldsymbol{\Sigma}$ is the covariance matrix of the residuals. In the linear regression model for the eggshell thickness that we fitted in the previous section, $\boldsymbol{\Sigma}$ has the following form.

$$\boldsymbol{\Sigma} = \sigma^2 \times \begin{pmatrix} 1 & 0 & \cdots & 0 \\ 0 & 1 & & \vdots \\ \vdots & & \ddots & 0 \\ 0 & \cdots & 0 & 1 \end{pmatrix}$$

The matrix with zeros and ones is the identity matrix with the dimensions 25 by 25. The first zero on the upper line represents the covariance between the residuals ε_1 and ε_2. The last zero on this line is the covariance between residuals ε_1 and ε_{25}. It is not that we estimated the values of these covariances as being 0. When we applied the linear regression model we implicitly *assumed* independence. In an ordinary linear regression model this means that all the covariances between different residuals are equal to 0.

In an ordinary linear regression model the covariance matrix of the residuals is equal to $\boldsymbol{\Sigma}$, where $\boldsymbol{\Sigma}$ is σ^2 times an identity matrix. Assuming independence, we can write this as $\boldsymbol{\varepsilon} \sim N(\mathbf{0}, \boldsymbol{\Sigma})$, where $\boldsymbol{\Sigma} = \sigma^2 \times \boldsymbol{I}$. This is the same as writing $\varepsilon_i \sim N(0, \sigma^2)$.

We can write the covariance matrix $\boldsymbol{\Sigma}$ in a more general notation as

$$\boldsymbol{\Sigma} = \begin{pmatrix} \sigma^2 & \phi_{1,2} & \phi_{1,3} & \phi_{1,4} & \cdots & \phi_{1,25} \\ & \sigma^2 & \phi_{2,3} & \phi_{2,4} & \cdots & \phi_{2,25} \\ & & \sigma^2 & \phi_{3,4} & & \phi_{3,25} \\ & & & \sigma^2 & \ddots & \vdots \\ & & & & \sigma^2 & \phi_{24,25} \\ & & & & & \sigma^2 \end{pmatrix} \tag{3.1}$$

A covariance matrix is symmetric, so we might as well omit the lower part of the matrix blank for notational convenience. The $\phi_{1,2}$ is the covariance between the residuals ε_1 and ε_2. The $\phi_{1,25}$ is the covariance between residuals ε_1 and ε_{25}. Expressed more generally, the $\phi_{i,j}$ term represents the covariance between the residuals ε_i and ε_j. In an ordinary linear regression model all the 300 (= 25 × 24 / 2) $\phi_{i,j}$ terms are assumed to be 0.

3.5 Dependency and the covariance matrix

To better understand the role of covariance in a linear regression model we carry out a small simulation study not related to the Osprey data. Suppose that we have two variables, for example the biomass of two species or the beak length of two chicks in a bird nest, eggshell thickness, and DDD, or temperature and salinity at two sites. Whatever we sample, just call the two variables z_1 and z_2 respectively. We are going to simulate 1,000 values for each of them. Let us assume that each variable is normally distributed and that there are no covariate effects. Suppose that the mean of z_1 is 10 and the mean of z_2 is 15. And suppose that the variances of both variables are 1. So we have $z_1 \sim N(10, 1)$ and $z_2 \sim N(15, 1)$. We can simulate such data in R using the following two lines of code. The `set.seed` function ensures that you have the same results.

```
> set.seed(12345)
> z1 <- rnorm(1000, mean = 10, sd = 1)
> z2 <- rnorm(1000, mean = 15, sd = 1)
```

This gives us two times 1,000 random observations from a normal distribution with the specified means and standard deviations. We can achieve the same task by using the `mvrnorm` function from the `MASS` package (Venables and Ripley 2002). You can also use the `mvtnorm` package (Genz et al. 2016). The required code is as follows. The `diag` function creates a diagonal matrix with the dimensions 2 by 2. The elements on the diagonal are equal to 1.

```
> library(MASS)
> SIGMA <- diag(2)
> SIGMA

1  0
0  1
```

The rest is a matter of using `mvrnorm` and storing the results in Z.

```
> Z <- mvrnorm(1000,
               mu = c(10, 15),
               Sigma = SIGMA)
> head(Z, 5)

[1,]  9.490108 14.39216
[2,] 10.715364 16.07622
[3,] 10.405522 14.42357
[4,] 10.012521 16.09863
[5,]  8.899119 16.40734
```

The `mvrnorm` function simulates data from a multivariate normal distribution. We can also write this as

$$\begin{pmatrix} z_1 \\ z_2 \end{pmatrix} \sim N\left(\begin{pmatrix} 10 \\ 15 \end{pmatrix}, \begin{pmatrix} 1 & 0 \\ 0 & 1 \end{pmatrix} \right)$$

The first column in the normal distribution represents the mean, and the matrix is the covariance $\mathbf{\Sigma}$. Figure 3.1A shows a scatterplot of the two columns of Z (these are the z_1 and z_2). Because we assume independence of z_1 and z_2 (and both have the same variance) we obtain a round cloud of simulated points. Knowing that z_1 is large does not give us any information on z_2.

Now let us change the scenario and suppose that z_1 and z_2 have a strong positive correlation. We can simulate such data using the following code.

```
> SIGMA        <- diag(2)
> SIGMA[1,2] <- 0.9
> SIGMA[2,1] <- 0.9
> SIGMA

1.0  0.9
0.9  1.0

> Z <- mvrnorm(1000,
              mu = c(10,15),
              Sigma = SIGMA)
```

We now use a different covariance matrix $\mathbf{\Sigma}$.

$$\begin{pmatrix} z_1 \\ z_2 \end{pmatrix} \sim N\left(\begin{pmatrix} 10 \\ 15 \end{pmatrix}, \begin{pmatrix} 1 & 0.9 \\ 0.9 & 1 \end{pmatrix} \right)$$

A scatterplot of the simulated data is presented in Figure 3.1B. A large value of z_1 nearly always goes hand in hand with a large value of z_2, and vice versa. We have generated highly correlated data. The simulation process uses a non-diagonal covariance matrix $\mathbf{\Sigma}$.

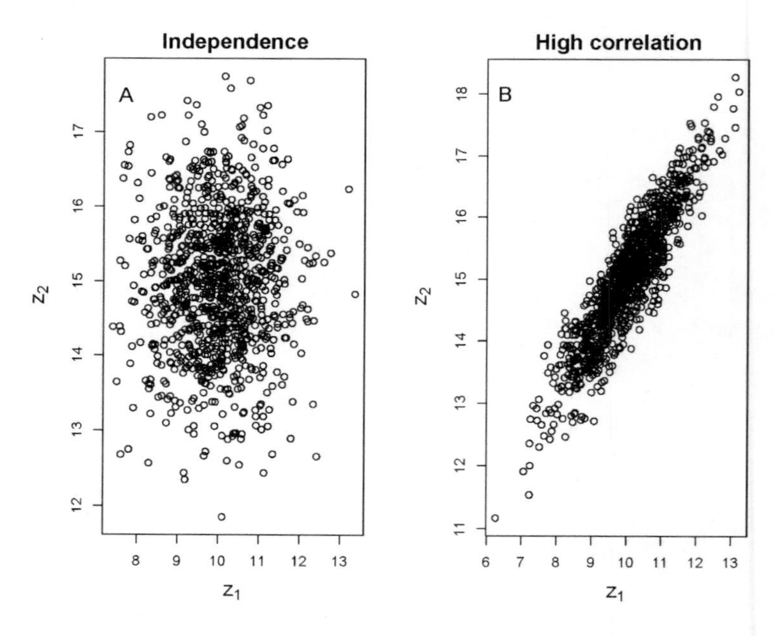

Figure 3.1. Simulated data from a normal distribution. A: Independence. B: A non-diagonal covariance matrix with a high covariance was used and as a result we have highly correlated data.

Because we use a variance of 1 for both variables in the covariance matrix $\mathbf{\Sigma}$, the covariance matrix equals the correlation matrix. In general this is not the case. The command `cov2cor(SIGMA)` can be helpful to convert a covariance matrix $\mathbf{\Sigma}$ into a correlation matrix as these are easier to understand.

High (absolute) values for the off-diagonal elements in a covariance matrix results in highly correlated values for the corresponding two variables. The same principle applies for the $\phi_{i,j}$s in the covariance matrix $\mathbf{\Sigma}$ in Equation (3.1). A non-zero value corresponds to dependent residuals.

3.6 GLS: Dealing with temporal dependency

This book is about dealing with dependency when we use regression models. This can be done (among various other ways) by allowing some of the $\phi_{i,j}$ terms in the residual covariance matrix to be non-zero. The problem is that we have 300 of these $\phi_{i,j}$s in the Osprey example. That is way too many to estimate for the size of the data set. Generalised least squares (GLS) allows for non-zero $\phi_{i,j}$s, but it imposes a mathematical model on them so that we only need to estimate a few of them. We will illustrate GLS using a penguin data set.

3.6.1 Adelie penguins

We will use data from Barbraud and Weimerskirch (2006), who analysed arrival and laying dates of Antarctic seabirds as part of a long-term study on Antarctic marine top predators. The data were also analysed in Zuur et al. (2009a). Here, we use the laying dates of Adelie penguins (*Pygoscelis adeliae*) as the response variable. We have one value per year for the colony (Figure 3.2A).

Adelie penguins use sea ice floes as resting platforms and forage within the pack ice. Therefore, we would like to use sea ice extent as an explanatory variable. Because the phenological data start in the early 1950s, and sea ice extent data derived from satellite observations are only available from the early 1970s, we use a proxy of sea ice extent developed for East Antarctica. Methanesulphonic acid (MSA) is a product of biological activity in surface ocean water whose production is heavily influenced by the presence of sea ice in the Southern Ocean. Figure 3.2B shows a scatterplot of annual laying dates versus MSA.

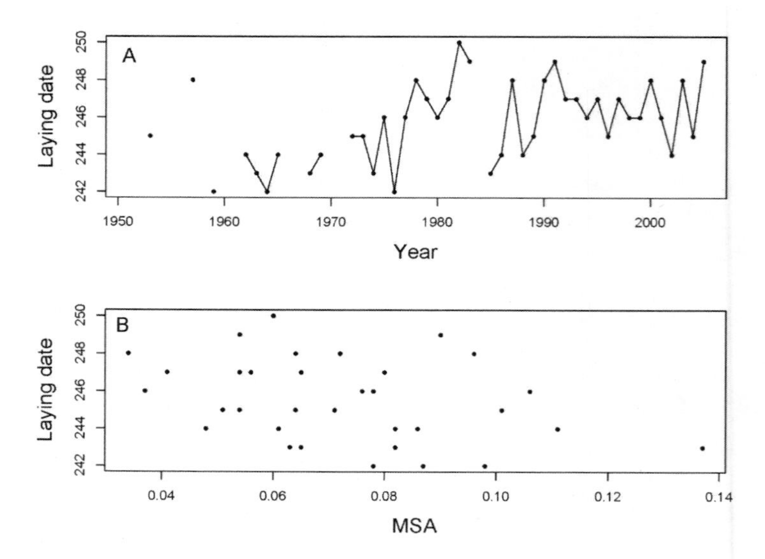

Figure 3.2. A: Scatterplot of laying date (response variable) versus year. Sequential years are connected with a line. If there is a missing value in a year then no line is drawn. B: Scatterplot of laying date versus MSA (sea ice extent proxy).

3.6.2 Do we have dependency?

We have one laying date value per year, and we have 55 years of data. The first question that we ask is whether we have independence in the response variable. If the laying date in year t is late, is it likely to be late in year $t + 1$ as well? The simple fact that the same penguin colony is observed in each year means that the answer is 'yes'. Perhaps in year t the colony consists mainly of relatively young birds that haven't reached sexual maturation, maybe there is a disease affecting the Adelie penguins in the colony, or maybe some physiological mechanisms (such as inappropriate prolactin and corticosterol levels) do not allow for a 'late in year t, early in year $t + 1$' breeding pattern. All these aspects may affect the egg-laying date in year $t + 1$.

It may be difficult to distinguish the effect of a covariate (MSA) and dependency in this example. Low values of MSA may imply a late laying date, but MSA itself may have a strong temporal trend. So, we may observe late laying dates in two or three sequential years, but this may be due to MSA having low values in all these years.

3.6.3 Formulation of the linear regression model

Because of the presence of dependency we should start with a regression model that deals with temporal dependency. However, at this stage of the book we don't know yet how to do this. Let us therefore start with a linear regression model for the laying dates without a dependency structure. Such a linear regression model is given by:

$$\begin{aligned} LD_t &= \beta_1 + MSA_t \times \beta_2 + \varepsilon_t \\ \varepsilon_t &\sim N(0, \sigma^2) \end{aligned} \tag{3.2}$$

LD_t is the laying date of Adelie penguins in year t (where t = 1, ..., 55 years), β_1 is the intercept, MSA_t is the methanesulphonic acid value in year t, β_2 is its slope, and ε_t is the residual. The residuals ε_1, ..., ε_{55} are assumed to be homogenous, normal, and independently distributed with mean 0 and variance σ^2.

Just as in the previous section we can write the linear regression model in Equation (3.2) as $\boldsymbol{y} = \boldsymbol{X} \times \boldsymbol{\beta} + \boldsymbol{\varepsilon}$, where the $\boldsymbol{y}$ vector contains all the LD_1, ..., LD_{55} values, and $\boldsymbol{X}$ is a matrix with the first column equal to 1 (for the intercept) and the second column contains the 55 MSA values. The vector $\boldsymbol{\varepsilon}$ contains all the residuals. Instead of writing that each individual residual ε_s has mean 0 and variance σ^2, and is independent, as in Equation (3.2), we can also write $\boldsymbol{\varepsilon} \sim N(\boldsymbol{0}, \sigma^2 \times \boldsymbol{I})$, where $\boldsymbol{I}$ is an identity matrix with the dimensions 55 by 55. See also Section 3.4.

In Chapter 2, we gave the mathematical expression for the estimated parameters and covariance matrix of the parameters of a linear regression model, and we reproduce these equations below. The regression parameters are estimated via the equation

$$\hat{\boldsymbol{\beta}} = (\boldsymbol{X}^t \times \boldsymbol{X})^{-1} \times \boldsymbol{X}^t \times \boldsymbol{y} \tag{3.3}$$

and the standard errors are given by the square root of the diagonal elements of the following matrix.

$$\text{cov}\left(\hat{\boldsymbol{\beta}}\right) = \sigma^2 \times (\boldsymbol{X}^t \times \boldsymbol{X})^{-1} \tag{3.4}$$

These expressions can be found in any book on linear regression models, for example Montgomery and Peck (1992).

A crucial fact is that the expressions in Equations (3.3) and (3.4) are based on the independence assumption!

3.6.4 Application of the linear regression model

Let us fit the model in Equation (3.2) in R. We use the `gls` function from the `nlme` package, though we could have used the `lm` function as well. First we import the data and load the `nlme` package.

```
> EP <- read.csv(file = "Antarcticbirds.csv",
                 header = TRUE)
> library(nlme)
```

Because we have missing values we use the `na.action` option in the `gls` function; it drops rows with missing values. So be careful when there are more covariates and when applying model selection. It is safer to remove all missing values manually before starting the analysis.

```
> B0 <- gls(LayingAP ~ MSA,
            data = AP,
            na.action = na.omit)
> summary(B0)
```

The numerical results of this model are as follows.

```
                 Value Std.Error   t-value p-value
(Intercept) 247.91825   1.25111 198.15867  0.0000
MSA         -33.01513  16.42976  -2.00947  0.0533
```

There is a weak MSA effect; the higher its value, the earlier the laying date. We have visualised the model fit in Figure 3.3. Before making any statements on the possible effects of climate change we should first address the dependency issue. As part of the model validation process we should plot residuals versus fitted values, versus each covariate in the model (Figure 3.4A), and we need to check the residuals for temporal dependency (Figure 3.4B).

In Figure 3.3 we want to see equal numbers of residuals above and below the line. We don't want to see too many residuals from sequential years all above or all below the fitted line. If this happens then we have temporal pseudoreplication. Looking at the graph, our first impression was that things were alright. However, a more detailed inspection indicates that there is (potentially) clustering of sequential years on one particular side of the fitted line. For example the residuals for 1962, 1963, 1964, and 1965 are all below the line. Figure 3.4A seems to confirm this; all the residuals from the 1960s are below zero.

Figure 3.5 is another attempt to visualise (and convince ourselves) whether there is temporal pseudoreplication. Again, we have plotted the fitted values against MSA. This time we have connected sequential years with a line. The thickness of the line is proportional to time (thicker lines are for the earlier sampling years and thinner lines for later sampling years). Note that the thick lines (representing the 1960s) are all on the right side, which is due to collinearity between MSA and Year. This aspect is not a violation of any assumption. But the fact that the residuals from the 1960s are *consistently* below the fitted line (and the residuals from the early 1990s are all above the line) is a problem. It means that we have pseudoreplication. The residuals from sequential years should be randomly scattered around the line.

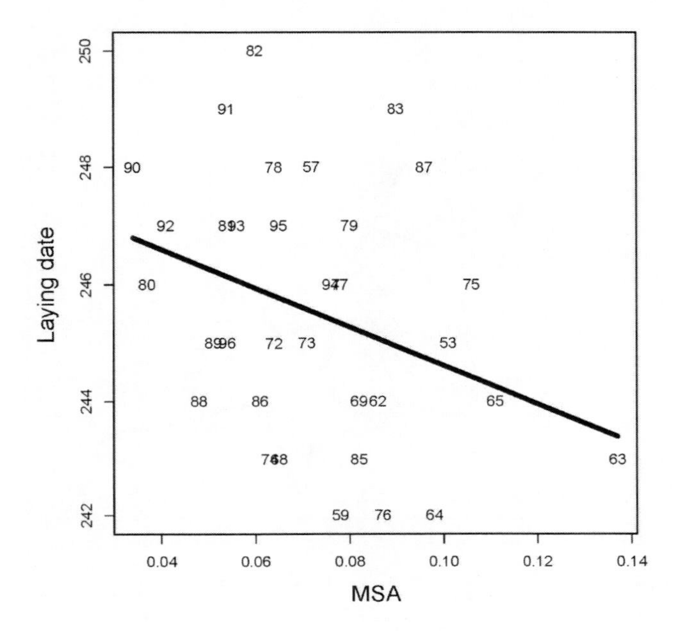

Figure 3.3. Model fit (thick line) of the linear regression model. Observed data are represented by the last two digits of the sampling year.

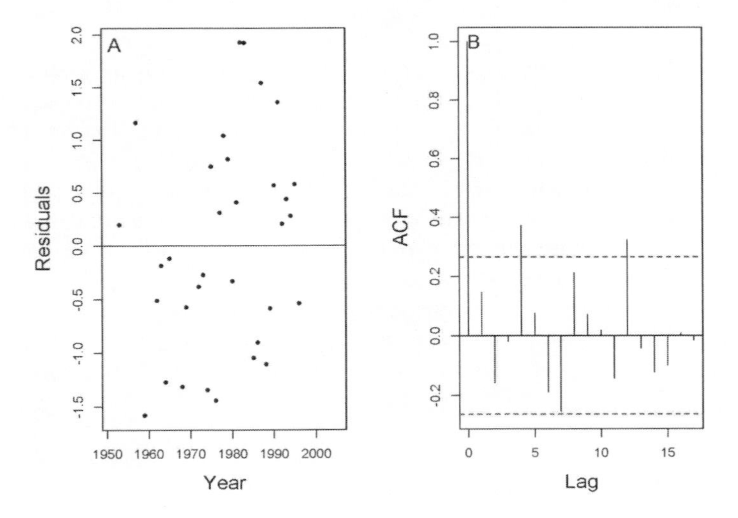

Figure 3.4. A: Residuals plotted versus year. B: Autocorrelation function of the residuals.

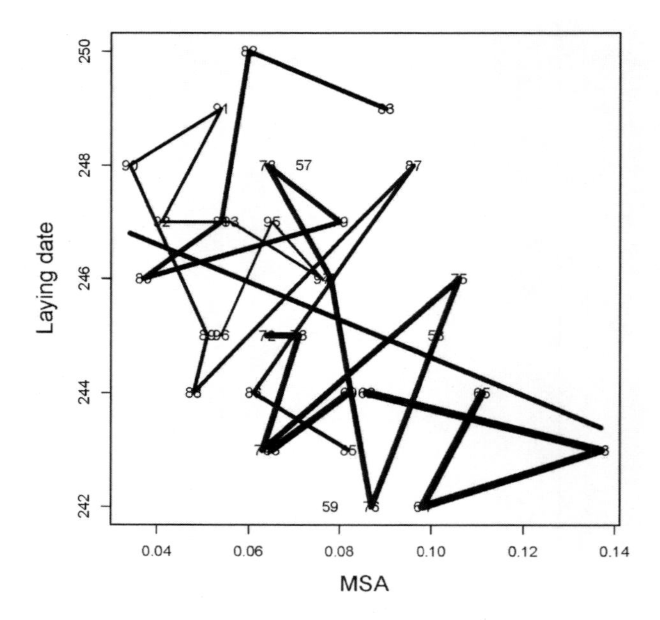

Figure 3.5. As Figure 3.3 but here sequential years are connected with a line. The thickness of a line is proportional to the sampling year (the thinner the line, the later the sampling year).

The autocorrelation function for the residuals (Figure 3.4B) shows that there are time lags (at 4, 7, and 12 years) with significant autocorrelation. Due to the large number of missing values, we also made a variogram of the residuals as it can cope better with gaps in time series and irregular spaced time series than the `acf` function. The variogram (Figure 3.6) of the residuals shows minor dependency at the same time lags.

The question is now how serious is the temporal pseudoreplication. Can we ignore it? To answer this question we should fit a model with a dependency structure and compare it (e.g. via the AIC) to a model without a dependency structure.

A word of warning (Schabenberger and Pierce 2002): Autocorrelation in residuals can also be due to a missing covariate or misspecification of the functional form of a covariate (e.g. linear instead of non-linear).

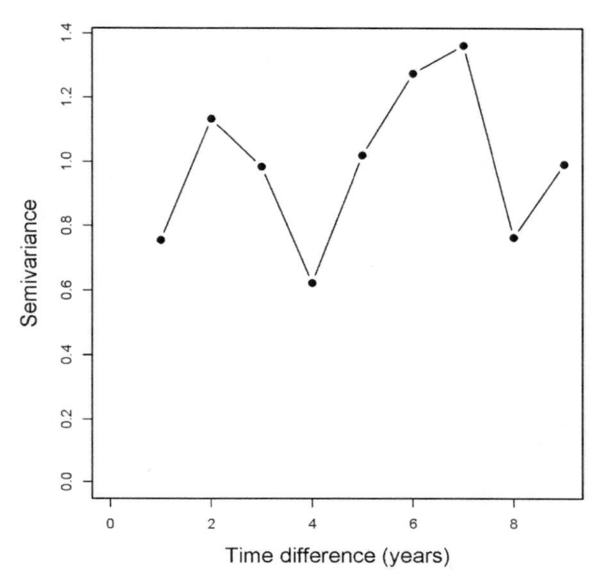

Figure 3.6. Sample variogram for the residuals of the linear regression model.

3.6.5 R code for acf and variogram

Before showing how to deal with temporal dependency structure in the regression model we catch up with the R code for the autocorrelation function and the variogram. First we extract the (normalised) residuals.

```
> E0 <- resid(B0, type = "n")
```

The main problem is that we have a considerable number of missing values. The gls function removes any row with a missing value for laying date or MSA. To avoid error messages with the plotting functions (due to variables having different lengths) we expand the vector with residuals so that they have the same length as the original variables. To do this we create a new variable E0Full consisting only of NAs (missing values), and wherever we have observed data for the laying dates and MSA we fill in the residuals from the model.

```
> E0Full      <- rep(NA, nrow(EP))
> I           <- is.na(AP$LayingAP)  | is.na(AP$MSA)
> E0Full[!I] <- E0
```

The E0Full variable has residuals at the correction positions. We can plot it versus year using simple plotting functions (see Figure 3.4A). And we can make an autocorrelation function using the acf function.

```
> acf(E0Full, na.action = na.pass)
```

The rows in E0full represent the years, and we instruct the acf function to take care of the NAs. The resulting graph is presented in Figure 3.4B.

The variogram function in the gstat (Pebesma 2004) package gave us some trouble so we use the variog function in geoR (Ribeiro and Diggle 2016). In Chapter 2 we explained the principle of a variogram. Recall that two coordinates were used to define geographical distance between locations. The Pythagorean theorem is used for this. But in the case of a time series we only have one variable that defines the distance between observations, namely the variable Year. However, the variog function in geoR (or variogram in gstat) still expects two variables to define distance. We can fool the function by supplying a dummy variable (e.g. a vector with only zeros). This is the variable with the name Zeros in the code below. To avoid an error message due to missing values we have to exclude the rows with NAs.

```
> library(sp)
> library(geoR)
> AP$Zeros   <- rep(0, nrow(AP))
> AP$E0Full <- E0Full
> AP2        <- na.exclude(AP[,c("E0Full", "Year",
                                 "Zeros")])
```

We define a vector with break points to control the range of years for which we want to calculate the experimental variogram. In this case we calculate the residual dependency for 10 distances between 0.5 and 9.5 years. Finally, we run the variog function on the residuals and plot the results (Figure 3.6).

```
> breaks <- seq(0.5, 9.5, length = 10)
> V1 <- variog(coords = AP2[,c("Year", "Zeros")],
               data = AP2[,"E0Full"],
               breaks = breaks)
> plot(V1,
       type = "b",
       xlab = "Time difference (years)",
       ylab = "Sample variogram",
       col = 1, pch = 16, cex = 1.5)
```

3.6.6 Formulation of the GLS model

Chapters 6 and 7 of Zuur et al. (2009a) and Chapter 5 of Pinheiro and Bates (2000) give detailed examples of how to include a residual correlation structure in linear regression models and additive models. In a linear regression model we assume

$$\begin{aligned} &LD_t = \beta_1 + MSA_t \times \beta_2 + \varepsilon_t \\ &\varepsilon_t \sim N(0, \sigma^2) \\ &cor(\varepsilon_t, \varepsilon_s) = \begin{cases} 0 & \text{if } t \neq s \\ 1 & \text{if } t = s \end{cases} \end{aligned} \tag{3.5}$$

The part that involves the *cor* function (which stands for correlation) is based on the independence assumption. It states that the correlation between the residuals from two different years is assumed to be 0.

To deal with a temporal dependency structure in a regression model, we simply allow for nonzero correlation between the residuals from two different years. This gives the following model.

$$\begin{aligned} &LD_t = \beta_1 + MSA_t \times \beta_2 + \varepsilon_t \\ &\varepsilon_t \sim N(0, \sigma^2) \\ &cor(\varepsilon_t, \varepsilon_s) = h(\phi, \varepsilon_t, \varepsilon_s) \end{aligned} \tag{3.6}$$

The *residuals* of this model are assumed to be normally distributed, but not independent of each other. We allow for a certain residual dependence structure using the function *h*(), which depends on an unknown parameter ϕ and on the residuals at time *s* and *t*. Another way to write this model is:

$$\boldsymbol{y} = \boldsymbol{X} \times \boldsymbol{\beta} + \boldsymbol{\varepsilon} \quad \text{where} \quad \boldsymbol{\varepsilon} \sim N(\boldsymbol{0}, \boldsymbol{\Sigma})$$

The $\boldsymbol{\Sigma}$ is a non-diagonal covariance matrix. Its off-diagonal elements define the covariance between residuals from two different years; see also Equation (3.1). As explained in Section 3.1, the covariance matrix $\boldsymbol{\Sigma}$ contains too many elements to estimate. We will use the function *h*() to impose a mathematical equation on the covariance elements of $\boldsymbol{\Sigma}$ and this reduces the number of parameters that we need to estimate. The challenge is to find an appropriate mathematical expression for the function *h*(). A popular choice to deal with dependency in a time series is the residual first-order autoregressive process (AR1):

$$\varepsilon_t = \phi \times \varepsilon_{t-1} + \nu_t$$

where ν_t is normal and independent distributed noise with mean 0 and variance σ_ν^2. The parameter ϕ is restricted to be between –1 and 1. In plain language: The residual AR1 process states that the residual in year *t* is equal to an unknown parameter ϕ times the residuals from year *t* – 1 plus pure noise.

It is relatively easy to show that the residual AR1 process results in the following function *h*():

$$h(\phi, \varepsilon_t, \varepsilon_s) = \phi^{|t-s|}$$

This correlation function $h()$ defines the correlation between two residuals ε_t and ε_s as a power function of the time difference. This choice assumes that the residuals are stationary. Without getting sidetracked, stationarity means (among other things) that the covariance (or correlation) only depends on the time difference. The parameter ϕ is unknown and we need to estimate it, together with the other regression parameters

Using the AR1 process the covariance matrix $\boldsymbol{\Sigma}$ has the following structure:

$$\boldsymbol{\Sigma} = \text{cov}(\boldsymbol{\varepsilon}) = \frac{\sigma_v^2}{1-\phi^2} \times \begin{pmatrix} 1 & \phi & \phi^2 & \phi^3 & \cdots & \phi^{54} \\ \phi & 1 & \phi & \phi^2 & \ddots & \vdots \\ \phi^2 & \phi & \ddots & \ddots & \ddots & \phi^3 \\ \phi^3 & \phi^2 & \ddots & \ddots & \ddots & \phi^2 \\ \vdots & \ddots & \ddots & \phi & 1 & \phi \\ \phi^{54} & \cdots & \phi^3 & \phi^2 & \phi & 1 \end{pmatrix} \tag{3.7}$$

Compare this to the expression in Equation (3.1) and note the huge decrease in number of parameters that we need to estimate!

If you find it hard to interpret a covariance matrix and prefer a correlation matrix, then use the following expression:

$$cor(\boldsymbol{\varepsilon}) = \begin{pmatrix} 1 & \phi & \phi^2 & \phi^3 & \cdots & \phi^{54} \\ \phi & 1 & \phi & \phi^2 & \ddots & \vdots \\ \phi^2 & \phi & \ddots & \ddots & \ddots & \phi^3 \\ \phi^3 & \phi^2 & \ddots & \ddots & \ddots & \phi^2 \\ \vdots & \ddots & \ddots & \phi & 1 & \phi \\ \phi^{54} & \cdots & \phi^3 & \phi^2 & \phi & 1 \end{pmatrix} \tag{3.8}$$

This correlation structure implies that the correlation between two sequential years is equal to ϕ, and if the time difference is 2 years then the correlation is ϕ^2, etc. Because ϕ is between –1 and 1 we have a correlation structure that is decaying for increasing time distances. As an example, suppose that the value of ϕ is 0.5. The correlation matrix is then equal to:

$$cor(\boldsymbol{\varepsilon}) = \begin{pmatrix} 1 & 0.5 & 0.25 & 0.125 & \cdots & 0 \\ 0.5 & 1 & 0.5 & 0.25 & \ddots & \vdots \\ 0.25 & 0.5 & \ddots & \ddots & \ddots & 0.125 \\ 0.125 & 0.25 & \ddots & \ddots & \ddots & 0.25 \\ \vdots & \ddots & \ddots & 0.5 & 1 & 0.5 \\ 0 & \cdots & 0.125 & 0.25 & 0.5 & 1 \end{pmatrix} \tag{3.9}$$

Two observations close in time have a higher correlation than two observations separated further in time.

So what is the purpose of using such a residual correlation matrix? The answer is that the covariance matrix Σ finds its way into the expression for the estimated regression parameters and its standard errors:

$$\hat{\boldsymbol{\beta}} = (\boldsymbol{X}^t \times \boldsymbol{\Sigma}^{-1} \times \boldsymbol{X})^{-1} \times \boldsymbol{X}^t \times \boldsymbol{\Sigma}^{-1} \times \boldsymbol{y} \tag{3.10}$$

and

$$\text{cov}\left(\hat{\boldsymbol{\beta}}\right) = (\boldsymbol{X}^t \times \boldsymbol{\Sigma}^{-1} \times \boldsymbol{X})^{-1} \tag{3.11}$$

In layman's terms: Using an AR1 process for the residuals results in a residual covariance matrix $\boldsymbol{\Sigma}$ that allows for a decaying temporal dependency pattern. This covariance matrix is substituted into the mathematical expressions for the estimated parameters and the covariance matrix of the standard errors. End result: Potentially different estimated parameters and quite often larger, but better, standard errors and p-values.

Instead of the GLS model with a residual AR1 process we can apply GLS models with other residual correlation structures. For example if we have a GLS with a residual autoregressive model of order 2, we can even apply correlation structures from spatial data analysis methods (see Chapter 4), which is especially useful if the time series are irregularly spaced. A correlation structure that we will see later in this book is the so-called exchangeable correlation. In that case we have the following correlation.

$$\text{cor}\left(\boldsymbol{\varepsilon}\right) = \begin{pmatrix} 1 & \phi & \phi & \phi & \cdots & \phi \\ \phi & 1 & \phi & \phi & \ddots & \vdots \\ \phi & \phi & \ddots & \ddots & \ddots & \phi \\ \phi & \phi & \ddots & \ddots & \ddots & \phi \\ \vdots & \ddots & \ddots & \phi & 1 & \phi \\ \phi & \cdots & \phi & \phi & \phi & 1 \end{pmatrix}$$

Whatever the time difference between two different points, the correlation is always ϕ in the model with exchangeable correlation. This may not be useful for data sets that contain long time series but it is an option for relatively short time series.

It should be stressed that a GLS model with residual AR1 process does not necessarily impose a temporal correlation structure on the response variable. The main purpose of the residual AR1 component in the GLS is to provide better estimators for the regression parameters and standard errors. These are in Equations (3.10) and (3.11). The ones in Equations (3.3) and (3.4) are faulty for time series.

3.6.7 Implementation using the `gls` function

The GLS model with the residual autoregressive correlation structure is given by

$$
\begin{aligned}
LD_t &= \beta_1 + MSA_t \times \beta_2 + \varepsilon_t \\
\varepsilon_t &\sim N\left(0, \sigma^2\right) \\
cor\left(\varepsilon_t, \varepsilon_s\right) &= \phi^{|t-s|}
\end{aligned}
\tag{3.12}
$$

To run this model in R we use the following code.

```
> B1 <- gls(LayingAP ~ MSA,
            correlation = corAR1(form =~ Year),
            data = AP, na.action = na.omit)
> summary(B1)
```

```
Correlation Structure: ARMA(1,0)
 Formula: ~Year
 Parameter estimate(s):
     Phi1
0.5236874

Coefficients:
                Value Std.Error   t-value p-value
(Intercept) 244.57591  1.443356 169.44949   0.000
MSA          10.33068 17.311974   0.59674   0.555
```

The new bit is the correlation option in the `gls` function. Because the data have gaps we need to specify the temporal positions of the observations via the `form` argument. It is always safer to use the form argument, otherwise the order of the observations is used. The estimated value for ϕ is 0.523 (the R output mentions ARMA(1, 0), but this is the same as AR1), which means that the dependency structure is very similar to the one in Equation (3.9). What is surprising is the change in the estimated slope; it has gone from negative in the linear regression model to positive in the model with the residual AR1 correlation, albeit the slope of MSA is not significant at the 5% level. The magnitude of the standard error is similar. The lower p-value is mainly because the estimate is closer to zero.

Figure 3.7 shows the fit of the model with and without the autoregressive correlation. We indeed go from a negative relationship to a positive relationship! That is a rather different biological message! We think that this change is due to the residual autocorrelation at both ends of the line in Figure 3.3. The model in Equation (3.12) has captured it with autocorrelation and as a result the fitted line is different.

We can compare both models with the AIC. Results indicate that the model with the autoregressive correlation is slightly better, though the difference in AICs is rather small. Due to the fact that the model without correlation has residual patterns we prefer to use the model with the temporal dependency structure. It is important not to have a model that violates the independence assumption. A model with a relevant dependence structure is more important than having the model with the lowest AIC. It is a valid option to use the AIC only to compare different correlation structures or apply model selection on a set of models with the same correlation structure.

```
> AIC(B0, B1)
   df      AIC
B0  3 139.7118  #Linear regression model
B1  4 138.8594  #Linear regression + AR1 model
```

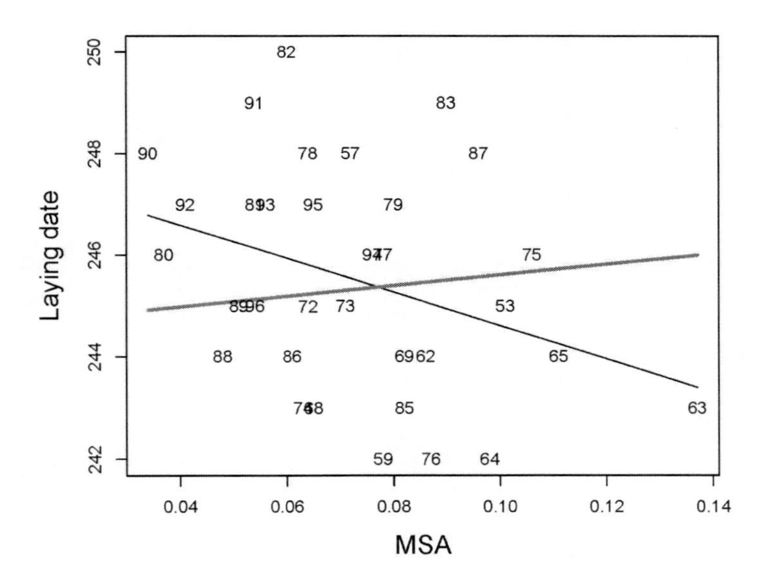

Figure 3.7. Fitted values of the model with (thick line moving upwards) and without (thinner line moving downwards) autoregressive correlation.

3.7 Multiple time series

Let us return to the subnivium average daily temperature time series that were discussed in Section 2.3. If we fit the model in Equation (1.3) or (2.5) then the residuals of each time series show a strong autoregressive correlation; see Figure 2.11. We extend the GAMM by allowing for autoregressive correlation inside a logger:

$$
\begin{aligned}
&T_{it} = Intercept + f_{habitat}\left(Day_t\right) + Habitat_i + a_i + \varepsilon_{it}\\
&a_i \sim N\left(0, \sigma^2_{Logger}\right)\\
&\varepsilon_{it} \sim N\left(\mu_{it}, \sigma^2_{\varepsilon}\right)\\
&cor\left(\varepsilon_{it}, \varepsilon_{is}\right) = \phi^{|t-s|}
\end{aligned}
\qquad (3.13)
$$

The R code for such a model is as follows.

```
> M3 <- gamm(Temp ~ s(TimeNum, by = Type) + Type,
             random = list(Logger=~1),
             correlation = corAR1(form =~ TimeNum|
                                          Logger),
             data = SN2)
```

The R function `gamm` is only capable of estimating one ϕ for all 12 loggers. Hence we have to assume that the strength of the residual autocorrelation is the same for each logger.

If the model contains a random intercept (logger in this case), then the autoregressive correlation will automatically be applied on the data from the same logger. This means that we could also have coded the correlation as

```
correlation = corAR1(form =~ TimeNum)
```

The residual correlation that this model imposes is as follows. Suppose for notational convenience that all loggers have the same number of observations, namely 89. The residual correlation for the data from logger 1 is equal to $\boldsymbol{\Sigma}_1$, where $\boldsymbol{\Sigma}_1$ is given by:

$$
\boldsymbol{\Sigma}_1 = \mathrm{cor}\begin{pmatrix} \varepsilon_{1,1} \\ \varepsilon_{1,2} \\ \vdots \\ \varepsilon_{1,88} \\ \varepsilon_{1,89} \end{pmatrix} = \begin{pmatrix} 1 & \phi & \phi^2 & \phi^3 & \cdots & \phi^{89} \\ \phi & 1 & \phi & \phi^2 & \ddots & \vdots \\ \phi^2 & \phi & \ddots & \ddots & \ddots & \phi^3 \\ \phi^3 & \phi^2 & \ddots & \ddots & \ddots & \phi^2 \\ \vdots & \ddots & \ddots & \phi & 1 & \phi \\ \phi^{98} & \cdots & \phi^3 & \phi^2 & \phi & 1 \end{pmatrix} \qquad (3.14)
$$

This is exactly the same dependency structure as for the Adelie penguin model in Equations (3.6) and (3.7). The dependency structure for the data in logger 2 is identical to the one for logger 1, and the same holds for all other loggers. This means that we have the following residual dependency structure for all the loggers:

$$\text{cor}\begin{pmatrix} \begin{pmatrix} \varepsilon_{1,1} \\ \vdots \\ \varepsilon_{1,89} \end{pmatrix} \\ \begin{pmatrix} \varepsilon_{2,1} \\ \vdots \\ \varepsilon_{2,89} \end{pmatrix} \\ \vdots \\ \begin{pmatrix} \varepsilon_{12,1} \\ \vdots \\ \varepsilon_{12,89} \end{pmatrix} \end{pmatrix} = \begin{pmatrix} \boldsymbol{\Sigma}_1 & \mathbf{0} & \cdots & \mathbf{0} \\ \mathbf{0} & \boldsymbol{\Sigma}_2 & & \mathbf{0} \\ \vdots & & \ddots & \vdots \\ \mathbf{0} & \mathbf{0} & \cdots & \boldsymbol{\Sigma}_{12} \end{pmatrix} = \begin{pmatrix} \boldsymbol{\Sigma}_1 & \mathbf{0} & \cdots & \mathbf{0} \\ \mathbf{0} & \boldsymbol{\Sigma}_1 & & \mathbf{0} \\ \vdots & & \ddots & \vdots \\ \mathbf{0} & \mathbf{0} & \cdots & \boldsymbol{\Sigma}_1 \end{pmatrix}$$

Simply put: The correlation between residuals from the same logger is autoregressive, this correlation is the same for each logger, and the residuals from different loggers are assumed to be independent (zero correlation).

In this case the `gamm` function estimates the value of ϕ as 0.5. The problem is that the normalised residuals of the GAMM in Equation (3.13) still show temporal correlation. Using one value ϕ for all 12 loggers did not work in this case.

Another problem that we frequently encounter with GAMs and GAMMs that contain a smoothing function of time and a residual AR1 process is that these two components compete for the same temporal patterns. Quite often the smoother ends up as a straight line and the value of ϕ is close to one. The solution is to fix one of the components. This problem may also appear in a linear regression model if time is used as a covariate and an AR1 residual process is used to deal with the temporal dependency.

Fox (2016) provides a series of cautionary notes. For example, if the temporal process is not AR1, then including a residual AR1 process may do more harm than good.

3.8 Discussion

In this section we discussed how to allow for temporal dependency in a regression model by using a non-diagonal covariance matrix. The main problem is to limit the number of parameters that we need to estimate in such a covariance matrix. The solution is to choose a mathematical equation that defines the covariance elements. The mathematical equation contains only a few parameters that we need to estimate.

Although the approach sketched above sounds good, in practice it does not always work very well due to software implementations. This is our main reason for adopting Bayesian tools later in this book, especially the use of INLA.

4 Spatial data and GLS

In Chapter 3 we used generalised least squares (GLS) to analyse time series data. It required the specification of the residual covariance matrix using an AR1 process, which eventually resulted in modified expressions for the estimated parameters and standard errors. In this chapter we will do exactly the same, but now for spatial data.

Just as in Chapter 3, our experience with GLS and spatial data are not positive. But this chapter allows us to set the scene for things to come.

Prerequisite for this chapter: You need to be familiar with the material in Chapter 3.

4.1 Variogram models for spatial dependency

In Chapter 3 we discussed how to extend the linear regression model with a temporal dependency structure. Recall that the idea was to allow for a non-zero correlation between the residuals from two different time points using a function $h()$:

$$cor(\varepsilon_s, \varepsilon_t) = h(\phi, \varepsilon_s, \varepsilon_t)$$

The task of the researcher is to choose a function $h(\phi, \varepsilon_s, \varepsilon_t)$ that adequately models the residual correlation. The option we illustrated for times-series models was the residual AR1 correlation, which is given by

$$h(\phi, \varepsilon_s, \varepsilon_t) = \phi^{|t-s|}$$

For any two time points t and s, the function $h()$ defines the residual correlation as a power function. The larger the time difference, the smaller the correlation. The values of $h()$ are then substituted into the residual covariance matrix $\boldsymbol{\Sigma}$, and this matrix is used in the expressions for the estimated parameters and standard errors; see Equations (3.10) and (3.11). The resulting method was called GLS.

For spatial data we can do the same thing. We only need a different $h()$ function. Depending on the shape of the sample variogram of the residuals we choose a mathematical model (also called a variogram model) and use it to calculate the elements of the residual covariance matrix $\boldsymbol{\Sigma}$.

There are various variogram models available in the package `nlme` to model the spatial residual patterns, for example, `corExp`, `corSpher`, `corLin`, `corGaus`, and `corRatio`. Depending on the selected variogram model and the parameters in such a model, the shapes of the imposed spatial dependency may look rather similar. Figure 4.1 shows five different types of variogram models (i.e. the exponential, Gaussian, linear, rational quadratic, and spherical variogram models). Each variogram model has two parameters. See Dale and Fortin (2014) for the mathematical expressions for these models.

The mathematics behind these variogram models is surprisingly simple. For example, the exponential variogram model is defined by (Schabenberger and Pierce 2002):

$$h(s,\phi) = 1 - e^{-\frac{s}{\phi}}$$

where s is the distance between two sampling locations and ϕ is an unknown parameter called the range (note that correlation = 1 – $h()$ in this case). The distance between any two observations can be calculated using the Pythagorean theorem, and therefore $h()$ can be calculated, assuming we know ϕ. The variogram model can easily be extended to allow for a so-called nugget effect at distance near 0 (representing variance due to measurement errors and variation at a finer resolution than the available data).

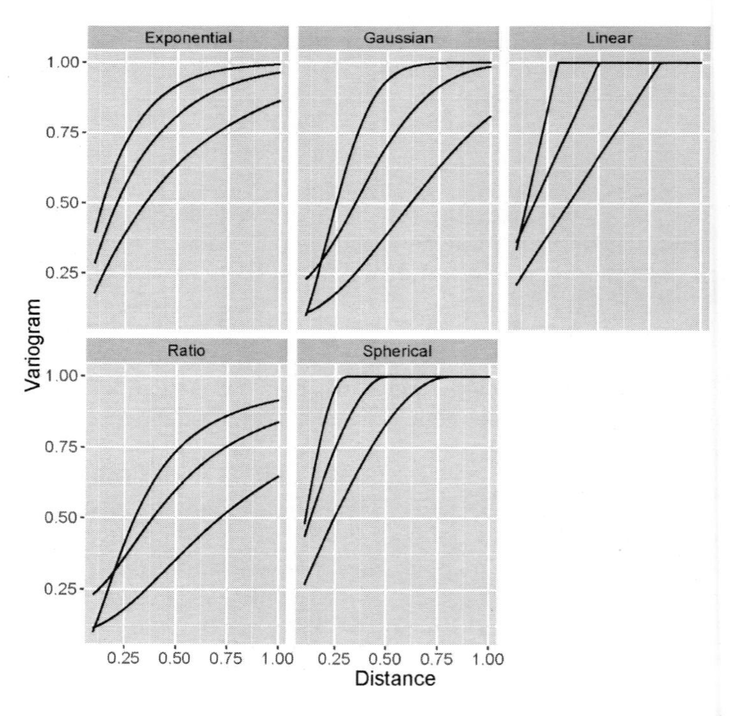

Figure 4.1. Variogram models. For each type of variogram model we simulated three different sets of parameters.

This results in the following five-step approach for the analysis.

1. Fit a linear regression model without a spatial dependency structure.
2. Make a sample variogram using the residuals of the model.
3. Based on the shape of the sample variogram (especially at smaller distances), pick a variogram model (e.g. one of the variogram models presented in Figure 4.1).
4. Fit a regression model that contains the variogram model for the residuals. The software will estimate the parameters of the variogram model and fills in the elements of $\mathbf{\Sigma}$. Estimated parameters and standard errors are obtained via Equations (3.10) and (3.11).
5. Check whether the GLS has solved the problem.

4.2 Application on the Irish pH data

As an example, we use the Irish pH data that was introduced in Chapter 2 to illustrate a sample variogram. Figure 1.8 showed some evidence of spatial dependency (so steps 1 and 2 were carried out in Chapter 2). We extend the multiple linear regression models with a residual spatial dependency structure using the `gls` function. This is step 3, except that we have no idea which variogram model to use. So, we fit all available variogram models in the `gls` function. Assuming that the Irish pH code has been loaded and altitude has been log-transformed, we use:

```
> M3 <- gls(pH ~ LOGAltitude + SDI + fForested +
                 LOGAltitude:fForested,
              method = "REML", data = iph)
> M4a <- update(M3, correlation =
          corExp(form=~ Xkm + Ykm, nugget = TRUE))
> M4b <- update(M3, correlation =
          corLin(form=~ Xkm + Ykm, nugget = TRUE))
> M4c <- update(M3, correlation =
          corGaus(form=~ Xkm + Ykm, nugget = TRUE))
> M4d <- update(M3, correlation =
          corSpher(form=~ Xkm + Ykm, nugget =TRUE))
> M4e <- update(M3, correlation =
          corRatio(form=~ Xkm + Ykm, nugget =TRUE))
```

The first model (`M3`) applies the linear regression model again, but this time we use (restricted) maximum likelihood to estimate the parameters (Zuur et al. 2009a). This allows us to compare the model with the spatial models. All the other models contain a spatial dependency structure. Each of them uses a specific mathematical form for *h*(). We compare models with AICs.

```
> AIC(M3, M4a, M4b, M4c, M4d, M4e)
    df      AIC
M3   6 212.7673
M4a  8 174.7766
M4b  8 216.7673
M4c  8 216.7673
M4d  8 216.7673
M4e  8 172.1448
```

It is suspect that various models have the same AIC. This may be an indication that the estimated range is 0, rendering the spatial correlation useless. The models with spatial correlation have two extra parameters and therefore the penalty in the AIC is four higher.

When doing this type of analysis it may be wise to add a good starting value for the range, which can be eyeballed from the sample variogram. If you do not provide a starting value then the `gls` function will use the default value, which is 90% of the smallest distance in the data set. For some data sets the optimisation process gets stuck near such a small value.

We inspected the estimated range of each model, and the range is indeed close to 0. The solution is to rerun all the models and specify better starting values for the range. To specify a starting value for the range and nugget (in this order), add `value = c(50, 0.1)` to the `corExp`, `corLin`, and other functions and rerun the code.

It is also important to specify the coordinates in sensible units; use kilometres for larger distances and metres for data sets in which the sampling locations are close to each other.

The AICs obtained with the starting values `value = c(50, 0.1)` are as follows:

```
    df      AIC
M3   6 212.7673
M4a  8 174.7766
M4b  8 167.9705
M4c  8 170.1817
M4d  8 172.0474
M4e  8 172.1448
```

Model `M4b` has the lowest AIC, which is the multiple linear regression model with the `corLin` dependency. The numerical output of this model is as follows:

```
Correlation Structure: Linear spatial correlation
 Formula: ~Xkm + Ykm
 Parameter estimate(s):
     range     nugget
67.7172373  0.1437146
```

```
Coefficients:
                         Value   SE     t-val  p-val
(Intercept)              10.039 0.185  54.152     0
LOGAltitude              -0.786 0.080  -9.774     0
SDI                      -0.021 0.001 -11.662     0
fForestedNo              -1.447 0.272  -5.303     0
LOGAltitude:fForestedNo  0.722 0.136   5.284     0
```

The estimated parameters and standard errors of the model with a spatial dependency structure may differ from those for a model without a dependency structure. In general the standard errors of the model with the spatial dependency are larger due to pseudoreplication.

Once the optimal model has been selected, a model validation has to be applied. We need to take the (normalised) residuals and check whether the variogram model has taken adequate account of the residual spatial dependency.

Our experience with adding spatial correlation structures to regression type models is that quite often the estimated variogram parameters do not make sense, and most of the time the residuals of the GLS model with the spatial dependency still has pseudoreplication.

4.3 Matérn correlation function

Before moving away from GLS models, we explain one more model to define the elements of $\boldsymbol{\Sigma}$. The reason for this is that this model will be used extensively in this book.

The mathematical function underlying the AR1 process was reasonably easy on the eye, and the same can be said for the exponential variogram model. The other variogram models available in the `gls` function are not terribly difficult to understand either; see Dale and Fortin (2014). We are now going to introduce a mathematical function for defining the elements of $\boldsymbol{\Sigma}$ that is not so easy on the eye. However, its underlying principle is identical to the AR1 or exponential variogram models.

The Matérn correlation function, which is named after a Swedish statistician, is defined as:

$$cor_{Matern}\left(s_i, s_j\right) = \frac{2^{1-\nu}}{\Gamma(\nu)} \times \left(\kappa \times \left\|s_i - s_j\right\|\right)^{\nu} \times K_{\nu}(\kappa \times \left\|s_i - s_j\right\|)$$

The terms s_i and s_j define the spatial positions for observations i and j. K_ν is a so-called Bessel function of the second kind. Don't ask what a Bessel function is unless you have a degree in mathematics (it has to do with the solution of a second-order differential equation). At this stage we are not interested in its mathematical form. The term || || stands for Euclidean distance and can be calculated with the `dist` function or simply with the Pythagorean theorem. The term κ (called kappa) is an unknown parameter, very much like the AR1 correlation parameter ϕ for the time-series correlation or the range parameter in the exponential variogram model. The Γ is the gamma function. Let us try to simplify the expression a little bit by using $\nu = 1$, in which case $\Gamma(1) = 1$. Let $Dist_{ij}$ be the distance between the two locations s_i and s_j.

$$cor_{Matern}\left(s_i, s_j\right) = \kappa \times Dist_{ij} \times K_1\left(\kappa \times Dist_{ij}\right)$$

R has existing functions to calculate the Matérn correlation function for specific κ values. Figure 4.2 shows various Matérn correlation functions. Although its mathematical expression is intimidating it follows the same principle as the residual AR1 process or the variogram models; points close in space are more correlated than points further separated.

Just as for the AR1 and the variogram models, the elements of a residual covariance matrix are given by the Matérn correlation function. The Matérn correlation function is not standard implemented in the `gls` function. The Matérn correlation function is used by the software (INLA) that we will use in later chapters.

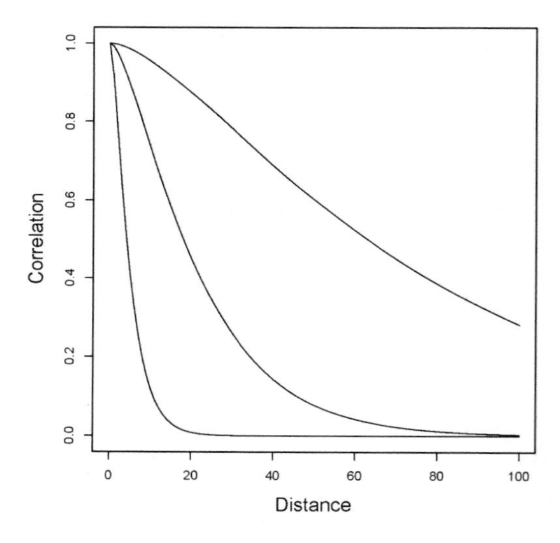

Figure 4.2. Three Matérn correlation functions for different κ values and $\nu = 1$. The online code can easily be adjusted to investigate the effect of using other values for ν.

The Matérn correlation function is a mathematical tool to define correlation as a function of distance. Its general principle is the closer in space, the larger the correlation. The value of the Matérn correlation function goes into a residual covariance matrix. As a result we avoid pseudoreplication.

5 Linear mixed-effects models and dependency

In this chapter we show how a linear mixed effects model deals with dependency. A mixed-effects model can be used when you have multiple observations from the same site, or from the same animal, or from the same person. These are repeated measurements. Other names are clustered data, hierarchical data, or panel data.

Useful references for mixed models are Pinheiro and Bates (2000), Bolker (2008), or Zuur et al. (2009a; 2013), among many others.

Prerequisite for this chapter: You need to be familiar with R and multiple linear regression.

5.1 White Storks

We use a small part of a data set provided by Boudjéma Samraoui, who studied factors affecting growth parameters of White Stork (*Ciconia ciconia*) nestlings in eastern Algeria; see also Bouriach et al. (2015). A large number of nests in a White Stork breeding colony were sampled. In each nest various chicks were sampled multiple times during the growth period from the first day of life to the maximum age of 54 days. In this section we use beak length measurements sampled in 2012. Figure 5.1 shows a scatterplot of beak length versus age.

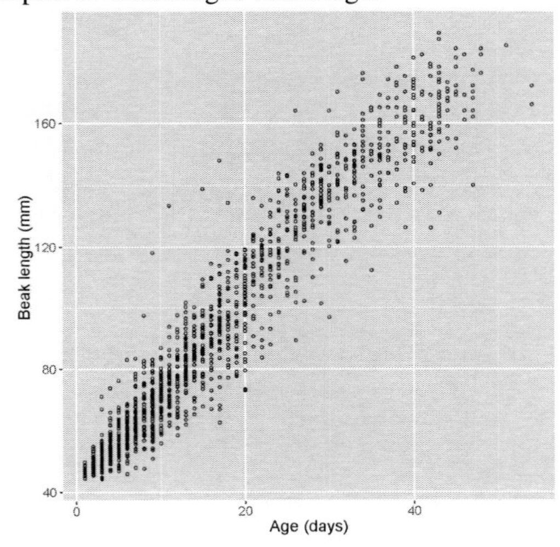

Figure 5.1. Scatterplot of beak length (mm) of White Stork chicks versus age (in days).

5.2 Considering the data (wrongly) as one-way nested

To model the growth pattern of beak length we naively specify a multiple linear regression model of the form

$$
\begin{aligned}
&BL_i = Intercept + Age_i + Nest_i + Chick_i + \varepsilon_i \\
&\varepsilon_i \sim N\left(0, \sigma^2\right)
\end{aligned}
\tag{5.1}
$$

BL_i is the beak length for observation i, Age_i is the age in days, $Chick_i$ is the identity of the chick, and $Nest_i$ is the nest from which the sample is taken. We have 1438 observations. Age is fitted as a continuous covariate, and Nest (73 levels) and Chick (261 levels) are fitted as categorical covariates.

We only use this model to build up a pedagogical explanation of linear mixed-effects models. For a full analysis we should also consider adding sex of the chick (male, female, and unknown) plus an interaction between age and sex. If we were to do that in this section, the numerical and graphical output would be much larger, too much for a simple explanation of a statistical technique.

Before starting the analysis, we should think carefully what we are about to do. The model has two major problems. Firstly, due to the categorical covariates Nest and Chick we have a large number of regression parameters in the model. These two variables alone consume 334 regression parameters! And image how many more parameters we would have if we also include an interaction between the covariates Age and Chick! Secondly, the model wrongly assumes that we have independent observations; the same chick is repeatedly sampled, and chicks from the same nests are likely to have similar beak lengths. We have pseudoreplication! These two problems are reasons enough not to fit the multiple linear regression model.

One solution is to take the average per nest, but this will reduce the sample size considerably. And it is not always possible to take averages of a covariate (e.g. a categorical covariate that differs for the observations within a nest). And what is the biological relevance of modelling the average beak length as a function of the average age per nest? A mixed-effects model solves both problems.

Model formulation

We will start simple and (wrongly) ignore the chick effect for the moment (this is for pedagogical reasons; we will add the chick effect later in this chapter). A linear mixed-effects model, in which nest is used as a random intercept, is given by

$$\begin{aligned} BL_{ij} &= Intercept + Age_{ij} + a_i + \varepsilon_{ij} \\ a_i &\sim N\left(0, \sigma^2_{Nest}\right) \\ \varepsilon_{ij} &\sim N\left(0, \sigma^2\right) \end{aligned} \tag{5.2}$$

We changed the index notation; BL_{ij} is now the jth observation from nest i, where $i = 1, \ldots, 73$, and $j = 1, .., n_i$, where n_i is the number of observations in nest i.

Changing the indices is a trivial change. The crucial change is the addition of the new term a_i, which is called a random intercept. It represents the nest effect, and in this example we have 73 of them. We assume that the a_is are normally distributed with mean 0 and variance σ^2_{Nest}. The term ε_{ij} is residual noise, and just as in the linear regression model we assume that it is normal, homogenous, and independently distributed. We can write these assumptions as $\varepsilon_{ij} \sim N(0, \sigma^2)$ or in matrix notation as $\boldsymbol{\varepsilon} \sim N(\mathbf{0}, \boldsymbol{\Sigma})$, where $\boldsymbol{\Sigma} = \sigma^2 \times \boldsymbol{I}_{1438\times1438}$. The a_is are also residuals, and this explains why some people do not consider them as parameters (which affects the degrees of freedom).

Although we assume that the ε_{ij}s and a_is are independent, they sort of dance the tango together. A model without the a_is would result in residuals ε_{ij}s with a clear nest effect.

An alternative mathematical formulation for the same model is as follows.

$$\begin{aligned} BL_{ij} &\sim N\left(\mu_{ij}, \sigma^2\right) \\ E\left(BL_{ij}\right) &= \mu_{ij} \quad \text{and} \quad \text{var}\left(BL_{ij}\right) = \sigma^2 \\ \mu_{ij} &= Intercept + Age_{ij} + a_i \\ a_i &\sim N\left(0, \sigma^2_{Nest}\right) \end{aligned} \tag{5.3}$$

The mean and variance are in fact the conditional mean and the conditional variance (Hanke et al. 2015). The random intercept does what its name indicates: it applies a random correction to the intercept. And it does this for each nest.

In the ecological literature the model in Equation (5.2) is also called a linear mixed-effects model for one-way nested models. In econometric applications it is sometimes referred to as a two-way nested model.

So what is the main purpose of a mixed-effects model? Chicks from the same nest are genetically linked (they are brothers and sisters), they get food from the same parents, and they also share the same environmental conditions (they grow up in the same nest). These are all aspects that affect the growth (and therefore the beak length) of the chicks from the same nest. So, we have dependency between multiple observations from the same nest. It is not too difficult to show that the linear mixed-effects model with a random intercept for nests imposes the following correlation on any two beak length observations BL_{ij} and BL_{ik} from the same nest i:

$$correlation(BL_{ij}, BL_{ik}) = \phi = \frac{\sigma^2_{Nest}}{\sigma^2_{Nest} + \sigma^2} \tag{5.4}$$

The model implies that beak length observations from different nests have a correlation of 0. The correlation in Equation (5.4) is also called the intraclass correlation (*ICC*). In normal words this means that beak length values of chicks from the same nests are allowed to be more similar than observations from different nests.

Equation (5.5) shows the correlation between any two beak length observations for nest i = 1 in matrix notation. This nest has six observations.

$$\boldsymbol{\Sigma}_1 = \text{cor}\begin{pmatrix} BL_{1,1} \\ BL_{1,2} \\ BL_{1,3} \\ BL_{1,4} \\ BL_{1,5} \\ BL_{1,6} \end{pmatrix} = \begin{pmatrix} 1 & \phi & \phi & \phi & \phi & \phi \\ \phi & 1 & \phi & \phi & \phi & \phi \\ \phi & \phi & 1 & \phi & \phi & \phi \\ \phi & \phi & \phi & 1 & \phi & \phi \\ \phi & \phi & \phi & \phi & 1 & \phi \\ \phi & \phi & \phi & \phi & \phi & 1 \end{pmatrix} \tag{5.5}$$

The software for linear mixed models will provide the estimated values for σ_{Nest} and σ so that we can easily calculate the intraclass correlation ϕ.

The linear mixed model with a random intercept assumes that the dependency in nest 2 is the same as in nest 1. Because nest 2 has 22 observations, we need to adjust the dimensions of the correlation matrix:

$$\boldsymbol{\Sigma}_2 = \text{cor}\begin{pmatrix} BL_{2,1} \\ BL_{2,2} \\ \vdots \\ BL_{2,22} \end{pmatrix} = \begin{pmatrix} 1 & \phi & \cdots & \phi \\ \phi & 1 & & \vdots \\ \vdots & & \ddots & \phi \\ \phi & \cdots & \phi & 1 \end{pmatrix}$$

This matrix has the same ϕ value as for nest 1, but instead of using 6 rows and 6 columns, the correlation matrix now has the dimensions 22 by 22. This shows that analysing a data set in which the nests have different numbers of observations is not an immediate problem for a mixed-effects model.

The correlation matrices for nests 3 through 73 have the same value of ϕ. The only thing that changes between these correlation matrices is the dimensions of each matrix. As a consequence we have the following dependency structure for the beak length data of all nests.

$$\text{cor}\begin{pmatrix} \begin{pmatrix} BL_{1,1} \\ \vdots \\ BL_{1,n_1} \end{pmatrix} \\ \begin{pmatrix} BL_{2,1} \\ \vdots \\ BL_{2,n_2} \end{pmatrix} \\ \vdots \\ \begin{pmatrix} BL_{73,1} \\ \vdots \\ BL_{73,n_{73}} \end{pmatrix} \end{pmatrix} = \begin{pmatrix} \boldsymbol{\Sigma}_1 & \mathbf{0} & \cdots & \mathbf{0} \\ \mathbf{0} & \boldsymbol{\Sigma}_2 & & \mathbf{0} \\ \vdots & & \ddots & \vdots \\ \mathbf{0} & \mathbf{0} & \cdots & \boldsymbol{\Sigma}_{73} \end{pmatrix}$$

Simply put: The correlation between beak length observations from the same nest is equal to ϕ. The value of this correlation is the same in each nest, and the correlation between beak length observations from different nests is assumed to be 0.

The mathematical expression in Equation (5.4) only holds for a linear mixed-effects model with a random intercept. A different expression is used for GLMMs, mixed models with multiple random intercepts, or models with random intercepts and slopes.

Recall from Chapters 3 and 4 that the GLS model corrects the standard errors. The linear mixed-effects model does the same thing (although the math is different); it produces 'better' standard errors as compared to the linear regression model that ignores dependency. Note that 'better' does not mean 'smaller'. It means better in a statistical sense. The mixed modelling standard errors tend to be larger than their linear regression equivalents.

In order to obtain reliable estimates for the variance of the random effects it is recommended to have a minimum number of 5 clusters (though 10 is probably better as a lower limit). For the stork data we have 73 nests, which is plenty!

A mixed-effects model imposes a dependency structure between all observations from the same cluster. This is how it deals with pseudoreplication.

5.3 Fitting the one-way nested model using `lmer`

The model in Equation (5.3) can be fitted in R using the `lmer` function from the `lme4` package (Bates et al. 2015) with the following R code. First we import the data and load the required packages and support file.

```
> WS <- read.csv(file = "Storks.csv", header=TRUE)
```

```
> library(lme4)
> library(ggplot2)
> library(plyr)
> source("HighstatLibV10.R")
```

We only use the 2012 data. To avoid problems with NAs we omit a few observations and we also drop the levels with no observations from the categorical variables (this is needed because we analyse a subset of the data).

```
> WS1 <- subset(WS, Year == 2012)
> WS2 <- na.exclude(WS1[, c("Beak", "Age", "Nest",
                            "Chick")])
> WS2$fNest  <- factor(WS2$Nest)
> WS2$fChick <- factor(WS2$Chick)
```

We are now ready to run the linear mixed-effects model and present its numerical output.

```
> MM1 <- lmer(Beak ~ Age + (1 | fNest),
              data = WS2)
> summary(MM1)
```

```
Random effects:
 Groups   Name        Variance Std.Dev.
 fNest    (Intercept) 57.23    7.565
 Residual             62.50    7.905
Number of obs: 1438, groups:  fNest, 73

Fixed effects:
            Estimate Std. Error t value
(Intercept)  44.4417     0.9651   46.05
Age           2.9925     0.0169  177.08
```

Note that the `summary` output does not contain p-values for the regression parameters. There is an ongoing debate within this field whether the t-value follows a t-distribution. It is relatively easy to get p-values; see Zuur et al. (2013) for the R code. Whether you can trust these p-values is a different matter.

The `summary` output indicates that we have the following estimated model.

$$\begin{aligned} &BL_{ij} \sim N\left(\mu_{ij}, 7.90^2\right) \\ &\mu_{ij} = 44.44 + 2.99 \times Age_{ij} + a_i \\ &a_i \sim N\left(0, 7.56^2\right) \end{aligned} \quad (5.6)$$

The fixed part of the model is given by the term μ_{ij} = 44.44 + 2.99 × Age_{ij}. This expression holds for all (comparable) nests. We also say: 'The fixed part represents the relationship between the expected values of beak length and age for a typical nest'. We have visualised the fixed part of the model in Figure 5.2. Note that there is a strong and positive relationship between age and beak length.

The random effects part in the output shows that we have σ_{Nest} = 7.56 and σ = 7.90 (from a statistical point of view we should have put a hat on σ_{Nest} and σ to indicate that these are estimators). These can be used to calculate the intraclass correlation.

$$\phi = \frac{7.56^2}{7.56^2 + 7.90^2} = 0.47$$

The value of ϕ = 0.47 means that the correlation between any two observations from the same nest is 0.47. It is difficult to state what constitutes a small enough value to omit the random effects all together. However, 0.47 is relatively high! The current consensus is always to include the random intercepts based on the design of the experiment, even if it turns out that they are not important.

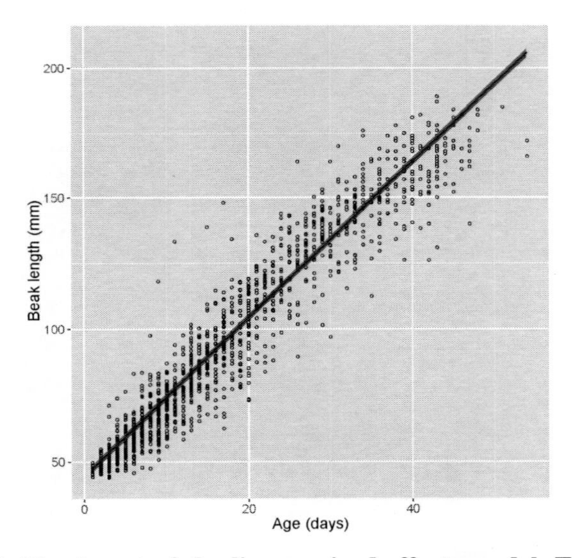

Figure 5.2. Fixed part of the linear mixed-effects model. The shaded area is a 95% confidence interval for the mean; it is very difficult to see as the 95% CI is rather small.

5.4 Model validation

Just as in multiple linear regression, we need to extract the residuals and plot them against the fitted values, each covariate in the model, each covariate not in the model, assess them for temporal and spatial correlation (if relevant), and see whether there are any outliers in the residuals. However, the model in Equation (5.4) is incorrect as we are ignoring the chick effect (causing pseudoreplication) and we also omitted the interaction between age and sex. Hence there is no point in presenting the results of a model validation here. Later chapters contain plenty of model validation examples.

5.5 Sketching the fitted values

Despite the fact that we are missing the chick effect and therefore use a wrong model (with pseudoreplication), for pedagogical reasons we present the visual presentation of the model as it allows us to explain the principles of a linear mixed-effects model. Figure 5.2 shows the fixed part of the model. The line represents the fitted values for a typical nest. We suggest that if you present the results of a mixed-effects model, you include the visual presentation of the fixed part of the model (Figure 5.2 in this case).

Figure 5.3 shows the fixed part plus the random part. This graph contains 73 lines. Each line is obtained by shifting the line in Figure 5.2 up or down with the value of a random intercept. The further apart these lines are from one another (in vertical direction), the more similar are the observations within a nest, and the larger the intraclass correlation.

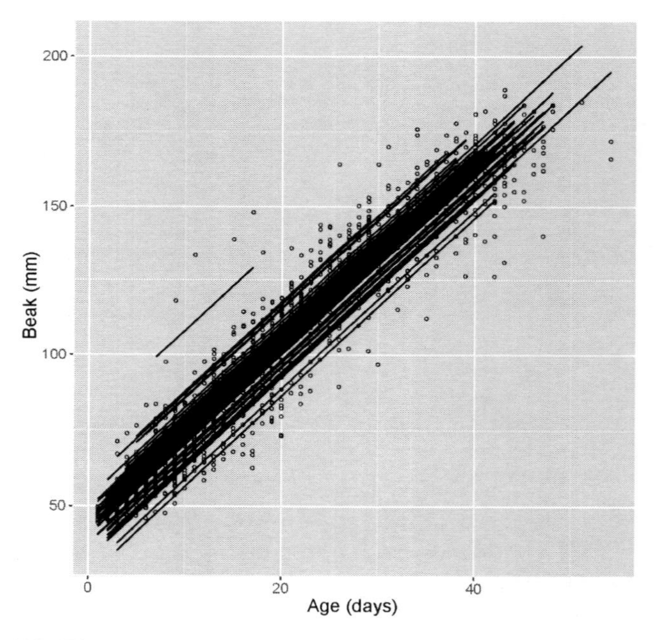

Figure 5.3. Fixed part plus the random effects for the linear mixed-effects model.

Figure 5.4 shows yet another explanation of what the random effects do. The thick line in the middle represents the fitted values of the fixed part of the model. It is the same line as in Figure 5.2. The black points are the observations from nest 6. The estimated random intercept for this nest is a_6 = 13.39. This value was added to the thick line, giving the thinner line. This line represents the fitted values for the beak length data in nest 6. And what about the residuals? These are obtained by subtracting the observed data and the thin line.

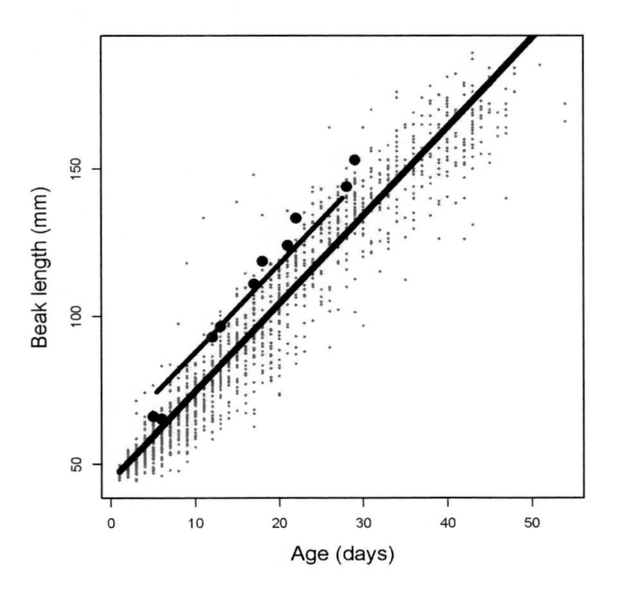

Figure 5.4. Fixed part of the model (thick line in the middle) and the fixed part + the random intercept for nest 6. The large black dots are the observed data for nest 6.

5.6 Considering the data (correctly) as two-way nested

The analysis of the stork data presented so far is wrong. We have ignored the chick effect, and therefore we still have pseudoreplication. We have multiple observations from the same nest, but we also have multiple observations from the same chick within a nest. It is likely that observations from the same chick are more similar than observations from different chicks within the same nest. To take the chick effect into account we formulate a two-way nested linear mixed-effects model.

$$
\begin{aligned}
& BL_{ijk} \sim N\left(\mu_{ijk}, \sigma^2\right) \\
& E\left(BL_{ijk}\right) = \mu_{ijk} \quad \text{and} \quad \text{var}\left(BL_{ij}\right) = \sigma^2 \\
& \mu_{ijk} = Intercept + Age_{ijk} + a_i + b_{ij} \\
& a_i \sim N\left(0, \sigma^2_{Nest}\right) \\
& b_{ij} \sim N\left(0, \sigma^2_{Chick}\right)
\end{aligned}
\qquad (5.7)
$$

We have again modified the subscripts. BL_{ijk} is now the kth observation for chick j in nest i. We have also added a new random intercept b_{ij}; it is the random intercept for chick. The linear mixed-effects model in Equation (5.7) imposes a correlation between two beak length observations from the same chick, and also between two observations from the same nest. As before, we assume that beak length observations from different nests are independent (although in later chapters we will relax this assumption by allowing for spatial dependency).

The correlation between any two observations from the same chick (obviously within the same nest) is given by

$$\phi_{Chick} = \frac{\sigma^2_{Nest} + \sigma^2_{Chick}}{\sigma^2_{Nest} + \sigma^2_{Chick} + \sigma^2} \tag{5.8}$$

And the correlation between any two beak length observations from the same nest (but from different chicks) is given by

$$\phi_{Nest} = \frac{\sigma^2_{Nest}}{\sigma^2_{Nest} + \sigma^2_{Chick} + \sigma^2} \tag{5.9}$$

The R code to fit the two-way nested mixed model and present the numerical output is as follows.

```
> MM2 <- lmer(Beak ~ Age + (1 | fNest / fChick),
              data = WS2)
> summary(MM2)

Random effects:
 Groups        Name        Variance Std.Dev.
 fChick:fNest (Intercept) 29.70    5.449
 fNest        (Intercept) 47.81    6.915
 Residual                 40.81    6.388
Number of obs: 1438, groups:  fChick:fNest, 262;
fNest, 73

Fixed effects:
            Estimate Std. Error t value
(Intercept) 44.41110    0.93656   47.42
Age          2.97532    0.01406  211.63
```

The estimated variances can be used to calculate the two intraclass correlations: $\phi_{Chick} = 0.65$ and $\phi_{Nest} = 0.40$. The fact that ϕ_{Chick} is larger than ϕ_{Nest} does not mean that it is more important. In fact, due to the definition of these two intraclass correlations ϕ_{Chick} will always be larger than ϕ_{Nest}. Let us try to understand what these values mean. Figure 5.5 shows three sets of fitted values. The thick line in the center of the cloud with observations represents the fitted values for the fixed part of the model. We then added the random intercept for nest 6, just as we did in Figure 5.4. This nest has multiple observations from two chicks. These are the black circles and the black squares in Figure 5.5. There are two more lines. These are the fitted values given by $\mu_{ijk} + a_i + b_{ij}$. We just added the values of $b_{6,1}$ and $b_{6,2}$. This process can be repeated for each nest and each chick. The results (not shown here) show that the lines for the chicks are all quite close to the lines for the nests. So the nest effect does a lot, and the chick effect within a nest is much smaller. This is reflected in the intraclass correlations; the smaller the difference between ϕ_{Chick} and ϕ_{Nest}, the less important the chick effect. It is also reflected by the difference in the variances of the random effects

The estimated parameters and standard errors obtained by the one-way and two-way nested mixed-effects models are slightly different. That raises the question of which model is better. We can use an AIC as a quick and dirty tool to compare them (though there are some technical issues with such a comparison; see Zuur et al. 2009a for 'testing on the boundary'). However, by design of the experiment we suggest going for the two-way nested model and refraining from comparing the one-way and two-way nested models.

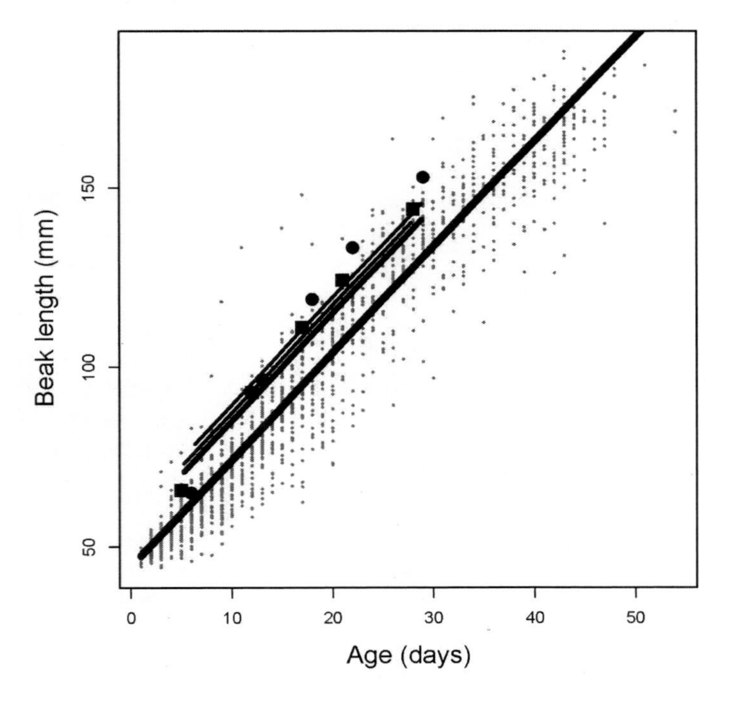

Figure 5.5. Fitted values due to the fixed part, fixed part + the random intercept Nest, and fixed part + the random intercept Nest + the random intercept Chick.

5.7 Applications to spatial and temporal data

Zuur et al. (2014) analysed seabird counts that were sampled from a boat that was moving along transects. Measurements were taken on a large number of surveys. To avoid applying complicated spatial-temporal models, a GLMM with two-way nested random effects was used. To be more precise, random effect 'survey' was used to model dependency between all observations made during the same survey, and the random effect 'transect' within a survey captured the small-scale spatial dependency. Such a model takes care of small-scale spatial dependency via the random intercept transect, and short-term temporal dependency via the random intercept survey.

Zuur et al. (2016A) presented the results of a negative binomial GLMM applied on a butterfly data set. Multiple observations of butterfly eggs on oak trees from the same area were available. The random effect area captured the small-scale spatial dependency.

Zuur et al. (2016A) also used a data set in which reproductive measures were sampled from plants in a greenhouse experiment. Three sets of random effects were used to model dependency due to the genetic origins, observations from the same bed, and multiple beds per garden.

Zuur et al. (2013) analysed kiwi pollen sampled in front on beehives. The measurements consisted of 16 short daily time series (4 days per time series). Random effects were used to deal with the short-term temporal dependency.

Zuur et al. (2013) analysed a beaver data set. Multiple observations per loch on morphometric variables of lilies (food for beavers) were analysed. The random effect loch took care of the small-scale spatial dependency within a loch.

5.8 Difference with the AR1 process approach

In Chapter 3 we used a GLS with a residual AR1 process to deal with temporal dependency. This resulted in modified expressions for the estimated parameters and standard errors. As a result we have dealt with pseudoreplication. But the GLS model with residual AR1 process does not necessarily impose a temporal dependency on the response variable.

The mixed model solves the pseudoreplication problem in a different way, namely by imposing a dependency structure on the response variable. This will also result in better standard errors (and potentially also different estimated parameters).

6 Modelling space explicitly

In Chapters 3 and 4 we dealt with temporal and spatial pseudoreplication by using generalised least squares (GLS). In the GLS a residual covariance matrix $\boldsymbol{\Sigma}$ is used that modifies the expressions for the parameter estimates and standard errors. In Chapter 5 we used a different approach, namely mixed-effects models. In the mixed modelling approach an extra residual term is added to the model, the random intercepts. As explained in Chapter 5, this term adds a dependency structure on the response variable, solving the pseudoreplication problem.

In this chapter we follow the mixed modelling concept, but instead of using random intercepts that are normal and independent distributed, we will use a random intercept that is normal and spatially *dependent*. In the last section of this chapter we discuss spatial-temporal dependency.

The frequentist software tools for the models that we discuss in this chapter are rather limited. For this reason we only present the underlying statistical mechanism behind the models. Once we have explained Bayesian analysis in Chapter 7 we will present a large number of spatial, temporal, and spatial-temporal case studies.

Prerequisite for this chapter: You need to be familiar with Chapters 1 through 5.

6.1 Model formulation

We will specify a series of models that increase in complexity using the White Stork data (see Chapter 5). Let us start with a linear regression model. In Section 5.2 we specified a multiple linear regression model for the beak length data. Let us for the moment ignore the nest and chick effects and only use age as the covariate.

$$\begin{aligned} BL_i &= Intercept + Age_i \times \beta + \varepsilon_i \\ \varepsilon_i &\sim N\left(0,\sigma^2\right) \end{aligned} \tag{6.1}$$

where BL_i is the beak length of specimen i, and Age_i is its age. We have 1438 specimens, so the index i runs from 1 through 1438. We can also write this model as:

$$\begin{aligned} BL_i &= z_i \times \beta + \varepsilon_i \\ \varepsilon_i &\sim N\left(0,\sigma^2\right) \end{aligned} \tag{6.2}$$

where z_i = (1, Age_i) contains the covariates. The first value of z_i is for the intercept. If there are more covariates (e.g. sex of the specimen) just add them as extra columns to z_i. In both models we assume that the residuals are independent.

The model in Equation (6.2) is a simple linear regression model. For the residuals $\boldsymbol{\varepsilon} = (\varepsilon_1, \ldots, \varepsilon_{1438})$ we can write $\boldsymbol{\varepsilon} \sim N(\mathbf{0}, \boldsymbol{\Sigma})$, where the residual covariance matrix $\boldsymbol{\Sigma}$ is equal to $\sigma^2 \times \boldsymbol{I}$, and $\boldsymbol{I}$ is an identity matrix with 1438 rows and 1438 columns. $\boldsymbol{\Sigma}$ has the dimensions 1438 by 1438 as well. There is no dependency in the model because $\boldsymbol{\Sigma}$ is diagonal.

To deal with spatial pseudoreplication, we add a new component to the model that represents a latent (which means unknown) spatial term. Diggle and Ribeiro (2007) use the notation $S(x_i)$ for this term, and Blangiardo and Cameletti (2015) use ξ_i. This term is pronounced 'ksi'. In this book we will use u_i as it looks friendlier.

$$\begin{aligned} BL_i &= z_i \times \beta + u_i + \varepsilon_i \\ \varepsilon_i &\sim N\left(0, \sigma^2\right) \end{aligned} \tag{6.3}$$

So, what is the purpose of the u_i term? How does this model deal with pseudoreplication? The answer is that we are going to assume that the u_is are random intercepts that are normal distributed with mean 0 and covariance matrix $\boldsymbol{\Omega}$ (pronounced 'omega'). The crucial point is that $\boldsymbol{\Omega}$ is *not* a diagonal matrix. The off-diagonal elements of $\boldsymbol{\Omega}$ will be used to model spatial dependency in the u_i terms. Because the u_is are random effects they impose a dependency structure on the response variable.

Let us recap. The model in Equation (6.3) can be formulated in words as:

$$\begin{aligned} BL_i &= \text{Fixed part} + \text{Spatial term} + \text{real noise} \\ &= \text{Fixed part} + \text{correlated random effects } u_i + \text{residuals } \varepsilon_i \end{aligned} \tag{6.4}$$

The residuals ε_i are our usual linear regression residuals (i.e. homogenous, normal, and independent), and the random effects u_i are spatially correlated. The mechanism that drives the spatial correlation is the non-diagonal covariance matrix $\boldsymbol{\Omega}$.

Table 6.1 summarises the crucial points.

In regression type models we can deal with spatial pseudoreplication by adding a random intercept that is spatially correlated

Table 6.1. Difference between the residual terms in Equation (6.4).

Residual	Type of noise	Covariance	Remark
ε_i	Real noise	$\boldsymbol{\Sigma} = \sigma^2 \times \boldsymbol{I}$	σ is estimated
u_i	Latent spatial term	$\boldsymbol{\Omega}$	Matérn correlation function is used

Before discussing the spatial regression model in more detail we present the full model specification.

$$\begin{aligned} BL_i &= \boldsymbol{z}_i \times \beta + u_i + \varepsilon_i \\ \boldsymbol{\varepsilon} &\sim N\left(0, \sigma^2 \times \boldsymbol{I}\right) \\ \boldsymbol{u} &\sim N\left(0, \boldsymbol{\Omega}\right) \end{aligned} \tag{6.5}$$

The $\boldsymbol{\varepsilon} = (\varepsilon_1, \ldots, \varepsilon_{1438})$ contains the independently distributed residuals and $\boldsymbol{u} = (u_1, \ldots, u_{1438})$ are the spatially correlated random effects.

For this specific data set, the spatial positions are given in terms of latitude and longitude. In general it is better not to use longitude and latitude to quantify distance with the Pythagorean theorem, but the study area for the White Storks is rather small. Using latitude and longitude instead of UTM coordinates will only marginally distort the relative spatial positions for this data set.[1]

6.2 Covariance matrix of the spatial random effect

The covariance matrix $\boldsymbol{\Omega}$ of the u_is has the dimensions 1438 by 1438. If we were to estimate all its elements then this means that we need to estimate a whopping 1438 × (1438 – 1) / 2 = 1033203 parameters! That is way too much. To drastically reduce the number of parameters in this covariance matrix we use a mathematical model that specifies the elements of $\boldsymbol{\Omega}$. Just as in Chapters 3 and 4 with the autoregressive correlation function and the variogram models, such a mathematical model has parameters that we need to estimate. Once we have *estimated* these parameters we can *calculate* the elements of $\boldsymbol{\Omega}$. Note the use of the words 'estimated' and 'calculate' in the previous sentence!

We will use the Matérn correlation function (which itself has a few parameters that we need to estimate) to calculate the elements of $\boldsymbol{\Omega}$. That means that instead of 1438 × (1438 – 1) / 2 parameters we only have to estimate two Matérn correlation parameters.

---

[1] In general, if a sampling area spans two UTM zones, it is better to use a different projection. If the area spans more than 3 UTM zones, then it is better to use great circle distances (also called orthodromic distances) based on latitude and longitude.

Simulation study

We will use a small simulation study with artificial data to explain how the Matérn correlation function determines the values of the covariate matrix $\boldsymbol{\Omega}$. Figure 6.1 shows the spatial positions of five sampling locations for the simulation study. These coordinates were obtained as follows and Figure 6.1 was obtained with a simple `plot` command.

```
> set.seed(123)
> Xloc <- runif(5, 0, 1)
> Yloc <- runif(5, 0, 1)
> Loc  <- cbind(Xloc, Yloc)
```

Suppose that at each location we sample the beak length of a bird. The model in Equation (6.5) dictates that besides a covariate, we also need five spatial random intercepts; these are the u_1, ..., u_5. We impose a normal distribution with mean 0 and covariance matrix $\boldsymbol{\Omega}$ on the us, where $\boldsymbol{\Omega}$ is given by

$$\boldsymbol{\Omega} = \sigma_u^2 \times \begin{pmatrix} 1 & \omega_{12} & \omega_{13} & \omega_{14} & \omega_{15} \\ & 1 & \omega_{23} & \omega_{24} & \omega_{25} \\ & & 1 & \omega_{34} & \omega_{35} \\ & & & 1 & \omega_{45} \\ & & & & 1 \end{pmatrix}$$

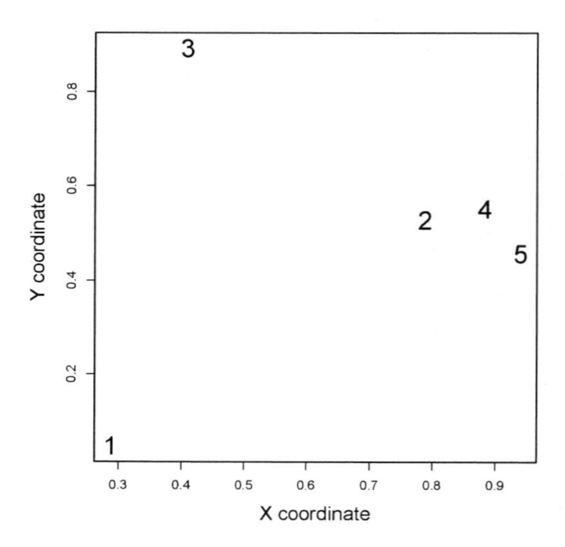

Figure 6.1. Position of five sampling locations in our simulation study. The graph was generated with simple plotting code.

For this simulation example the covariance matrix $\boldsymbol{\Omega}$ consists of $5 \times 4 / 2 + 1 = 11$ unknown parameters. As explained above that is too much for the sample size of these simulated data and we will use the Matérn correlation function to define the elements of $\boldsymbol{\Omega}$:

$$\omega_{ij} = \text{cov}\left(u_i, u_j\right) = \sigma_u^2 \times \text{Matérn correlation sites } i \text{ and } j \qquad (6.6)$$

Recall from Chapter 4 that the Matérn correlation is a function of distance between the two sites and two unknown parameters. In plain language, the expression in Equation (6.6) reads: The covariance between the two random intercepts u_i and u_j is a function of the distance between two sampling locations and two unknown parameters. Once we have these unknown parameters we can calculate every element in $\boldsymbol{\Omega}$ using Equation (6.6). We show this with some R code. First, we calculate the distance between each site in Figure 6.1. We use the `dist` function for this.

```
> Dist <- dist(Loc)
> Dist <- as.matrix(Dist)
> Dist

     1    2    3    4    5
1 0.00 0.69 0.86 0.78 0.77
2 0.70 0.00 0.53 0.09 0.17
3 0.86 0.52 0.00 0.58 0.69
4 0.78 0.09 0.58 0.00 0.11
5 0.77 0.16 0.69 0.11 0.00
```

These are the distances between each of the five sampling locations in Figure 6.1. Next we need to choose some Matérn correlation–specific parameter values.

```
> kappa   <- 5 #Choosen arbitrary
> Sigma_u <- 1 #Choosen arbitrary
```

The `kappa` was discussed in Chapter 4 (it determines the distance at which dependency diminishes) and to convert a correlation matrix into a covariance matrix we need the σ_u. The Matérn correlation values can be calculated as follows (see also Chapter 4).

```
> d.vec <- as.vector(Dist)
> Cor.M <- (kappa * d.vec)*besselK(kappa*d.vec, 1)
> Cor.M[1] <- 1
```

We can plot the Matérn correlation values versus distance; see Figure 6.2. Sites close to one another have a high correlation and sites further separated from one another have a lower correlation. This figure is for any distance, but we need the covariance for specific combinations. To do this we define a matrix with 5 rows and 5 columns, and start a double loop (this is very inefficient coding but simple to understand and explain). For each i and j combination we take the distance between sites i and j, and calculate the matching covariance following Equation (6.6).

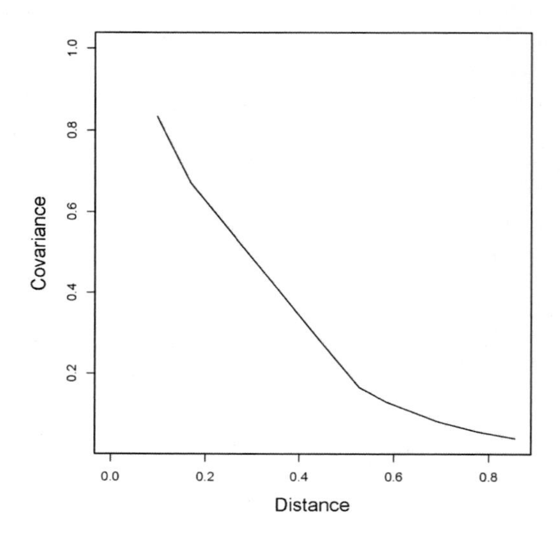

Figure 6.2. Matérn correlation function versus distance.

```
> OMEGA    <- matrix(nrow = 5, ncol = 5)
> for (i in 1:5){
   for (j in 1:5){
    Distance <- Dist[i,j]
    OMEGA[i,j] <- Sigma_u^2 * (kappa * Distance) *
                        besselK(kappa * Distance,1)
  }}
> diag(OMEGA) <- Sigma_u^2
```

The resulting values are as follows.

```
> OMEGA
```

	1	2	3	4	5
1	1.00	0.07	0.03	0.05	0.05
2	0.07	1.00	0.16	**0.83**	**0.67**
3	0.03	0.16	1.00	0.12	0.08
4	0.05	**0.83**	0.12	1.00	**0.80**
5	0.05	**0.67**	0.08	0.80	1.00

A high value in this matrix $\boldsymbol{\Omega}$ means a high covariance between the two corresponding sites. And this means strong spatial dependency. Sites 2 and 4, 2 and 5, and 4 and 5 (all in bold) are physically close to one another. Equation (6.5) has ensured that the corresponding u_is are correlated.

Let us recap what we have done. We have discussed the following points:

1. Instead of choosing the values of $\boldsymbol{\Omega}$ ourselves, we use the Matérn correlation function to define the elements in $\boldsymbol{\Omega}$.
2. The general principle is as follows: The closer in space, the larger the values of the covariance matrix $\boldsymbol{\Omega}$, and the more dependent the u_i values.

In Chapter 9 and onwards we will use the software package INLA to apply the model in Equation (6.5). Figure 6.3 shows which parameters are associated with which term in the model.

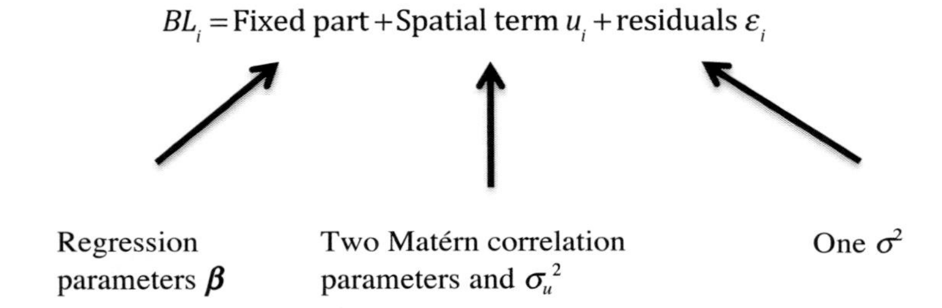

Figure 6.3. Parameters per component. For the spatial term two Matérn correlation parameters need to be estimated. Once they are available $\boldsymbol{\Omega}$ can be calculated.

6.3 Spatial-temporal correlation*

The material covered in this section is only required if you are going to work with spatial-temporal models later in this book. We suggest that you skip this section on first reading.

So far we have only discussed time-series examples and spatial examples. Many data sets consist of measurements that are sampled at multiple spatial positions and multiple moments in time. For example, the beak length of the White Storks was sampled at a large number of nests for at least 4 years. The owl data that we used in Chapter 2 were sampled at 27 nests (which are all within a 10 kilometre radius) during a period of 3 months. The subnivium snow data that we used in Chapters 2 and 3 consisted of data sampled daily at 12 locations.

We will discuss the spatial-temporal extension of the spatial model that we presented for the White Stork data in the previous section. Let BL_{it} be the sampled beak length at location i at time t. The covariates for the White Stork data will also change in space and in time. In this case we only use age of the specimens, which we denote by Age_{it}. Just as in the previous section, we put all covariates in a vector z_{it} = (1, Age_{it}). This is a vector with the dimensions 1 by 2. If there are more covariates then just add them to $\mathbf{z}_{it}$.

Using our new time subscript t, the Equation (6.5) model becomes

$$BL_{it} = \mathbf{z}_{it} \times \beta + w_{it} + \varepsilon_{it} \tag{6.7}$$

To add a temporal dependency element to the model we will do something special with the w_{it}s; we will assume that they change in space *and* in time with the following construction.

$$w_{it} = \phi \times w_{i,t-1} + u_{it} \tag{6.8}$$

The expression in Equation (6.8) is rather intimidating, but it looks vaguely like the autoregressive model that we used for the time-series models in Chapter 3. We will explain the general idea behind Equation (6.8) with a simulation study.

Simulation study (continued)

We continue the simulation study with the same five sampling locations that we used in the previous section. The autoregressive parameter ϕ in Equation (6.8) is between –1 and 1, and it determines the temporal strength of the dependency. If $\phi = 0$ then there is no autoregressive correlation (of order 1). It doesn't mean that there is no correlation; perhaps another mathematical construction should be used.

We start with the u_{it}s. Just as in the previous section we let the u_{it}s be spatially correlated by using a non-diagonal covariance matrix $\boldsymbol{\Omega}$. To emphasise that they play the same role we used the same notation.

$$\begin{pmatrix} u_{1t} \\ \vdots \\ u_{5t} \end{pmatrix} \sim N(\mathbf{0}, \boldsymbol{\Omega})$$

Just as before, we will *calculate* the values of Ω using the Matérn correlation function.

$$\boldsymbol{\Omega} = \sigma_u^2 \times \text{Matérn correlation sites } i \text{ and } j$$

There is no temporal dependency for the u_{it} terms.

The u_{it}s and Ω are playing exactly the same roles as in the previous section; they model the spatial dependency.

Then what about the w_{it} with the time index t? Suppose that ϕ is large, say 0.9. That means that the random intercept w_{it} at time t is similar to $w_{i,t-1}$ at time $t - 1$. Imagine that you are walking on a footpath. The position of your left foot (at time t) depends very much on where your right foot was just a second ago (at time $t - 1$). Will it be straight in front of you or slightly to the left? Formulated differently, we have something like this: The left foot position ($= w_{it}$) is equal to ϕ times the previous right foot position ($= w_{i,t-1}$) plus a random spatially correlated wind effect ($= u_{it}$).

Let us demonstrate this with some R code; perhaps it clarifies things. Let us use a process with a strong temporal dependency: $\phi = 0.9$.

```
> phi <- 0.9
```

We want to calculate the w_{it} for times t = 1, …, 100. This means that at each time point t we end up with five values for w, one value at each site in Figure 6.1. We have a little problem at t = 1. Actually, we always have a problem with autoregressive models at time 1 because we don't have measurements at time t = 0. A little bit of mathematical theory that we don't explain here shows that we can get $w_{i,1}$s as follows.

$$\begin{pmatrix} w_{1,1} \\ \vdots \\ w_{5,1} \end{pmatrix} \sim N\left(\mathbf{0}, \frac{\sigma_u^2}{1-\phi^2} \times \boldsymbol{\Omega}\right)$$

In R code we can simulate the five w_{i1}s as follows.

```
> Sigma_u   <- 1                  #Chosen arbitrary
> Cov.w1    <-(Sigma_u^2 / (1 - phi^2)) * OMEGA
> Zeros     <- rep(0, 5)          #= five zeros
> w.1       <- mvrnorm(1, mu = Zeros, Sigma=Cov.w1)
```

We now have five values $w_{1,1}$, …, $w_{5,1}$ at time t = 1. We can run the expression for w_{it} in Equation (6.8) for t = 2 to 100. The R code is as follows.

```
> w <- matrix(nrow = 5, ncol = 100)
> w[,1] <- w.1
> for (t in 2:100){
    u <- mvrnorm(1, mu = Zeros, Sigma = OMEGA)
    w[,t] <- phi * w[,t-1] + u
  }
```

The `w[,t]` term in the code is the w_{it} in Equation (6.8). It is obtained by multiplying ϕ with the previous `w` values and adding some spatially correlated noise. The object `w` in R has five rows (for the sites) and 100 columns (for time). Each row is a time series. Figure 6.4 shows a plot of the w_{1t}, .., w_{5t} time series. The whole point of this process is to generate a latent variable w_{it} that is slowly changing over time and that is spatially correlated. We can clearly see that the sites that are close in space (sites 2, 4, and 5) behave similarly (due to spatial correlation). There is also autocorrelation in these time series (which is obvious from the trends).

In this example we used ϕ = 0.9, which represents strong temporal dependency. If we use ϕ = 0, then as mentioned above, w_{it} is not temporally correlated. In that case w_{it} is only spatially correlated, and the w_{1t}, .., w_{5t} at time point t are a different realisation from the spatial process.

In this section we simulated spatially correlated random intercepts u_{it} and plugged these into an equation to get spatial-temporal correlated random intercepts w_{it} values. In one sense, the role of the w_{it} components is very similar to that of the random intercepts in a mixed-effects model; add dependency to the response variables and ensure that the residuals ε_{it} are independent.

The w_{it} term is a latent variable that changes in space and time. As such it captures any spatial and temporal patterns that are not already modelled by the covariates. It is like a random intercept that ensures that the residuals ε_{it} are independent, and it imposes a dependency structure on the response variable.

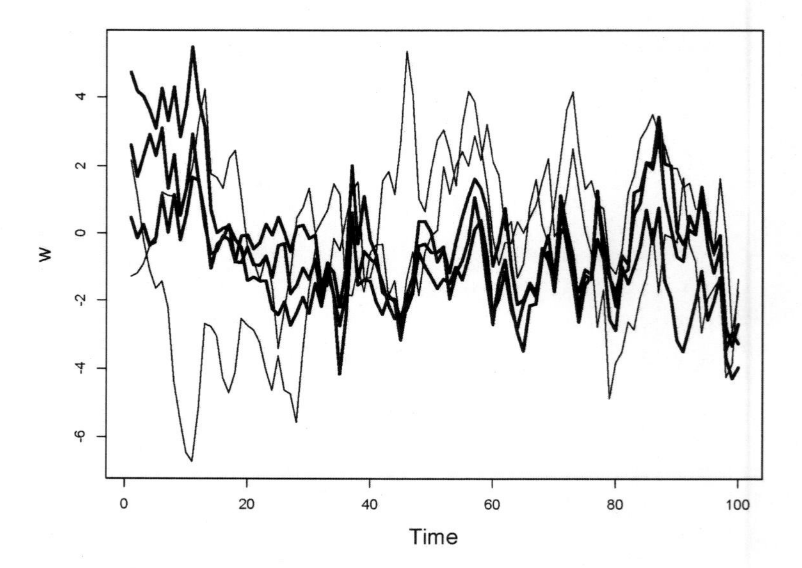

Figure 6.4. Time-series plot for each element of w. The five time series correspond to the five locations. The three lines in bold are for sites 2, 4, and 5 (which are close to one another).

The reality is that we don't simulate u_{it} and w_{it} values, and we don't choose the values of the parameters in the Matérn correlation function. Instead we need to estimate all the parameters that we conveniently selected in the simulation process. Diggle and Riberio (2007) give R code to fit the model in Equation (6.3), but it is not as user friendly as the `lm`, `glm`, or `lme` functions. Estimating these parameters is a rather ugly process, especially if we use GLMs and GLMMs. Luckily there is software available (INLA), but it is Bayesian in nature. We explain the underlying principles of Bayesian analysis in Chapter 7.

7 Introduction to Bayesian statistics

This chapter introduces Bayesian statistics, Markov chain Monte Carlo (MCMC) techniques, and integrated nested Laplace approximations (INLA). We will keep the explanations simple and conceptual. However, some sections do contain some mathematics. We marked them with an asterisk '*'. If you are not interested in the underlying mathematics, then you can skip these sections, or just read the summaries in the 'owl notes'.

Prerequisite for this chapter: A working knowledge of R and linear regression is required.

7.1 Why go Bayesian?

The use of Bayesian techniques can be motivated in different ways. Some scientists start by arguing that they have prior knowledge, and that this prior knowledge should be incorporated into the models. In Chapter 3 we analysed osprey data; eggshell thickness was modelled as a function of DDD. You don't need to be a scientist in order to know that a breakdown product of a pesticide is bad for ospreys. Why would you not use this knowledge (which essentially translates as a negative slope for DDD) in the models? Using prior knowledge during the analysis then immediately leads to other scientists criticising Bayesian-based approaches because (according to them) using prior knowledge means that the models are not objective anymore.

Another angle of motivating Bayesian approaches is to first criticise frequentist approaches and then show how useful the output from Bayesian techniques is. The criticism is about the interpretation of frequentist confidence intervals and *p*-values. Their interpretation goes via a statement like 'if we were to repeat the experiment a large number of times, then in 5% of the cases we would expect to find an even larger *t*-value'. This is a statement based on fictive data. In reality we are not repeating an experiment. What we would like to say from our results is that there is a 95% probability that a regression parameter is between a and b. This is wishful thinking for a frequentist scientist, but it is reality for a Bayesian analyst.

Figure 7.1 contains the so-called posterior distribution of a regression parameter. Pay special attention to the word 'distribution' in the previous sentence. In a Bayesian analysis we assume that the parameters are unknown stochastic quantities, and we estimate their distribution using the data (resulting in a picture like Figure 7.1). In a Bayesian analysis we end up with the probability of a parameter given the data (written as $P(\beta \mid \text{data})$), whereas in a frequentist analysis we look at the probability of the data given the parameters (written as $P(\text{data} \mid \beta)$). In other words, in a frequentist analysis we assume there is only one β and the analysis gives us an estimate plus a 95% confidence interval. The $P(\text{data} \mid \beta)$ tells us how likely (or unlikely) the data are, given the betas. In a Bayesian analysis there is no fixed value for β.

Both arguments for using Bayesian statistics will probably not convince the scientist with a frequentist background, who has limited time available, doesn't like programming, and sees colleagues publishing papers with *p*-values smaller than 0.05. And Bayesian analysis doesn't even give you *p*-values!

We decided to base our motivation to convince the reader to adopt Bayesian techniques on another argument: You have no choice. The reason that you are reading this book is most likely because you have data with a spatial and / or temporal dependency structure. The packages in R that can cope with such data are rather limited. To take full advantage of spatial and temporal models we need tools that allow us to fit such models. At the time of this writing the majority of these tools require Bayesian statistics.

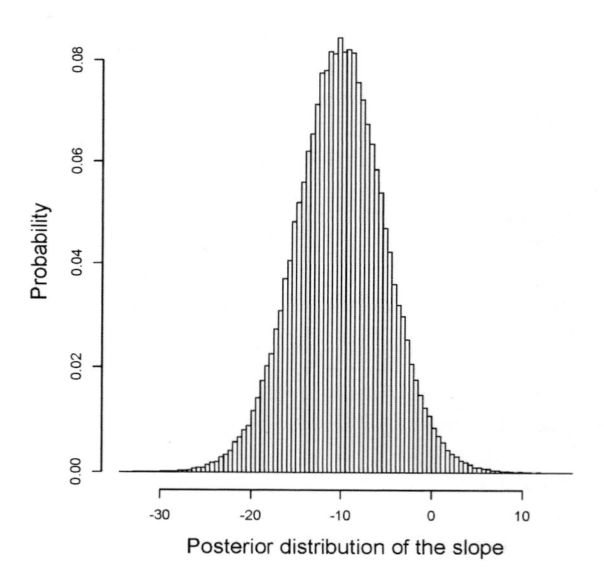

Figure 7.1. Posterior distribution of the regression parameter β.

7.2 General probability rules

We begin by reviewing some basic probability rules. Let $P(A)$ be the probability of an event A and $P(B)$ the probability of an event B. Define the joint probability $P(A \text{ and } B)$ as the probability that events A and B both occur. The following basic probability rule holds.

$$P(A \text{ and } B) = P(B \text{ and } A) \tag{7.1}$$

If A and B are independent then we can write

$$P(A \text{ and } B) = P(A) \times P(B) \tag{7.2}$$

If A and B are not independent of each other, then we use the conditional probability $P(A \mid B)$, which reads as the probability of A *given* B. The symbol | reads as 'given that'. If A and B are not independent then we have

$$P(A \text{ and } B) = P(A \mid B) \times P(B) \tag{7.3}$$

We now present Bayes' theorem. Many textbooks on Bayesian statistics state this theorem without explaining how it is derived. This is rather simple, so we show it here. The following two equations hold (see above).

$$P(A \text{ and } B) = P(A \mid B) \times P(B)$$
$$P(B \text{ and } A) = P(B \mid A) \times P(A)$$

Because $P(A \text{ and } B) = P(B \text{ and } A)$ we can rewrite these two equations as

$$P(A \mid B) = \frac{P(B \mid A) \times P(A)}{P(B)} \tag{7.4}$$

This expression is also called Bayes' theorem.

7.3 The mean of a distribution*

You probably know that the expected value of a variable Y that is Poisson distributed with parameter μ is equal to μ. In mathematical notation we write this as

$$Y \sim Poisson(\mu)$$
$$E(Y) = \mu$$

But why is the mean of Y equal to μ? To answer this question we present the general statistical expression for calculating the mean of a variable Y that takes discrete values. This expression can be found in any statistical textbook that explains basic statistics.

$$E(Y) = \sum_y y \times p(y) \tag{7.5}$$

The expected value of Y is equal to the sum of each possible value of Y (these are the y_is) times the corresponding value of the probability function $p(y)$. The expression in Equation (7.5) can be used to calculate the mean of a variable that follows a Poisson, negative binomial, Bernoulli, binomial, or zero-inflated Poisson distribution as these are all distributions for discrete data. For example, if we assume that Y is Poisson distributed, then we can write Equation (7.5) as follows.

$$E(Y) = \sum_{y=0}^{\infty} y \times \frac{e^{-\mu} \times \mu^{y}}{y!} \tag{7.6}$$

We substituted the Poisson probability function. So the question 'why is the mean of a Poisson distributed variable equal to μ' simplifies to: Why is the right side of Equation (7.6) equal to μ? It is actually very simple to answer this; see Equation (7.7). First we write out the definition of the expected value, then we get rid of the $y = 0$ component, use some basic mathematics rules,[1] and rearrange things a little bit by using $z = y - 1$. The last step comes from the fact the sum of all probabilities for a probability function adds up to 1.

$$\begin{aligned} E(Y) &= \sum_{y=0}^{\infty} y \times \frac{e^{-\mu} \times \mu^{y}}{y!} \\ &= 0 \times \frac{e^{-\mu} \times \mu^{0}}{0!} + \sum_{y=1}^{\infty} y \times \frac{e^{-\mu} \times \mu^{y}}{y!} \\ &= 0 + \sum_{y=1}^{\infty} y \times \frac{e^{-\mu} \times \mu^{y}}{y!} \\ &= \sum_{y=1}^{\infty} y \times \frac{e^{-\mu} \times \mu^{1} \times \mu^{y-1}}{y!} \\ &= \mu \times \sum_{y=1}^{\infty} y \times \frac{e^{-\mu} \times \mu^{y-1}}{y \times (y-1)!} \\ &= \mu \times \sum_{y=1}^{\infty} \frac{e^{-\mu} \times \mu^{y-1}}{(y-1)!} \\ &= \mu \times \sum_{z=0}^{\infty} \frac{e^{-\mu} \times \mu^{z}}{z!} = \mu \end{aligned} \tag{7.7}$$

This is an example in which high school mathematics is used to calculate the mean of a discrete distribution. In the case that the variable is continuous, we are not working with a sum but with an integral. This leads to the following expression for the mean:

$$E(Y) = \int_{-\infty}^{\infty} y \times p(y) dy \tag{7.8}$$

The $p(y)$ is now called a density function. Equation (7.8) is the continuous equivalent of Equation (7.5), and can also be found in any

[1] Recall from high school mathematics that 5! is called '5 factorial' and equals $5 \times 4 \times 3 \times 2 \times 1$. We also use $x^3 = x^2 \times x$ and $x! = x \times (x - 1)!$

basic statistics book. It can be used to calculate the mean of a variable that follows for example a normal, gamma, beta, or lognormal distribution. If Y is normal distributed we can substitute the normal density function with parameters μ and σ^2 for $p(y)$ in Equation (7.8) and it is relatively simple to show that the mean of Y is equal to μ. There are similar expressions for the variance.

In the next section we will explain why we presented these details from high school mathematics.

To calculate the mean of a discrete variable we use Equation (7.5). To calculate the mean of a continuous variable we use Equation (7.8).

7.4 Bayes' theorem again

In Section 7.2 we presented Bayes' theorem. Instead of abstract variables like A and B we will fill in something more relevant.

$$P(\beta \mid \text{data}) = \frac{P(\text{data} \mid \beta) \times P(\beta)}{P(\text{data})} \tag{7.9}$$

The term $P(\beta \mid \text{data})$ is called the posterior distribution of β, given the data. It is the part of primary interest. It is the part that goes into the paper. The $P(\text{data} \mid \beta)$ is the likelihood of the data given the parameters. The $P(\beta)$ is called the prior of β; it is information on the β that is known a priori. Finally, $P(\text{data})$ is the probability of the data, and it is a scaling factor so that the distribution on the left adds up to 1. The term $P(\text{data})$ is difficult to calculate and quite often it is dropped from the equation, resulting in

$$P(\beta \mid \text{data}) \propto P(\text{data} \mid \beta) \times P(\beta) \tag{7.10}$$

The $\propto$ symbol stands for 'proportional to'. This expression states that if we choose the likelihood of the data, say a normal distribution, and also choose a distribution for the prior of the regression parameter, then we can take a pen and paper and work out the posterior distribution. The good news is that for some likelihood functions and some prior distributions we can indeed do this by hand. We will show an example of this in the next section.

We now answer the question why we showed Equations (7.6) and (especially) (7.8). We presented the expression for the posterior distribution in Equation (7.10). Having a posterior distribution is nice, but the crucial thing we want is the expected value of this distribution. And because β is continuous we are going to see things like

$$E\left(\beta \mid data\right) = \int_{-\infty}^{\infty} \beta \times p(\beta \mid data) d\beta \tag{7.11}$$

in later sections. This expression follows from Equation (7.10).

Using Bayes' theorem we can specify the posterior distribution of a parameter, say β. However, we also want to know the expected value of β. That means that we need to solve Equation (7.11), which follows immediately from Equation (7.10).

In the next three sections we will present three approaches to estimate the posterior distribution and its mean. In approach 1 (Section 7.5) we will use special priors, also called conjugate priors, and derive an expression by hand. In Section 7.6 we will use Markov chain Monte Carlo simulation. Finally, in Section 7.8 we will use numerical approximation (integrated nested Laplace approximation).

7.5 Conjugate priors

Suppose that your colleagues carried out an osprey field study and they used the data to model eggshell thickness as a function of DDD. Just like we did in Chapter 3, they fitted the following model.

$$\begin{aligned} &Thickness_i = \beta_1 + DDD_i \times \beta_2 + \varepsilon_i \\ &\varepsilon_i \sim N(0, \sigma^2) \end{aligned} \tag{7.12}$$

They found a negative relationship; let us suppose their estimated slope was –9 with a standard error of 1. This is the β_2 in Equation (7.12). Why would you ignore their results in your analysis? With Bayesian analysis you can include these results.

We are going to use the expression in Equation (7.10), but we will keep the presentation as simple as possible and only focus on the β_2 part of the story, and we will pretend that we know β_1 and σ. On the left-hand side in Equation (7.10) is the posterior distribution $P(\beta_2 \mid \text{data})$. That is the one that we need for the paper. To get this distribution we need to specify the other two components, namely $P(\text{data} \mid \beta_2)$ and $P(\beta_2)$.

Likelihood function

The $P(\text{data} \mid \beta_2)$ is the likelihood of the data given the parameters (and keep in mind that we only focus on β_2 to keep the discussion simple). Eggshell thickness is continuous so a normal distribution is a potential candidate. A gamma distribution may be better as eggshell thickness cannot be negative, but again, let's keep things simple and stick to the normal distribution. For count data the Poisson or negative binomial distributions are good candidates and for presence / absence data the Bernoulli distribution should be used.

Using the normal distribution, the likelihood of the data, given the parameters, is given by

$$P(data|\beta_2)=\prod_{i=1}^{n}\frac{1}{\sigma\sqrt{2\pi}}\times\exp\left(-\frac{(Thickness_i-\beta_1-DDD_i\times\beta_2)^2}{2\sigma^2}\right) \quad (7.13)$$

The roman pillar $\prod$ stands for 'product' and the rest of the equation is based on the normal density function.[2] In frequentist statistics we obtain parameter estimates by taking the logarithm of the likelihood function (so that the product becomes a sum), obtaining derivatives with respect to the parameters, setting these to 0, and solving them. This is called maximum likelihood estimation, and is it is implemented in the `glm` function in R.

Priors

Now we focus on the $P(\beta_2)$ in Equation (7.10). This is the prior distribution. We could use the results from your colleagues, or in the absence of other field studies we can use common sense to come up with a sensible range of likely values. We can use a normal distribution to quantify such a range of likely values for β_2.

$$\beta_2\sim N\left(\beta_2^0,\sigma_0^2\right) \quad (7.14)$$

We used the superscript 0 to indicate that β_2^0 and σ^2_0 are for the prior distribution. Instead of a 0, we could have used 'p' for prior. Because we are using a normal distribution for the prior, its density function has the following form.

$$P(\beta_2)=\frac{1}{\sigma_0\sqrt{2\pi}}\times\exp\left(-\frac{(\beta_2-\beta_2^0)^2}{2\sigma_0^2}\right) \quad (7.15)$$

This is the same normal density function as in Equation (7.13) or in the footnote, but this time it is for the prior distribution.

Posterior distribution

To get the posterior distribution we need to multiply the two normal density functions in Equations (7.13) and (7.15). If you take a pen and paper and work this out using the mathematical rule $e^a\times e^b = e^{a+b}$, then 20 pages and half a day later you end up with the following expression for the posterior distribution.

$$P(\beta_2|data)\propto \text{Ugly stuff}\times e^{-\frac{1}{\text{Ugly stuff}}\times\left(\beta_2-\text{Ugly stuff}\right)^2} \quad (7.16)$$

---

[2] The normal density function is given by $f(y)=\frac{1}{\sigma\sqrt{2\pi}}\times e^{-\frac{(y-\mu)^2}{2\sigma^2}}$, where μ is the mean and σ^2 the variance. The product in Equation (7.13) comes from the independence assumption: $P(A \text{ and } B) = P(A)\times P(B)$.

Each 'ugly stuff' component is about half a page with $Thickness_i$, DDD_i, cross products, sums of squares, and σ^2 and σ^2_0 terms. The key aspect of Equation (7.16) is that the posterior distribution belongs to the same family of distributions as the prior distribution; both are normal density functions. This is called conjugacy in Bayesian analysis. We also say that the prior and posterior are conjugated distributions, or that the prior is a conjugated prior to the likelihood function. There are only a few specific combinations for which this is happening; see Ntzoufras (2009) for examples.

The expression in Equation (7.16) allows us to write an expression for the mean and the variance of the posterior distribution $P(\beta_2 \mid \text{Data})$. This means that there is no need to try and solve the integral in Equation (7.11); we can avoid that! The mean and variance are given by

$$\begin{aligned} E\left(\beta_2 \mid data\right) &= \hat{\beta}_2 \times w + (1-w) \times \beta_2^0 + \text{Ugly stuff} \\ \text{var}\left(\beta_2 \mid data\right) &= \sigma^2 \times \frac{w}{\text{Ugly stuff}} \\ w &= \frac{\sigma_0^2 \times \text{Something}}{\sigma_0^2 \times \text{Something} + \sigma^2} \end{aligned} \tag{7.17}$$

The 'Ugly stuff' component on the first line is there because we pretended that the intercept is known. In a more general situation where all regression parameters are unknown the 'Ugly stuff' on the first line is not there. The $\hat{\beta}_2$ is the maximum likelihood estimator, and it is also the frequentist solution. So, if you just use the `glm` function with the Gaussian distribution, then that is the one you get.

Let us go back one step and consider two extreme prior distributions for β_2. For the one extreme we don't know anything; for the other extreme we have very precise knowledge.

Diffuse prior

Perhaps we do not have any prior knowledge at all. We can quantify this as

$$\beta_2 \sim N(0, 100^2) \qquad \Longleftrightarrow \qquad \beta_2^0 = 0 \text{ and } \sigma_0 = 100$$

Essentially we are stating that β_2 is somewhere around 0, but it can be anywhere between –200 and +200, and it can even be smaller or larger than that. This is called a *non-informative prior* or *diffuse* prior. Choosing the variance of the posterior distribution is a matter of picking a value in such a way that the normal distribution covers the likely values with high probability. There is no major difference between a standard deviation of 100 or 101.

So, what happens with w in Equation (7.17) when we use a diffuse prior? A diffuse prior is quantified via a large σ_0, say 100 or 1,000. In that case the w in Equation (7.17) is close to 1, and this gives:

$$E(\beta_2 | data) \approx \widehat{\beta}_2$$

The Bayesian results are nearly identical to the frequentist results when we use a diffuse prior. The same holds for the variance. We have visualised the (minimal) effect of a diffuse prior distribution on the posterior distribution in Figure 7.2. The right panel is the posterior distribution and it is obtained by multiplying the likelihood of the data with the diffuse $N(0, 100^2)$ prior. The shape of the posterior is nearly identical to the likelihood function.

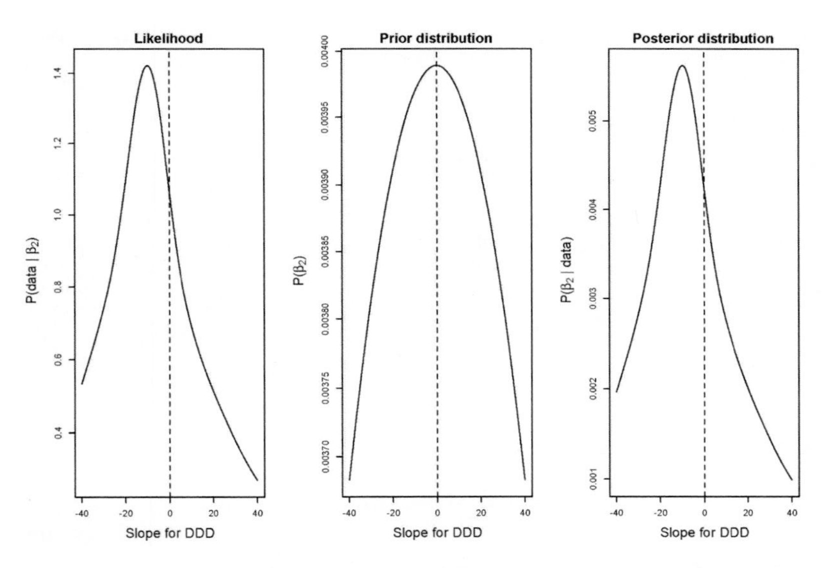

Figure 7.2. Left: Likelihood of the data as a function of β_2. Middle: Diffuse normal prior. Right: Posterior distribution which is obtained by multiplying the left and middle curves.

Informative prior

On the other hand, perhaps we do have a good idea about possible values for β_2. In that case we can use an informative prior. Again, we can use the normal distribution for this; we just need to modify the range of possible values.

$$\beta_2 \sim N(-9, 1^2)\ ) \qquad \Longleftrightarrow \qquad \beta_2^0 = -9 \text{ and } \sigma_0 = 1$$

What happens with the w in Equation (7.17) now? The analysis in Chapter 3 showed that the value of σ is around 5. This means that the more informative the prior is, the smaller the σ_0, the closer w is to 0, and the closer the expected value of the posterior distribution is to the prior specified value β_2^{0}, which is –9 in this example. The formula is:

$$E\left(\beta_2 \mid data\right) \approx \beta_2^0$$

We have visualised the effect of an informative $N(-9, 1^2)$ prior in Figure 7.3. In this case the posterior distribution has nearly the same shape as the prior distribution.

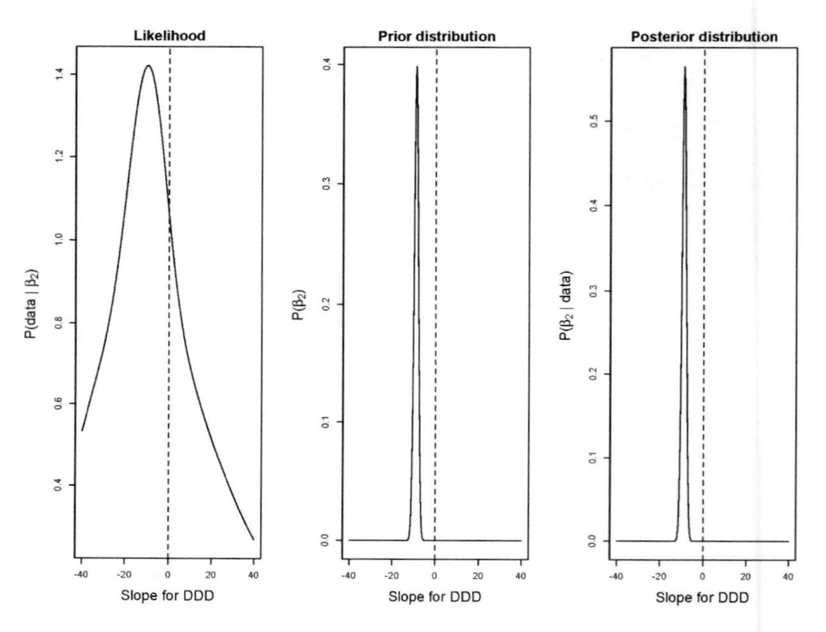

Figure 7.3. Left: Likelihood of the data as a function of β_2. Middle: Informative normal prior $N(-9, 1^2)$. Right: Posterior distribution which is obtained by multiplying the left and middle curves.

If we use non-informative (also called diffuse) priors then the expected value of the posterior distribution is close to the frequentist results. The more informative the prior, the more influential the prior.

Diffuse normal priors are commonly used for regression parameters that can be positive or negative. If we use diffuse priors then the Bayesian results tend to be similar to the frequentist results. In the Osprey example we only focused on the parameter β_2. In reality we put priors on all regression parameters, for example,

$$\beta_1 \sim N(0, 100^2) \text{ and } \beta_2 \sim N(0, 100^2)$$

Priors for standard deviations need to take into account that these terms are strictly positive. Hence a normal distribution is not a good candidate for a prior of a standard deviation σ. It can be a major problem to choose a prior for the standard deviation term σ, especially when we introduce multiple random effects later on in this book. Each of those random effects will come with its own sigma! For simple models (i.e. linear regression, linear mixed models) the diffuse uniform distribution for a standard deviation performs well.

Ntzoufras (2009) shows that within the context of a linear regression model a normal prior on the betas and an inverse gamma prior for σ^2 leads to conjugate priors. An inverse gamma distribution for σ^2 is the same as assuming a gamma prior for $\tau = 1 / \sigma^2$. The latter term is also called *precision*.

In a regression model it is common to assume diffuse normal priors for the betas and a uniform or gamma prior for the standard deviation parameters.

In this section we used special priors that allowed us to calculate the posterior distribution, and its mean and variance, by hand. Hence, there is no need to calculate the integral in Equation (7.11). The reality is that this is only possible for the simplest models. For more complicated models we need a different mechanism to obtain the posterior distribution and its mean.

7.6 Markov chain Monte Carlo simulation

7.6.1 Underlying idea

Instead of deriving a mathematical expression for the posterior distribution in Equation (7.10) by hand, Markov chain Monte Carlo (MCMC) simulates parameter values that converge to the posterior distribution. And MCMC does it for all parameters simultaneously. For the Osprey data we have $\boldsymbol{\theta} = (\beta_1, \beta_2, \sigma)$. The 'Markov Chain' in MCMC refers to simulating vectors $\boldsymbol{\theta}^{(1)}, \boldsymbol{\theta}^{(2)}, \boldsymbol{\theta}^{(3)}, \ldots, \boldsymbol{\theta}^{(T)}$ in such a way that the density function for $\boldsymbol{\theta}^{(t+1)}$ depends only on $\boldsymbol{\theta}^{(t)}$ and not on any of the other samples. In mathematical notation this is written as:

$$f(\boldsymbol{\theta}^{(t+1)} \mid \boldsymbol{\theta}^{(t)} \ldots, \boldsymbol{\theta}^{(1)}) = f(\boldsymbol{\theta}^{(t+1)} \mid \boldsymbol{\theta}^{(t)})$$

We need a mechanism that will provide many random samples for $\boldsymbol{\theta}$ and that will do this in such a way that each draw depends only on the previous draw. This is an iterative process. A formal statistical explanation of MCMC is given in Zuur et al. (2012), and a cartoon explanation was presented in Zuur et al. (2016a). MCMC is essentially an iterative simulation process. The bottom line is that MCMC gives a large number of simulated values (also called iterations) for each parameter, say 10,000 per parameter. We can make a histogram of these 10,000 iterations. So, if you have a model with 10 parameters, you end up with 10 histograms. The underlying statistical theory dictates that the iterations converge to the posterior distributions of the parameters. Instead of a histogram per parameter, we can summarise the iterations with the mean, median, standard deviation, or with quantiles. The underlying theory also dictates that the mean of the iterations converges to the posterior mean.

MCMC is a simulation technique that produces a large number of iterations for each parameter. These iterations approximate the posterior distribution in Equation (7.10) and the mean of the iterations per parameter converges to the posterior mean; hence there is no reason to calculate the integral in Equation (7.11).

You can use existing software packages to run MCMC from within R. One of the first software packages that was developed for MCMC was WinBUGS (Lunn et al. 2000). The BUGS in WinBUGS stands for Bayesian inference Using Gibbs Sampling. Development of WinBUGS has ceased. Its successors are OpenBUGS (Lunn et al. 2009), JAGS (Plummer 2003), and STAN (Gelman et al. 2015), among others. The abbreviation JAGS stands for Just Another Gibbs Sampler. The syntax of JAGS and OpenBUGS is 99% identical. We will use JAGS in this chapter as it is cross-platform.

7.6.2 Installing JAGS and R2jags

Installing JAGS requires two steps. The first step involves installing a file from http://mcmc-jags.sourceforge.net/. Note that this has to be done from outside R. JAGS can be executed from within R using special packages, for example `R2jags`.

7.6.3 Flowchart for running a model in JAGS

Figure 7.4 shows a flowchart that illustrates the required steps to carry out an analysis using JAGS. As an example we will use the Irish pH data that we discussed in Chapter 4. We will fit the following linear regression model in JAGS.

$$
\begin{aligned}
pH_i &= \beta_1 + SDI_i \times \beta_2 + \varepsilon_i \\
\varepsilon_i &\sim N\left(0, \sigma^2\right)
\end{aligned}
\tag{7.18}
$$

To keep the code and model simple, we only use one covariate and ignore the pseudoreplication for the moment.

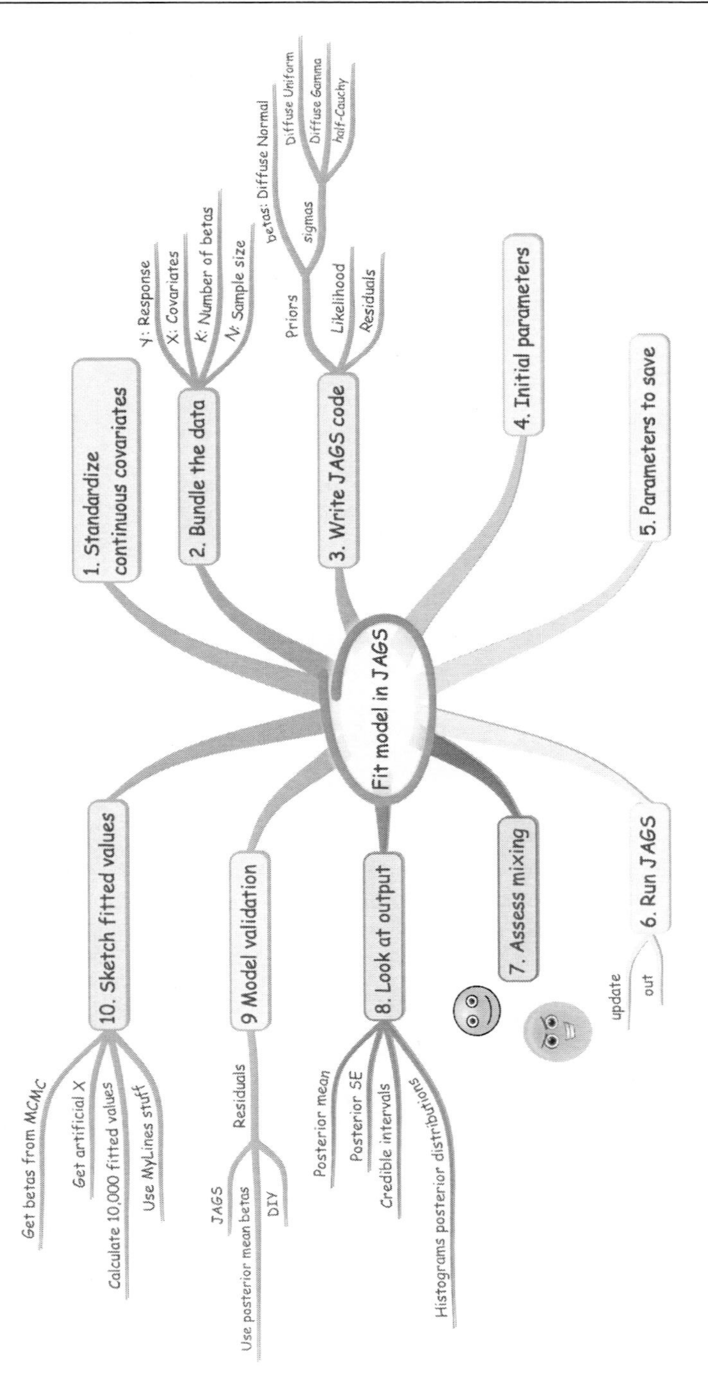

Figure 7.4. Flowchart showing essential JAGS steps.

7.6.4 Preparing the data for JAGS

In step 1 of the flowchart we need to standardise the continuous covariates. The reason for this is that keeping all covariates within the same range makes it easier for the computer to perform numerical optimisation tasks. If you don't standardise, then most likely the algorithm needs to run much longer.

The following code imports the Irish pH data, loads the required packages and our support file, and standardises the SDI covariate using our function Mystd.

```
> iph <- read.table(file = "IrishPh.txt",
                    header = TRUE,
                    dec = ".")
> library(R2jags)
> library(lattice)
> source("MCMCSupportHighstatV4.R")
> iph$SDI.std <- Mystd(iph$SDI)
```

We first fit the model in a frequentist setting so that we can compare the results with the Bayesian approach in a moment.

```
> M1 <- lm(pH ~ SDI.std, data = iph)
> summary(M1)

              Estimate Std. Error t value Pr(>|t|)
(Intercept)    7.43219    0.02626  283.07   <0.001
SDI.std       -0.41367    0.02632  -15.72   <0.001
```

In step 2 in the flowchart we need to put all the required data that we will need in JAGS in a list. We will make the code as generic as possible so that for another data set only minimal changes to the code are required.

The code below stores the pH data (Y), the standardised SDI variable (via the X matrix), the number of parameters (K), and the sample size (N) in a list with the name JAGS.data.

```
> X <- model.matrix(~ 1 + SDI.std, data = iph)
> K <- ncol(X)
> JAGS.data <- list(Y = iph$pH,     #Response
                    X = X,          #Covariate
                    K = K,          #Num. parameters
                    N = nrow(iph))#Sample size
```

We recommend that you check that all data are indeed in the list. It is easy to make mistakes typing the names in the list; R will not give an error message.

7.6.5 JAGS code

We have now reached step 3, the JAGS code. We present the code below and explain it in detail.

```
> sink("iphJAGS.txt")
> cat("
  model{
    #1A. Diffuse normal priors beta
    for (i in 1:K) { beta[i] ~ dnorm(0, 0.0001)}

    #Diffuse uniform prior for sigma
    tau  <- 1 / (sigma * sigma)
    sigma ~ dunif(0, 20)

    #2. Likelihood
    for (i in 1:N) {
      Y[i]   ~ dnorm(mu[i], tau)
      mu[i] <- inprod(beta, X[i,])
    }}
    ",fill = TRUE)
  > sink()
```

The first `sink`, `cat`, `fill = TRUE` and the second `sink` function are used to create a text file in your working directory with the name `iphAGS.txt`.

The JAGS code is between the curly brackets in `model{ }`. Keep in mind this is JAGS code, not R code. JAGS code looks like R code but it is different. JAGS will compile the code, so you do not have to worry about the order in which you type things.

The JAGS code for this model consists of two blocks of code: something for the priors and something for the likelihood.

Priors

We need priors for β_1 and β_2, and we need a prior for σ. We discussed the concept of a prior in the previous section. Recall that for a regression parameter β_i we have to specify, a priori, a distribution that specifies the possible values of this parameter. For example, are all values equally likely, or are some values more likely than others? So what do we know, a priori, about β_1 and β_2? Perhaps you are a specialist on pH data, and you therefore have a reasonably good idea about possible values for β_1 and β_2. For example, based on your experience or other research you expect that β_1 is somewhere between 5 and 10, and β_2 is somewhere between –1 and –0.5, though other values may be possible. Based on these numbers, we could use the following informative priors:

$$\beta_1 \sim \text{normal}(7.5, 1^2) \text{ and } \beta_2 \sim \text{normal}(-0.75, 0.5^2)$$

On the other hand, perhaps we don't have any idea about possible values. In that case we can use diffuse or non-informative priors. Again, we can use a normal distributions for this; we just need to modify the range of likely values.

$$\beta_1 \sim \text{normal}(0, 100^2) \text{ and } \beta_2 \sim \text{normal}(0, 100^2)$$

We will use diffuse normal priors for β_1 and β_2. At this point we have a small coding issue. JAGS does not want us to specify the variance of the prior; instead it uses 'precision', which is defined as 1 divided by the variance. Hence in JAGS we need to write

$$\beta_1 \sim N(0, 1 / 100^2) \text{ and } \beta_2 \sim N(0, 1 / 100^2)$$

for diffuse normal priors. This explains the line

```
for (i in 1:K) { beta[i] ~ dnorm(0, 0.0001)}
```

in the JAGS code above. The 0.0001 is equal to $1 / 100^2$.

We also have a variance term, namely σ^2. This term is not related to the variance of the priors! It is the variance that is being used in the normal distribution for pH. The JAGS code above uses a diffuse uniform prior for the standard deviation of the form

$$\sigma \sim \text{uniform}(0, 20)$$

It is diffuse because the range 0 to 20 is relatively large. It is an option to make the interval even larger, say between 0 and 100.

Likelihood

In Equation (7.18) we gave the mathematical expression for the linear regression model. It is reproduced below, in a slightly different form.

$$
\begin{aligned}
& pH_i \sim N\left(\mu_i, \sigma^2\right) \\
& E\left(pH_i\right) = \mu_i \quad \text{and} \quad \text{var}\left(pH_i\right) = \sigma^2 \\
& \mu_i = \beta_1 + SDI_i \times \beta_2
\end{aligned}
$$

In this notation we state that each pH value is normally distributed with mean μ_i and variance σ^2. This is translated into JAGS code as

```
for (i in 1:N) {
   Y[i]    ~ dnorm(mu[i], tau)
   mu[i] <- inprod(beta, X[i,])
}
```

Just as before, the JAGS function `dnorm` expects the precision and not the variance. That is the reason that we calculated `tau` as `1/sigma`^{2}. Note that the JAGS code follows the mathematical expression of the model.

You need to remember what is precision and what are priors.

7.6.6 Initial values and parameters to save

The rest of the code is much easier. We need to specify initial values and tell JAGS which parameters to save. These are steps 4 and 5 in the flowchart in Figure 7.4.

```
> inits  <- function () {
    list(beta  = rnorm(K, 0, 0.1),
         sigma = runif(0, 20))}
> params <- c("beta", "sigma")
```

This is R code, not JAGS code. We specify a vector of length K = 2 (intercept and slope) for `beta`. Its values are drawn from a normal distribution with mean 0 and a small degree of uncertainty around it (we deliberately did not mention the term *standard deviation* here as it may cause further standard deviation confusion). For more complicated models it may save some time to choose sensible starting values. For `sigma` we choose a value from a uniform distribution. We need to ensure that the starting values comply with the prior distributions. So don't use `sigma = 250` as a starting value because it does not comply with the diffuse prior for sigma in the JAGS code.

Initial values are used by the algorithm to start the iteration process.

7.6.7 Running JAGS

To run the model (step 7 in the flowchart in Figure 7.4) we execute the following R code.

```
> library(R2jags)
> J1 <- jags(data        = JAGS.data,
             inits       = inits,
             parameters  = params,
             model       = "iphJAGS.txt",
             n.thin      = 10,
             n.chains    = 3,
             n.burnin    = 4000,
             n.iter      = 5000)
> J2  <- update(J1, n.iter = 50000, n.thin = 10)
> out <- J2$BUGSoutput
```

We specify the data, initial values, parameters to save, and the name of the file with the JAGS modelling code. There are four arguments that require some additional explanation: the thinning rate, number of chains, burn-in, and number of iterations. We specify 5,000 iterations with a burn-in of 4,000 iterations. The simulation process takes a while to reach a situation in which the iterations converge to the posterior distribution. We had best throw away the iterations that are generated by the simulation process before convergence. This is called the burn-in.

Hence, we have 5,000 – 4,000 = 1,000 iterations for each posterior distribution. But instead of doing the MCMC iteration process only once, we do it three times (this is the number of chains), each with a different initial starting value. This would suggest that we have 3 × (5,000 – 4,000) iterations for each posterior distribution. But instead of storing all iterations we only store every 10th iteration. This is called the *thinning rate*. The motivation for using thinning (or not) is to improve the quality of the simulated values (i.e. remove correlation in the simulated values).

Hence, we have 3 × (5,000 – 4,000) / 10 = 300 iterations for each posterior distribution. This is obviously not enough, and therefore we run the `update` function. This function considers all iterations before the `update` as burn-in, but it will continue where previous chains ended. In the case of poor mixing we can run the `update` function a couple of times.

In this case we have 3 × 50,000 / 10 = 15,000 iterations for each posterior distribution. That is good enough for a simple model like this, but for more advanced models you may want to take a higher number of iterations for the final posterior distributions.

At this point common questions from our students are: How many iterations should we use? How many chains should we use? Why not one chain instead of three? Why not five chains? And how big should the burn-in be? These are all valid questions but we will not address them in this book. See Zuur et al. (2016a) for a full discussion.

We store the output in an object with the name `out`.

7.6.8 Accessing numerical output from JAGS

When you type `print(out)` in R you get rather cumbersome output. Instead of relying on existing functions to present the output (and ending up with a bunch of numbers that we won't look at anyway), we programmed a series of utility functions to present the numerical and graphical outputs in a nicer format. These utility functions are available on the website for this book. To access them we need to source the file in which they are stored.

```
> source(file = "MCMCSupportHighstatV4.R")
```

All the support files access the MCMC iterations for the parameters that we specified in step 5 of the flowchart in Figure 7.4.

7.6.9 Assess mixing

Before investigating whether SDI has an effect on pH we need to assess whether we can trust the MCMC iterations. We can only use the MCMC iterations for a posterior distribution when mixing is good. To assess mixing (step 7 of the flowchart) we will plot each chain versus its iteration number (Figure 7.5). If the iterations do flip around an equilibrium then the underlying Bayesian theory dictates that the iterations converge to the real distribution. The main problem with MCMC is to get chains that flip around an equilibrium. If that is the case we say: 'We have good mixing'. This essentially means that we can use the iterations to calculate the posterior distribution and trust the results. Judging whether we have good mixing is a challenge and requires some expertise. Based on the results in Figure 7.5 we can state that mixing is excellent.

To create this graph we used the following R code.

```
> MyNames <- c("Intercept", "Slope", "sigma")
> MyBUGSChains(out, c("beta[1]", "beta[2]",
                      "sigma"),
               PanelNames = MyNames)
```

First we specify the names of the variables that we want to see above the panels. Then we execute our utility routine MyBUGSChains, which takes care of the plotting of the chains.

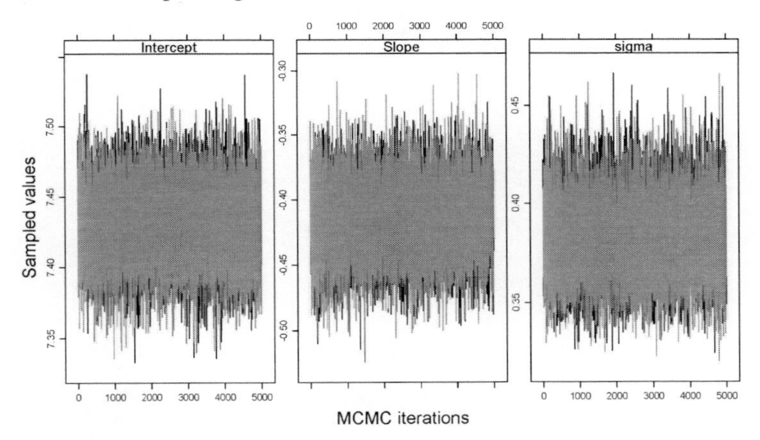

Figure 7.5. Iterations for each chain per parameter. We have three chains per parameter.

7.6.10 Posterior information

We can easily calculate numerical summary statistics (step 8 of the flowchart) of the MCMC iterations per parameter, for example the average and standard deviation, and the 2.5% and 97.5% quantiles. We use one of our utility routines for this. First we define the names that we would like to print in front of the numerical output (and in the graph panels). Then we execute our MyBUGSOutput function.

```
> MyNames <- c("Intercept", "Slope", "sigma")
> OUT1 <- MyBUGSOutput(out,
                       c("beta[1]", "beta[2]",
                         "sigma"),
                       VarNames = MyNames)
```

```
> print(OUT1, digits = 5)

              mean        se      2.5%     97.5%
Intercept  7.43211 0.026419  7.37996   7.48325
Slope     -0.41415 0.026450 -0.46557  -0.36198
sigma      0.38290 0.019009  0.34773   0.42268
```

The posterior mean of the slope is –0.41 and its 95% credible interval (this is the Bayesian equivalent of a confidence interval and it is obtained by taking the 2.5% and 97.5% quantiles of the MCMC iterations) runs from –0.46 to –0.36. This does not contain 0. We can't use the word 'significant' in the same way as we did in the frequentist world because there is no *p*-value that we can compare to 0.05. Based on the Bayesian results you can state that the probability is 0.95 that the slope is between –0.46 and –0.36. Bayesians call this the 95% credible interval. This interval does not contain 0; hence we consider SDI an important variable. If you really want to use the word 'significant' then associate it with the word 'important'.

Note that the frequentist and Bayesian results are nearly identical!

Instead of the numerical summary statistics we can also make histograms of the 15,000 MCMC iterations for each parameter. These are the posterior distributions. They are presented in Figure 7.6.

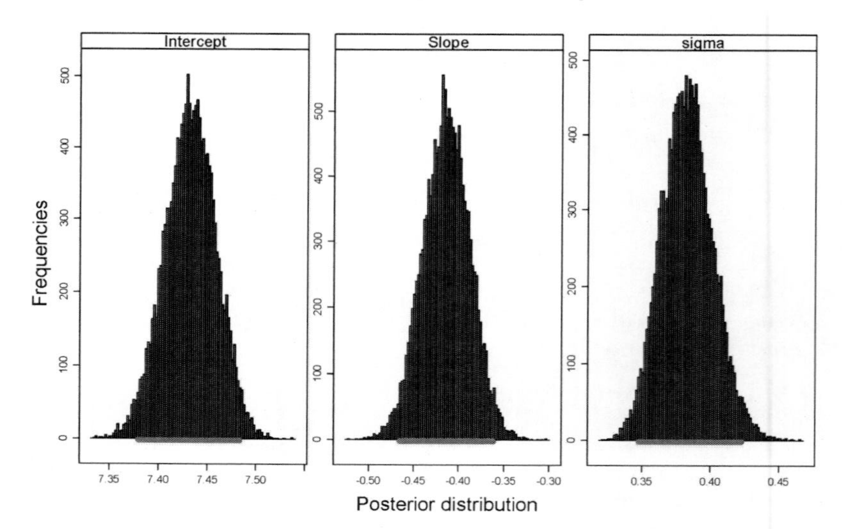

Figure 7.6. Histogram of the iterations for each parameter.

We used the following code to create this figure.

```
> MyBUGSHist(out,
             c("beta[1]", "beta[10]", "sigma"),
             PanelNames = MyNames)
```

After carrying out a frequentist analysis we need to verify that all model assumptions (independence, homogeneity, normality) hold. We need to do exactly the same for a Bayesian analysis. Examples of this process are given in later chapters.

7.7 Integrated nested Laplace approximation*

The JAGS code presented in Section 7.6 can easily be extended to linear mixed-effects models, generalised linear mixed-effects models, zero-inflated models, and even generalised additive mixed-effects models. See Zuur et al. (2013, 2014, 2016) for a large number of examples. Misery comes with models that contain spatial, temporal, and spatial-temporal dependencies. Zuur et al. (2013) contains an example in which a generalised linear model is combined with an autoregressive correlation structure. The example required some extensive JAGS coding, and a large number of MCMC iterations were used. In Zuur et al. (2012) we used WinBUGS and fitted zero-inflated models with a so-called spatial CAR correlation. Mixing was a serious problem, even after running the computer for 2 days.

In this section we show how numerical approximation can be used to obtain a posterior distribution. The title of this section contains an asterisk, *, indicating it is more difficult than other sections and that it may be skipped without losing track of the main storyline. To be honest, the underlying mathematics of the approach that we will discuss in this section merits at least five stars for complexity! We will try to keep it as conceptual as possible. Readers interested in the actual mathematics are referred to Rue et al. (2009).

7.7.1 Joint posterior distribution

In the previous section we used a linear regression model for the Irish pH data. To avoid browsing too much through this book we present it here again.

$$\begin{aligned} pH_i &= \beta_1 + SDI_i \times \beta_2 + \varepsilon_i \\ \varepsilon_i &\sim N\left(0, \sigma^2\right) \end{aligned} \tag{7.19}$$

This model has three parameters, namely β_1, β_2, and σ. In the INLA literature a distinction is made between betas and variance parameters. Variance parameters are also called hyperparameters. These can be the σ from the residual noise term in Equation (7.19), but also the dispersion parameter in a negative binomial GLM, the variance parameter of random effects in a mixed-effects model, the ϕ from the autoregressive correlation term in a time-series model, or the parameters for the spatial correlation function. In the model in Equation (7.19) for the Irish pH data we only have one hyperparameter, namely σ.

Using Bayes' theorem we can write the following expression:

$$
\begin{aligned}
P(\beta_1,\beta_2,\sigma \mid Data) &= \frac{P(Data \mid \beta_1,\beta_2,\sigma) \times P(\beta_1,\beta_2,\sigma)}{P(Data)} \\
&\propto P(Data \mid \beta_1,\beta_2,\sigma) \times P(\beta_1,\beta_2,\sigma)
\end{aligned} \tag{7.20}
$$

The $P(\beta_1, \beta_2, \sigma \mid \text{Data})$ is called the joint distribution of the parameters given the data. Just as in Section 7.2, we dropped $P(\text{Data})$ as it is only a scaling factor. Recall that '$\propto$' means 'proportional to'. The $P(\text{Data} \mid \beta_1, \beta_2, \sigma)$ is the likelihood of the data. Because we are analysing pH data, the Gaussian distribution is an option (though the gamma distribution would be better as pH cannot be negative). The $P(\beta_1, \beta_2, \sigma)$ is the prior. The problem is that this prior contains two betas and a sigma, and we cannot use the same type of prior for these components because σ has to be positive, whereas the betas can take any value. So, we have to rewrite the prior $P(\beta_1, \beta_2, \sigma)$ as something that contains two components, one for the betas and one for the sigma.

In Section (7.2) we gave the definition of conditional probability: $P(A \mid B) = P(A \text{ and } B) / P(B)$. We can easily rearrange the terms and write this as $P(A \text{ and } B) = P(A \mid B) \times P(B)$. We can use this to rewrite the expression of the joint probability of the prior: $P(\beta_1, \beta_2, \sigma) = P(\beta_1, \beta_2 \mid \sigma) \times P(\sigma)$. All we did was to set A = 'β_1 and β_2', and $B = \sigma$. That is the beauty of mathematics. This means that we now have:

$$
\begin{aligned}
P(\beta_1,\beta_2,\sigma \mid Data) &\propto P(Data \mid \beta_1,\beta_2,\sigma) \times P(\beta_1,\beta_2,\sigma) \\
&= P(Data \mid \beta_1,\beta_2,\sigma) \times P(\beta_1,\beta_2 \mid \sigma) \times P(\sigma)
\end{aligned} \tag{7.21}
$$

This reads as: The joint probability for all parameters (given the data) is equal to the likelihood of the data (given all the parameters) times the prior for all the betas (given the hyperparameters) times the prior for all the hyperparameters. The advantage of doing this is that we have split the priors into two groups: priors related to the betas and priors related to the hyperparameters. This means that we can specify something for the betas (say: any possible value) and something else for the sigmas (say: something positive).

The likelihood of the data is an easy one: Gaussian, gamma, Poisson, Bernoulli, or whichever distribution is relevant for the response variable. The prior for the betas (given the hyperparameters) is a trickier one. For most data sets and models the posterior distributions of the betas look unimodal and symmetric. Formulated differently, they look normal (as in: normal distributed). We won't do much damage if we use a multivariate normal prior for the betas. And computing time is going to be much shorter if we also add the phrase 'independent' to this. The reason for shorter computing time is that the covariance matrix in such a prior distribution will contain lots of zeros.[3]

The expression in Equation (7.21) gives the joint distribution as a function of the likelihood of the data, (conditional) priors of betas, and priors of hyperparameters. In a Gaussian Markov random field we assume that the priors of the betas are multivariate normal and independent. That simplifies the numerical calculations.

7.7.2 Marginal distributions

Before going any further with the expression in Equation (7.21), we have to confess that this equation is not the one we actually are interested in. But it allowed us to set the scene for things to come in this subsection. Recall from the previous section that MCMC gives the posterior distribution for *each* parameter. These are the *marginal* distributions $P(\beta_1 \mid \text{Data})$, $P(\beta_2 \mid \text{Data})$, $P(\sigma \mid \text{Data})$ that were presented in Figure 7.6. We never focussed on the joint distribution $P(\beta_1, \beta_2, \sigma \mid \text{Data})$.

The joint distribution is $P(\beta_1, \beta_2, \sigma \mid \text{Data})$. The marginal distributions are $P(\beta_1 \mid \text{Data})$, $P(\beta_2 \mid \text{Data})$, $P(\sigma \mid \text{Data})$, and these are the ones we want.

Maybe we should explain a bit more clearly the difference between a joint distribution and a marginal distribution because it is rather essential to understand the difference between them. Zuur and Ieno (2016a) used an oystercatcher data set to explain Bayesian statistics. These data are part of a 3-year study on the feeding ecology of the American oystercatcher (*Haematopus palliates*), which inhabits coastal areas of Argentina (Ieno, unpublished data). For our present purpose, we use a subset of data consisting of observations of the shell length of clams eaten by oystercatchers at three sites during December and January. These data were also used in Zuur et al. (2012) and Ieno and Zuur (2015).

To break open a shell, an oystercatcher uses its beak to either hammer or stab the shell. Depending on the technique used, the bird is called a 'hammerer' or a 'stabber'. An underlying question is whether clams eaten by hammerers are larger than those consumed by stabbers. Table 7.1 shows the number of clams eaten by hammering oystercatchers observed at three feeding sites and records whether the shells were small or large. We convert these numbers into proportions by dividing each number by 172; see Table 7.2.

The six values of P(Shell size **and** Feeding site) define the joint probability.

---

[3] If you have ever seen the phrase 'Gaussian Markov random field' (GRMF), that is what this is.

Table 7.1. Number of clams eaten by hammering oystercatchers per shell size (small versus large) and feeding site.

	Feeding site			
Shell size	1	2	3	**Total**
Small	46	35	57	**138**
Large	14	10	10	**34**
Total	**60**	**45**	**67**	**172**

Table 7.2. Proportions of clams eaten by hammering oystercatchers per shell size (small versus large) and feeding site.

	Feeding site			
Shell size	1	2	3	**Total**
Small	46 / 172	35 / 172	57 / 172	**138 / 172**
Large	14 / 172	10 / 172	10 / 172	**34 / 172**
Total	**60 / 172**	**45 / 172**	**67 / 172**	**172 / 172**

The probabilities *P*(Feeding site) and *P*(Shell size) are marginal distributions. The marginal distribution *P*(Feeding site) is the lower row with probabilities and it is obtained by factoring out shell size.

P(Feeding site = 1) = *P*(Feeding site = 1 and Shell size = Small) +
P(Feeding site = 1 and Shell size = Large)
= 46 / 172 + 14 / 172 = 60 / 172

P(Feeding site = 2) = *P*(Feeding site = 2 and Shell size = Small) +
P(Feeding site = 2 and Shell size = Large)
= 35 / 172 + 10 / 172 = 45 / 172

P(Feeding site = 3) = *P*(Feeding site = 3 and Shell size = Small) +
P(Feeding site = 3 and Shell size = Large)
= 57 / 172 + 10 / 172 = 67 / 172

To factor out the shell size effect we just add up all shell size probabilities. We can do the same with shell length:

P(Shell size = Small) = *P*(Feeding site = 1 and Shell size = Small) +
P(Feeding site = 2 and Shell size = Small) +
P(Feeding site = 3 and Shell size = Small)
= 46 / 172 + 35 / 172 + 57 / 172 = 138 / 172

and

P(Shell size = Large) = *P*(Feeding site = 1 and Shell size = Large) +
P(Feeding site = 2 and Shell size = Large) +
P(Feeding site = 3 and Shell size = Large)
= 14 / 172 + 10 / 172 + 10 / 172 = 34 / 172

In a more general notation this can be written as

$$P(X=x)=\sum_{y} P(X=x \text{ and } Y=y)$$

However, this expression is only for discrete variables. We are working with priors and posteriors of regression parameters and hyperparameters, and these are all continuous variables. If we have a joint distribution with continuous parameters, how do we factor out a term so that we obtain the marginal distribution? Here is where you will see an integral:

$$P(X=x)=\int_{y} P(X=x \text{ and } Y=y)dy$$

For a lot of people mathematics wasn't their favorite subject in high school. Unfortunately, when you try to understand the mathematics behind INLA, you will see plenty of integrals. If you still have an appetite to understand the mathematical principle behind INLA, please read on.

To go from a joint distribution to a marginal distribution we need to factor out parameters. If the variables are continuous (like our betas and sigmas) then this requires integrals.

7.7.3 Back to high school

Let us go back to your high school time. By this we don't mean the time that you met your first girl- or boyfriend. Instead we will take you back to high school math class.

What was an integral? Figure 7.7A shows a line and we want to know the size of the area under the curve between x = 0 and x = 1. Formulated differently, we want to know the size of the colored area. One option is to create multiple small rectangles between x = 0 and x = 1, and for each rectangle we determine the size (= width × height). In Figure 7.7B we use five rectangles. Adding up the five size values gives an approximation of the total surface under the line between 0 and 1. The more rectangles we use, the more accurate the answer. If you dive more deeply into the mathematics of INLA you will see summations being used as approximations of integrals.

We just mentioned the word integral. If we know the mathematical function $f(x)$ that was used to generate the line, then we can directly calculate the surface under the line between x = 0 and x = 1 with an integral. In this case we do know the mathematical function, namely, $f(x) = -0.2 \times x^2 + 1$. Let's see whether you remember the following from high school!

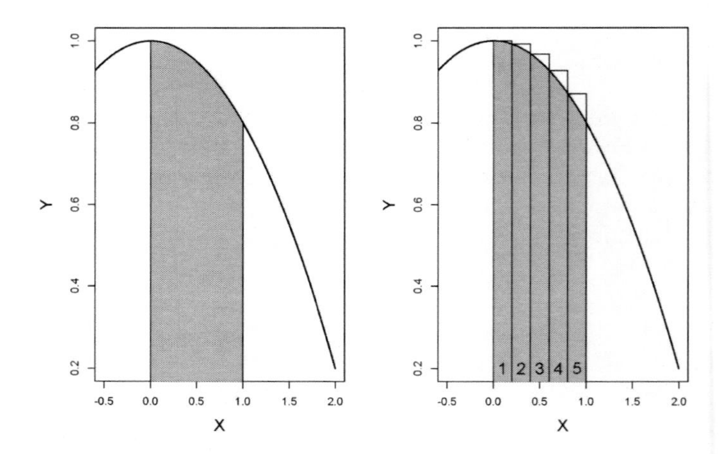

Figure 7.7. A: Line obtained by $f(x) = -0.2 \times x^2 + 1$. B: As panel A. To calculate the surface under the line between 0 and 1 we have created five rectangles, each with width 0.2. The total surface of the five rectangles is 0.952.

$$\begin{aligned}\int_0^1 f(x)dx &= \int_0^1 \left(-0.2 \times x^2 + 1\right)dx \\ &= -0.2 \times \int_0^1 x^2\,dx + \int_0^1 1 \times dx \\ &= -0.2 \times \frac{1}{3} \times x^3\Big|_0^1 + x\Big|_0^1 \\ &= -0.2 \times \frac{1}{3} + 1 = 0.93\end{aligned}$$

The integral[4] defines the area between 0 and 1 under the line. The answer is similar to adding up all the rectangles in Figure 7.7B.

In this case we were able to solve the integral because the function $f(x) = -0.2 \times x^2 + 1$ is rather simple. But what can you do if the function is very difficult and solving the integral analytically is impossible? Here is where Laplace approximation comes in. Instead of using the function $f(x)$, we use a mathematical approximation of $f(x)$. This approximation is based on a so-called Taylor series (which probably goes beyond high school math).

---

[4] When we solve an integral $\int_a^b f(x)dx$ we look for a function $H'(x) = f(x)$, and then we calculate the difference $H(b) - H(a)$. The difference is also written as $H(x)\Big|_a^b$. The notation H' stands for derivative.

If the integral contains a function that is too complicated, we can open a tin of math tricks and approximate it with an exponential function that looks like a Gaussian density function. Computing time will be much (!) shorter.

7.7.4 INLA

In this subsection we glue together all the pieces that we discussed so far in this section.

In Subsection 7.7.2 we explained that a marginal distribution is obtained by factoring out other terms. This means that we need to solve the following two integrals.

$$\begin{aligned} P(\beta_1 \mid Data) &= \int P(\beta_1, \sigma \mid Data) d\sigma \\ P(\beta_2 \mid Data) &= \int P(\beta_2, \sigma \mid Data) d\sigma \end{aligned} \qquad (7.22)$$

This is just integrating out a term from a joint distribution; see Section 7.7.2. In our linear regression model for the pH data we have only one hyperparameter (the sigma for the residuals) but for a mixed-effects model there will be multiple sigmas, say σ_1 and σ_2. For such a model we also need to calculate

$$\begin{aligned} P(\sigma_1 \mid Data) &= \int P(\sigma_1, \sigma_2 \mid Data) d\sigma_2 \\ P(\sigma_2 \mid Data) &= \int P(\sigma_1, \sigma_2 \mid Data) d\sigma_1 \end{aligned} \qquad (7.23)$$

Assuming for the moment that we only have one sigma and using the standard conditional probability rules,[5] we can write the expressions in Equation (7.22) as

$$\begin{aligned} P(\beta_1 \mid Data) &= \int P(\beta_1 \mid \sigma, Data) \times P(\sigma \mid Data) d\sigma \\ P(\beta_2 \mid Data) &= \int P(\beta_2 \mid \sigma, Data) \times P(\sigma \mid Data) d\sigma \end{aligned} \qquad (7.24)$$

In order to calculate $P(\beta_1 \mid \text{Data})$, we need two components: the $P(\beta_1 \mid \sigma, \text{Data})$ and the $P(\sigma \mid \text{Data})$. In this case the second component is easy to obtain as there is only one sigma. But if we have multiple sigmas (or a sigma and time-series or spatial parameters) then the second term is again a joint distribution, in which case we need to break it down again into easier terms. For the second component, Rue et al. (2009) used the notation $P(\theta \mid \text{Data})$ where the θ contains all the hyperparameters, and Blangiardo and Cameletti (2015) wrote the same term as $P(\psi \mid \text{Data})$,

[5] Conditional probability is defined as $P(A \mid B) = P(A \text{ and } B) / P(B)$. Refining A and B gives $P(A \text{ and } B \mid C) = P(A \text{ and } B \text{ and } C) / P(C)$. We can also write $P(A \text{ and } B \text{ and } C) = P(A \mid B \text{ and } C) \times P(B \text{ and } C)$. And $P(B \mid C) = P(B \text{ and } C) / P(C)$. Combining these expressions gives $P(A \text{ and } B \mid \text{C}) = P(A \mid B, C) \times P(B \mid \text{C})$. We can apply this on $P(\beta \text{ and } \sigma \mid Data)$.

where the ψ contains all the hyper-parameters. It doesn't really matter how we write it, the problem is that we need to estimate it. Using the probability rule in Footnote 5 on the previous page and Bayes' theorem these authors rewrite $P(\psi \mid \text{Data})$ as a ratio with multiple components. Some of these components can easily be calculated but for one of them the Laplace approximation is used.

To get the first term in Equation (7.24) Blangiardo and Cameletti (2015) discuss three approaches. In two of the approaches, including the one implemented in the software that we will use in the next section, Laplace approximation is used. A list of the required computer steps is presented in Section 4.7 of Blangiardo and Cameletti (2015), and a (reasonably) simple example is given in their Section 4.9.

To get the marginal distributions of the regression parameters in Equations (7.22) and the marginal distribution of hyperparameters in Equation (7.23) we need two components. Each of these components is obtained with Laplace approximation.

Because the solution to obtain a marginal distribution requires a series of approximations (which sort of are nested) using Laplace, the resulting methods were called 'Integrated Nested Laplace Approximation'. The software for this is called R-INLA.

7.8 Example using R-INLA

In this section we show how to fit the linear regression model in Equation (7.19) for the Irish pH data. First we need to install the R-INLA package. The easiest way to install R-INLA on your computer is to copy the following command into your R console and press enter.

```
> source("http://www.math.ntnu.no/inla/givemeINLA.R")
```

It is wise to always use the latest stable version. You can upgrade R-INLA at a later stage with `inla.upgrade(testing = FALSE)`.

Now that we have R-INLA installed we can fit our model.

```
> library(INLA)
> I1 <- inla(pH ~ SDI.std, data = iph,
             family = "gaussian")
```

The R code is nearly identical to that of fitting a linear regression model with the `glm` or `lm` functions. The `family` argument specifies the distribution, and the `data` argument tells `inla` the name of the object with the data. As to the `family` argument, typing

```
> names(inla.models()$likelihood)
```

gives a list of 50^+ other distributions that can be used in R-INLA. The usual candidates like the Gaussian, Poisson, negative binomial, gamma, and binomial distributions are all present. Default link functions are used by `inla`. In later chapters we will change the distribution and link functions.

Being a Bayesian method, obviously the terminology in R-INLA is slightly different compared to functions like `lm`, `glm`, and `lme`. It is all about fixed parameters and hyperparameters, posterior mean values and credible intervals, and posterior marginal distributions.

The fixed parameters are the regression parameters and the hyperparameters are variance-type parameters. In our linear regression model for the Irish pH data the fixed parameters are β_1 and β_2, and the hyperparameter is the variance of the residuals. Actually, R-INLA works with precision and not with the variance.

Posterior mean values

We get posterior means and other information for the parameters using

```
> summary(I1)
```

```
Fixed effects:
              mean    sd 0.025quan 0.5quan 0.975quan   mode kld
(Intercept) 7.432 0.026    7.380   7.432     7.483  7.432   0
SDI.std    -0.413 0.026   -0.465  -0.413    -0.362 -0.413   0

The model has no random effects

Model hyperparameters:
                                mean   sd 0.025quan 0.5quant
Precision for Gaussian observ. 6.98 0.62  5.83    6.95
                                           0.975quant  mode
Precision for Gaussian observ.             8.26        6.90

Expected number of effective parameters(std dev):
2.024(0.0023)
Number of equivalent replicates : 103.74
Marginal log-Likelihood:  -113.24
```

The numerical output is not very aesthetically appealing. It is perhaps more convenient to extract the relevant information and present it in the way we want. For example, the fixed effects can also be extracted with the `$summary.fixed` argument. We extract the posterior mean, standard deviation, and the 2.5% and 97.5% quartiles.

```
> Betas <- I1$summary.fixed[,c("mean", "sd",
                               "0.025quant",
                               "0.975quant")]
> print(Betas)
```

```
              mean    sd  0.025quant 0.975quant
(Intercept)  7.43  0.026        7.38       7.48
SDI.std     -0.41  0.026       -0.47      -0.36
```

These posterior means, standard errors, and 95% credible intervals are nearly identical to their frequentist equivalents obtained with the `lm` function, and also with the MCMC results.

The first column in the `Betas` object contains the posterior means of the regression parameters obtained by R-INLA, and these can be used to calculate fitted values:

```
> X   <- model.matrix(~ 1 + SDI.std, data = iph)
> Fit <- X %*% Betas[,1]
```

Subtracting the observed data and `Fit` gives residuals that can be used for model validation purposes.

```
> E1 <- iph$pH - Fit
```

Instead of calculating fitted values manually, it is also possible to let R-INLA do this by adding the `control.predictor` option; see the code below.

```
> I2 <- inla(pH ~ SDI.std, data = iph,
             family = "gaussian",
             control.predictor = list(
                                  compute = TRUE))
> Fit <- I2$summary.fitted.values[,"mean"]
```

The fitted values are now in the `summary.fitted.values` slot (together with a 95% credible interval). It is recommended to let R-INLA calculate the fitted values, especially for generalised linear models with random effects. Typing

```
> I2$summary.hyperpar
```

gives the posterior mean and standard error for each hyperparameter. The output was already printed with the `summary` function, and it is not presented again. The posterior mean of the 'Precision for the Gaussian observations' is 6.982. Just like JAGS, R-INLA employs precision where $\tau = 1 / \sigma^2$. In a moment we will show how to obtain the posterior mean of σ.

To get the posterior mean values and credible intervals of the regression parameters use `$summary.fixed`. Use `$summary.hyperpar` for the hyperparameters.

Posterior marginal distributions

In the previous section we explained the mathematical background of INLA and we showed the starting expression for posterior distributions in INLA. Because these are posterior distributions for *one* specific parameter, they are called posterior 'marginal' distributions.[6] Up to now we have been sloppy and called these 'posterior distributions'. In our model we have three marginal distributions: one for the intercept, one for the slope, and one for sigma. Actually, because R-INLA is using precision, we have four marginal distributions. We can obtain them using

```
> I2$marginals.fixed
```

The object contains two posterior marginal distributions, one for the intercept and one for the slope. We can extract the information.

```
> pmbeta1  <- I2$marginals.fixed$`(Intercept)`
> pmbeta2  <- I2$marginals.fixed$SDI.std
> pmtau    <- I2$marginals.hyperpar$`Precision for
                    the Gaussian observations`
```

The object pmbeta1 now contains the posterior marginal distribution for β_1. The pmtau contains the marginal distribution for the precision parameter. R-INLA has lots of support functions to manipulate the marginal distributions. For example we can calculate the 95% credible interval via:

```
> inla.qmarginal(c(0.025, 0.975), pmbeta2)
-0.4654632   -0.362281
```

We have already seen these numbers, namely in the summary output. Other useful commands are inla.zmarginal(pmbeta2) which gives the gives mean, standard deviation, 2.5%, 25%, 50%, 75%, and 97.5% quantiles of the marginal distribution, and inla.hpdmarginal(0.95, pmbeta2). The latter command gives the highest posterior density interval. This is an interval in which most of the distribution lies (which is useful for non-symmetric distributions).

The distribution of τ can easily be converted into the posterior marginal distribution for σ with another support function.

```
> MySqrt   <- function(x) sqrt(1/x)
> pm.sigma <- inla.tmarginal(MySqrt, pmtau)
```

Figure 7.8 shows the posterior marginal distribution of the two regression parameters, the precision parameter τ (which we are not really interested in) and σ.

To get the posterior mean value of σ, we use yet another support function.

```
> inla.emarginal(function(x) MySqrt, pmtau)
```

---

[6] The posterior distributions in MCMC are in fact 'marginal' distributions. Because we never talked about 'joint distributions' when we explained MCMC, we didn't bother using the phrase 'posterior marginal distribution'.

```
0.3799003
```

This value is nearly the same as what we obtained via the `lm` function and MCMC. We could also have used

```
> inla.zmarginal(pm.sigma)
```

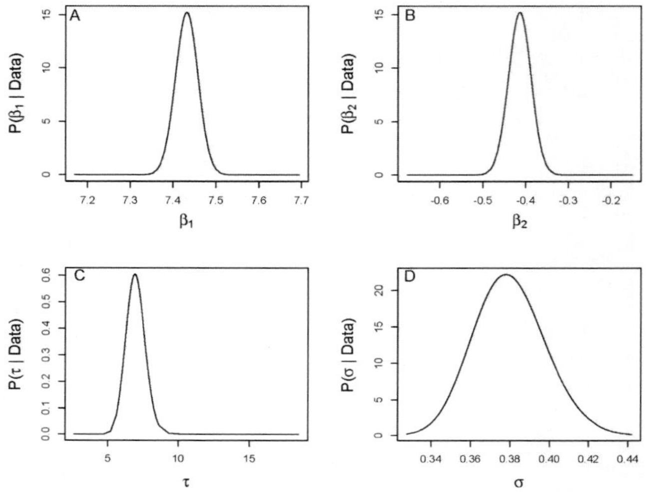

Figure 7.8. A: Posterior marginal distribution of the intercept. B: Posterior marginal distribution of the slope. C: Posterior marginal distribution of the precision parameter τ (= 1 / σ^2). D: Posterior marginal distribution of the standard deviation σ.

The R code to create Figure 7.8A is as follows.

```
> plot(x = pmbeta1[,1],
       y = pmbeta1[,2],
       type = "l",
       xlab = expression(beta[1]),
       ylab = expression(paste("P(", beta[1] ," |
                                Data)")))
```

The other three graphs are produced with similar code.

7.9 Discussion

In this chapter we presented three approaches for obtaining a posterior distribution and its mean value. The conjugate approach was reasonably easy to understand but it only works in the simplest models. The second approach, MCMC, simulates the posterior distribution but is time-consuming for complicated models and mixing may be a problem. In the third approach we use advanced numerical optimisation routines to obtain posterior distributions and corresponding mean values.

8 Multiple linear regression in R-INLA

In Chapter 7 we explained how to obtain posterior distributions of regression parameters in R-INLA. Before diving into models with dependency structures (e.g. repeated measurements, or spatial and temporal data) we will analyse a relatively simple data set in this chapter using multiple linear regression. It allows us to discuss topics like fitted values, residuals, model validation, model selection, model visualisation, and simulations. Once we have this knowledge we will deal with dependency in all remaining chapters.

Prerequisite for this chapter: Knowledge of R and multiple linear regression is required.

8.1 Introduction

The data that will be analysed in this chapter are taken from Hopkins et al. (2013), who studied tool use by captive chimpanzees (*Pan troglodytes*). An experiment was carried out to simulate termite fishing in wild chimpanzees. Small PVC pipes were filled with food and the animals were given thin sticks, which allowed them to eat the food (provided they could figure out how to put the stick into the PVC pipe). The underlying question that we will address in this chapter is whether tool-use skills differ by sex, age, and rearing experience.

In the Hopkins et al. paper, tool-use skill is quantified as the time required per successful dip. This variable is called 'latency'. For each chimpanzee a minimum of 50 successful attempts are recorded and average latency score for each of the 243 monkeys is provided in the online material of the paper. The provided latency scores are standardised. The larger the latency score, the longer it took (on average) the chimp to get food.

The covariates are sex (males vs. female), age (in years), rear (MR = 'mother rear', HR = 'human rear', WC = 'wild caught'), and colony (chimps came from two research units).

8.2 Data exploration

Figure 8.1 shows Cleveland dotplots of all the variables. We temporarily coded the categorical variables colony, rear, and sex as numerical, otherwise the `dotplot` function in R gives an error message. Each point in the Latency panel represents the average motor performance score for a specific chimpanzee. Note that there are two animals with a relative large score. This means that on average it took these two animals a long time to figure out how to get the food! These chimps were removed from the analysis in the Hopkins et al. (2013) paper, but we will keep them in. The Age panel shows the age (in years) of the chimpanzees; there are no animals that are considerably younger or older. Sample sizes differ per colony, rear, and sex.

There are no spatial or temporal dependency aspects in this data set. Because we have an average score per animal, we do not have repeated measurements from the same animal. Assuming that the animals don't learn from one another or are in any other way related there is no need for regression models with complicated dependency structures. Actually, within a colony the chimpanzees are genetically linked, which in principle causes pseudoreplication. If you have this then you may want to look into models with phylogenetic correlation. Lajeunesse and Fox (2015) provide an easy-to-understand starting point.

Further data exploration steps did not indicate any major problems.

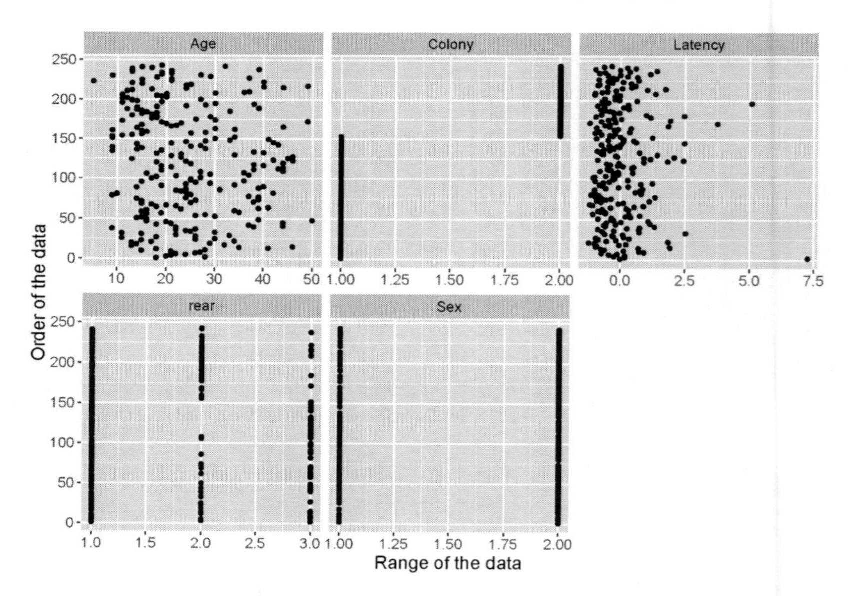

Figure 8.1. Cleveland dotplots of all the variables. The horizontal axes show the values of the variables and the vertical axes are the row numbers as imported from the data file.

8.3 Model formulation

To answer the underlying question formulated in Section 8.1 we will model the latency score as a function of age, sex, colony, and rear using a multiple linear regression model; see Equation (8.1). The categorical variables sex and colony each have two levels, and therefore each of them consumes one regression parameter. The categorical variable rear has three levels and consumes two regression parameters. To keep the expression simple we did not write a beta in front of the $Rear_i$ term in Equation (8.1).

$$
\begin{aligned}
&Latency_i \sim N(\mu_i, \sigma^2) \\
&E(Latency_i) = \mu_i \quad \text{and} \quad \text{var}(Latency_i) = \sigma^2 \\
&\mu_i = \beta_1 + \beta_2 \times Age_i + \beta_3 \times Sex_i + \beta_4 \times Colony_i + Rear_i
\end{aligned}
\tag{8.1}
$$

In case of numerical estimation problems it may be an option to standardise the continuous covariate age.

8.4 Linear regression results

8.4.1 Executing the model in R-INLA

We first import the data with the `read.table` function, define categorical variables, load the R-INLA package (see Chapter 7), and then run the linear regression model in Equation (8.1) in R-INLA.

```
> Chimp <- read.table(file = "Chimps.txt",
                      header = TRUE,
                      dec = ".")
> Chimp$fSex     <- factor(Chimp$Sex)
> Chimp$fColony <- factor(Chimp$Colony)
> Chimp$fRear    <- factor(Chimp$rear)
> library(INLA)
> I1 <- inla(Latency ~ Age + fSex + fColony +
                       fRear,
             family = "gaussian", data = Chimp)
```

8.4.2 Output for the betas

Numerical output for the betas

The numerical output of the linear regression model is stored in the list `I1`. It has various objects that we can inspect. The first two are `I1$summary.fixed` and `I1$summary.hyperpar` as these give us the posterior mean values and measures of variation of the posterior distributions. In this subsection we will focus on the information in `I1$summary.fixed`.

```
> Beta1 <- I1$summary.fixed[, c("mean", "sd",
                      "0.025quant", "0.975quant")]
```

```
> print(Beta1, digits = 3)

               mean      sd 0.025quant 0.975quant
(Intercept) -0.4458 0.19911   -0.83703    -0.0549
Age          0.0227 0.00818    0.00666     0.0388
fSex2       -0.2241 0.13316   -0.48573     0.0373
fColony2     0.0926 0.14409   -0.19050     0.3755
fRear2       0.0500 0.16464   -0.27354     0.3732
fRear3      -0.0348 0.20765   -0.44287     0.3728
```

The first column shows the posterior mean and the third and fourth columns show the 95% credible intervals. The 95% credible interval for β_2 (regression parameter for age) goes from 0.006 to 0.038. This means that there is a 95% probability that β_2 is in this interval. Because 0 is not in this interval we state that β_2 is 'important' as compared to the frequentist phrase 'significant'. Sex is not important and neither is colony.

The interpretation of fRear2 and fRear3 is identical as in the frequentist analysis; these are corrections for the baseline level.

Graphical output for the betas

It is also possible to obtain posterior (marginal) distributions for each fixed parameter. Recall from Chapter 7 that a marginal distribution refers to the distribution of one specific variable. To see the names of the variables for which we can make the posterior (marginal) distributions, type

```
> names(I1$marginals.fixed)

"(Intercept)"   "Age"     "fSex2"    "fColony2"
"fRear2"        "fRear3"
```

So we can make six posterior (marginal) distributions for the fixed parameters. Instead of presenting all six distributions we only plot the one for β_2, which is the slope for age; see Figure 8.2A. In the next subsection we show how you can improve this graph (e.g. make the curve smoother). This graph shows the probability distribution for β_2, very much like the Markov chain Monte Carlo results we showed in Chapter 7.

Figure 8.2A was created with the following R code. The object I1$marginals.fixed$Age contains two columns, each with 75 values; the first column contains a range of β_2 values and the second column the corresponding posterior (marginal) distribution values. Let us have a look at the first five rows.

```
> betaAge <- I1$marginals.fixed$Age
> head(betaAge, 3)   #Show first 3 rows

               x            y
 [1,] -0.059087082 1.580493e-18
 [2,] -0.042722370 9.514732e-12
 [3,] -0.026357658 1.900550e-06
```

Plotting these two columns against each other gives the posterior distribution in Figure 8.2A. When confronted for the first time with the two columns in `betaAge`, you may wonder what exactly these numbers are. It is in fact very simple. Figure 8.2B shows a standard normal distribution. To make this graph we need values for the x-axis, say 75 values between –4 and 4. And for each value in `x1` we calculate the density value of the normal distribution.

```
> x1 <- seq(from = -4, to = 4, length = 75)
> y1 <- dnorm(x1, mean = 0, sd = 1)
```

Figure 8.2B is just a scatterplot of `x1` (possible values) versus `y1` (corresponding normal density values). The `x1` and `y1` in Figure 8.2B are comparable to the `betaAge$x` and `betaAge$y` in Figure 8.2A.

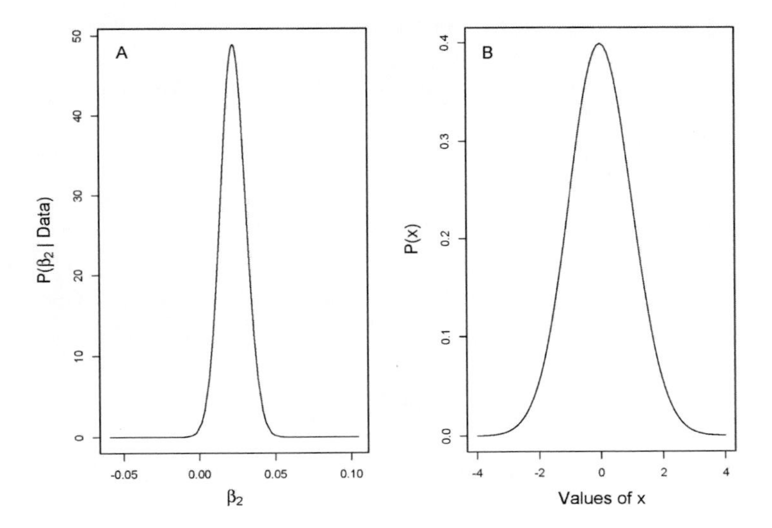

Figure 8.2. A: Posterior (marginal) distribution for the slope β_2 for age. B: Standard normal distribution.

8.4.3 Output for the hyperparameters

Numerical output for the hyperparameters

We now focus on the hyperparameter. The model contains a standard deviation parameter σ that is being used for the variance (σ^2) of the normal distribution for the latency score. As explained in Chapter 7, R-INLA works with precision (τ), which is defined as $\tau = 1 / \sigma^2$. The `$summary.hyperpar` output contains information on the precision parameter τ.

```
> I1$summary.hyperpar
```

```
                          mean     sd  0.025q  0.5q
Precision Gaussian obs. 1.055  0.089   0.892 1.052

                         0.975q    mode
Precision Gaussian obs.   1.237   1.045
```

The output does not fit on one line, but it shows the posterior mean, standard deviation, quantile information, and the mode for τ. This is all nice, but we are not really interested in the information on τ. We want to see the information on σ. Recall that $\tau = 1 / \sigma^2$ and this means that σ can be obtained via

$$\sigma = \sqrt{\frac{1}{\tau}} = \frac{1}{\sqrt{\tau}} \tag{8.2}$$

Here we have a little problem. The expected value of σ is ***not*** equal to 1 divided by the square root of the expected value of τ. In Chapter 7 we gave the mathematical expression for expected values. We will reproduce it here.

$$E(\tau) = \int_{-\infty}^{\infty} \tau \times p(\tau) d\tau \tag{8.3}$$

In this equation the τ is the value along the x-axis in Figure 8.3A and $p(\tau)$ is the density function along the y-axis. If σ is a function of τ, say $\sigma = h(\tau)$, then the following expression holds:

$$E(\sigma) = \int_{-\infty}^{\infty} h(\tau) \times p(\tau) d\tau = \int_{0}^{\infty} \frac{1}{\sqrt{\tau}} \times p(\tau) d\tau \tag{8.4}$$

Luckily, R-INLA has support functions to calculate this integral for us.

```
> tau <- I1$marginals.hyperpar$`Precision for the
         Gaussian observations`
```

Just as in the previous subsection, the object `tau` contains an `x` and a `y` variable, each of length 75. The `x` variable contains 75 possible values for τ and the `y` variable contains the corresponding values of the distribution (and these will be plotted against each other in a moment). The support function `inla.emarginal` applies Equation (8.4) to get the mean value of σ. To avoid having to type 1 / sqrt(x) all the time we create a small support function `MySqrt`.

```
> MySqrt <- function(x) { 1 / sqrt(x) }
> sigma  <- inla.emarginal(MySqrt, tau)
> sigma

0.9764446
```

If you just type `inla.emarginal` in R without any arguments then you can see the source code for this function. It first applies the spline smoothing (it makes the density function smoother; see Figure 8.3B) and then approximates the integral in Equation (8.4) with a sum of the product of the density function and 1 divided by the square root of τ.

Typing the posterior mean value of τ directly into the square root

```
> 1 / sqrt(1.055884)

0.9731771
```

gives a similar number but this approach is faulty! So, we really need to use the support functions from R-INLA to convert the posterior mean of τ into the posterior mean of σ. The same holds for other quantities of the distribution of σ. The function `inla.emarginal` is fine if you only want the expected value. Use `inla.tmarginal` in combination with e.g. `inla.zmarginal` for more statistics on the transformed marginal.

Use R-INLA support functions to obtain information on σ from τ.

Graphical output for the hyperparameters

Besides these numbers we can also visualise the posterior marginal distribution of τ; see Figure 8.3A. This graph was created with the following code.

```
> tau <- I1$marginals.hyperpar$`Precision for the
         Gaussian observations`
```

The object `tau` contains an `x` and a `y` variable. The `x` variable contains a range of τ values and the `y` variable contains the corresponding values of the distribution. Plotting them against each other gives us Figure 8.3A.

```
> plot(x = tau[,1], y = tau[,2],  type = "l",
       xlab = expression(tau),
       ylab = expression(paste("P(", tau ," |
                                 Data)")))
```

The line representing the posterior marginal distribution is slightly peaky because only 75 points are used. You can make it smoother by applying spline smoothing; see Figure 8.3B. It is essentially the same graph as in panel A, except the density curve is less peaky. The R code to make a smoother curve uses the R-INLA support function `inla.smarginal`.

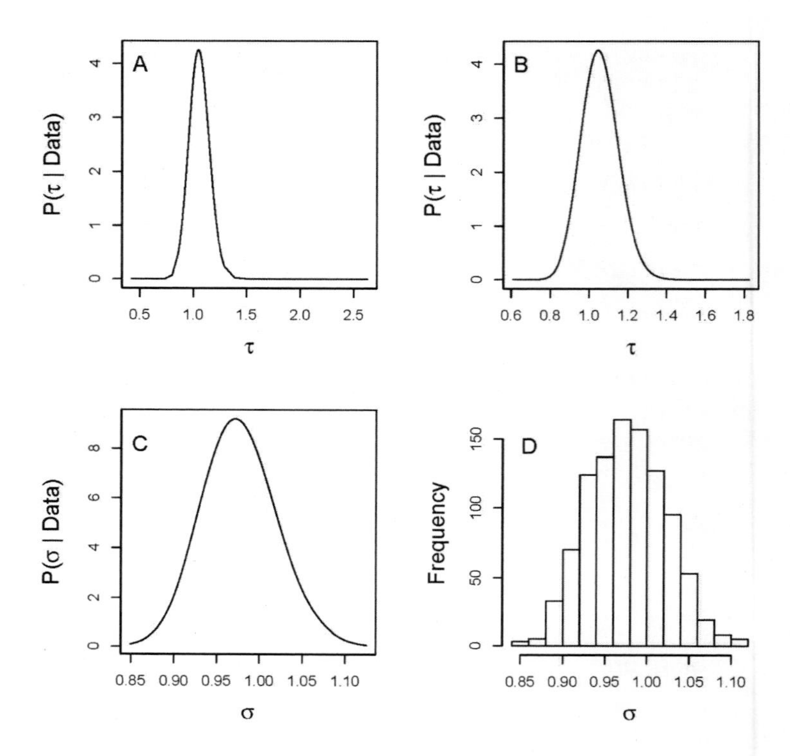

Figure 8.3. A: Posterior marginal distribution of τ. B: Spline smoothing applied on the posterior marginal distribution of τ. C: Posterior marginal distribution of σ. D: 1,000 random samples from the posterior marginal distribution of σ.

```
> tau.distr <- inla.smarginal(tau)
> plot(x = tau.distr$x,
       y = tau.distr$y,
       type = "l",
       xlab = expression(tau),
       ylab = expression(paste("P(", tau ," |
                                Data)")))
```

The function `inla.tmarginal` transforms the marginal distribution of τ into that of σ.

```
> sigma.distr <- inla.tmarginal(fun = MySqrt,
                                 marg = tau)
```

This function does things similar to what we just described for the posterior means of τ and σ, but instead uses quantiles. We can plot the posterior marginal distribution of σ (Figure 8.3C) via the following code.

```
> plot(x = sigma.distr[,1],
       y = sigma.distr[,2],
       type = "l",
       xlab = expression(sigma),
       ylab = expression(paste("P(", sigma ," |
                                       Data)")))
```

The distribution for sigma is not necessarily symmetric. There is also a function `inla.contrib.sd` that simulates 1,000 values from the posterior marginal distribution of σ. We can make a histogram of these 1,000 simulated values (Figure 8.3D), or we can take summary statistics like a mode, mean, or 95% credible intervals.

```
> sigma.info <- inla.contrib.sd(I1)
> hist(sigma.info$samples,
       xlab = expression(sigma),
       main = "")
```

8.4.4 Fitted model

Now that we have all the numerical output, let us try to understand what these results tell us. We have a model with four covariates; three of them are categorical variables. The concept of a categorical variable adjusting the intercept (and an interaction adjusting the slope) is the same as in frequentist analysis. This means that we can write the fitted model as:

$$Latency_i \sim N\left(\mu_i, 0.97^2\right)$$
$$E\left(Latency_i\right) = \mu_i \quad \text{and} \quad \text{var}\left(Latency_i\right) = 0.97^2$$

For a chimpanzee that has sex = 1, colony = 1, and rear = 1 we have

$$\mu_i = -0.44 - 0.02 \times Age_i$$

and when sex = 2, colony = 2, and rear = 2 the mean value takes the form

$$\begin{aligned}\mu_i &= -0.44 + 0.02 \times Age_i - 0.22 + 0.09 + 0.05 \\ &= -0.52 + 0.02 \times Age_i\end{aligned}$$

We have six different scenarios and therefore there are six equations for the fitted values.

8.5 Model validation

Now that we have the numerical results we can proceed to the next step of the analysis and apply model validation. We need to obtain residuals and plot these against fitted values (which we also need to obtain from R-INLA) to check for homogeneity. We must also plot the residuals versus each covariate in the model and each covariate not in the model to check for patterns. And finally, we need to inspect the residuals for spatial and temporal dependency, though that is not relevant in this specific example.

In a frequentist analysis we can use the `resid` function to obtain residuals, but there is no such function available to extract the residuals of a model fitted in R-INLA. Instead, we have to calculate the residuals ourselves. To do this we first calculate an X matrix containing the covariate values (see Chapter 2) with the `model.matrix` function. We then multiple the X matrix with the posterior mean values of the regression parameters to get the fitted values. Once we have the fitted values we can calculate the residuals. Here is the R code.

```
> X <- model.matrix(~ Age + fSex + fColony +fRear,
                    data = Chimp)
> F1 <- X %*% Beta1[,1]
> E1 <- Chimp$Latency - F1
```

We can use the residuals `E1` for model validation. Figure 8.4 shows the residuals plotted versus fitted values. Note that there are two observations with relatively large residual values.

It is also possible (and safer) to let R-INLA calculate the fitted values via the `control.predictor` option in R-INLA; see the R code below. The `compute = TRUE` argument in the `control.predictor` option ensures that fitted values are calculated inside R-INLA.

It is also possible to get the 95% credible intervals for each fitted value. This is what the quantiles argument achieves; if you omit it you get the default quantiles (0.025, 0.25, 0.5, 0.75, 0.975).

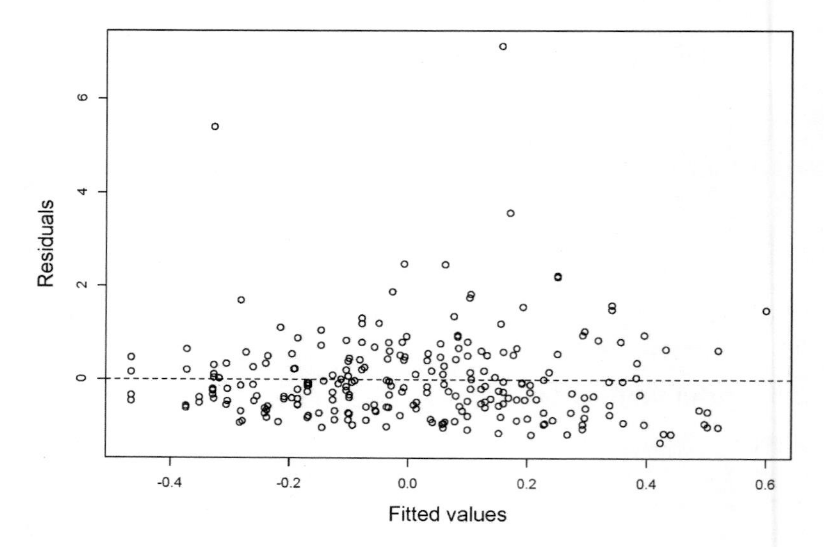

Figure 8.4. Residuals plotted versus fitted values.

```
> I2 <- inla(Latency ~ Age + fSex + fColony +
                       fRear,
             family = "gaussian",
             control.predictor = list(
                 compute = TRUE,
                 quantiles = c(0.025, 0.975)),
             data = Chimp)
> Fit2 <- I2$summary.fitted.values[,"mean"]
```

The object `I2$summary.fitted.values` contains the posterior mean fitted values of the model and the 95% credible intervals for the fitted values.

Summarising, we can obtain fitted values in two different ways. We can either use $\mathbf{X} \times \boldsymbol{\beta}$ (where $\boldsymbol{\beta}$ contains the posterior mean values) or via the `control.predictor` option). These two approaches give fitted values that are not necessarily identical, but the differences between them are typically small (<0.001). However, this only holds for a model with the identity link. As soon as we use a generalised linear (mixed) model with a link function that is not the identity link then the expected value of the function is not the same as the function of the expected value.

$$E\left(h(x)\right) \neq h(E\left(x\right))$$

See also Equations (8.3) and (8.4). Figure 8.5 shows a plot of the fitted values versus the observed latency scores. The fit is rather poor!

Though not relevant for a Gaussian model with the identity link, the `summary.linear.predictor` yields the fitted values at the link scale whereas `summary.fitted.values` yields them at the original scale. Both are provided with `compute = TRUE`.

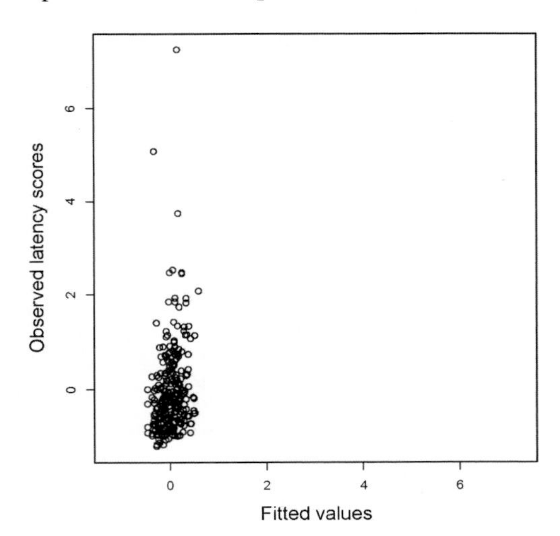

Figure 8.5. Observed latency scores plotted versus fitted values.

If possible, use R-INLA support functions to obtain information on fitted values. When using a linear regression model or linear mixed effects model with the identity link you can still calculate fitted values manually.

8.6 Model selection

8.6.1 Should we do it?

Most scientists fit a statistical model with a certain number of covariates and want to know which covariates are important and which are not. When you have a model with a large set of covariates it is tempting to drop the covariates that are not important because it makes the interpretation part of the analysis much easier.

The process of dropping covariates is called model selection. This is a highly controversial area (Johnson and Omland 2004; Whittingham et al. 2006; Murtaugh 2009; Mundry and Nunn 2009, among others).

In the frequentist world the process of model selection is not that difficult. Tools like the Akaike Information Criterion (Akaike 1973; AIC), the Bayesian Information Criteria (BIC), *p*-values (Halsey et al. 2015), and the Information Theoretic approach (Burnham and Anderson 2002) are all worked out and explained well in the literature.

In the Bayesian world model selection is more complicated. Zuur et al. (2016a) discuss various model selection tools that can be used for Markov chain Monte Carlo (MCMC) simulation, e.g. the Kuo and Mallick approach and Gibbs variable selection. But R-INLA does not do MCMC and therefore these approaches do not work with R-INLA. In this section we will discuss various model selection tools that can be used with R-INLA.

Before reading on, ask yourself the question whether you really want to do model selection; you formulated biological questions and you fitted a model to answer these questions. We would say: 'Job done'.

8.6.2 Using the DIC

In classical model selection the AIC is a popular tool. The AIC is defined as

$$AIC = -2 \times Log(L) + 2 \times p$$

where L is the likelihood of the data given the parameters and p is the number of parameters. The term $-2 \times$ log likelihood is called the deviance. The more covariates we add the better the fit of the model (higher L), but more covariates also means a larger p (penalty). We can calculate the AIC for models with different sets of covariates, and the model with the lowest AIC is deemed the 'best' model.

The AIC works fine as long as we can calculate the log likelihood and we know the number of parameters. When we use multiple linear regression or generalised linear models with diffuse priors we know the number of parameters. If we use informative priors then the regression parameters do not have the full flexibility anymore to take any value; their range is limited. This affects the degrees of freedom and therefore the p term in the AIC. In such a situation it may be better not to use the AIC or to be cautious with it. Up to now we have not used regression parameters with informative priors in INLA. But as soon as we use linear mixed-effects models or generalised linear mixed models in INLA we will use informative priors for the random effects, in which case it becomes rather difficult to define p in the AIC. In those cases we can use the deviance information criteria (DIC, Spiegelhalter et al. 2002).

Bayesian model selection with AIC creates problems if you use informative priors (which is the case for mixed models). Use the DIC in that case.

The DIC is defined rather similarly to the AIC, and we use it in the same way (i.e. to fit models with different sets of covariates and compare the DICs). To simplify notation, let θ contain all the regression parameters, and $f(y \mid \theta)$ is the likelihood. The DIC is defined by

$$DIC = D(\bar{\theta}) + 2 \times p_D$$

According to this expression the DIC is calculated as the deviance using the posterior mean of the parameters plus two times p_D, where the latter is the effective number of parameters. INLA uses the posterior mean values for the betas and the posterior mode for the hyperparameters to calculate the DIC. The reason for using the mode is that posterior marginal distributions for hyperparameters tend to be skewed.

Note the similarity between the AIC and DIC! If diffuse priors are used then p_D is approximately equal to p (number of parameters) and the posterior mean is approximately equal to the frequentist maximum likelihood estimators, and therefore DIC ≈ AIC. Seeing is believing, so let us verify this with some R code. First we rerun model `I2` in a frequentist setting using the `lm` function and get the log likelihood and the AIC.

```
> M1 <- lm(Latency ~ Age + fSex + fColony + fRear,
           data = Chimp)
> logLik(M1)

'log Lik.' -336.2983 (df=7)

> AIC(M1)

686.5965
```

If you calculate $-2 \times -336.2983 + 2 \times 7$ then you get indeed the 686.596. Next we run the same model in INLA but add the `dic = TRUE` option to `control.compute`; it tells INLA to calculate the DIC.

```
> I3 <- inla(Latency ~ Age + fSex + fColony +
```

```
                         fRear,
              family = "gaussian",
              control.compute = list(dic = TRUE),
              data = Chimp)
> I3$dic$dic

686.6012
```

The AIC and DIC are indeed similar.

Effective number of parameters

There are two ways to calculate the effective number of parameters. Spiegelhalter et al. (2002) used

$$p_D = \bar{D} - D(\bar{\theta})$$

This is the mean of the deviances minus the deviance calculated at the posterior means. The justification for this expression is rather technical and is not explained here. Using the expression of the p_D we can also write the DIC as

$$\begin{aligned} DIC &= D(\bar{\theta}) + 2 \times p_D \\ &= D(\bar{\theta}) + 2 \times \left(\bar{D} - D(\bar{\theta})\right) \\ &= 2 \times \bar{D} - D(\bar{\theta}) \\ &= \bar{D} + \bar{D} - D(\bar{\theta}) \\ &= \bar{D} + p_D \end{aligned}$$

Millar (2009) showed that for generalised linear mixed models applied on overdispersed count data, model selection based on DIC produced rather conflicting results.

Use DIC with care if GLMMs are used. See Millar (2009).

DIC-related output

DICs always confuse us a little bit, so let's calculate them ourselves. If we type `summary(I3)` we get the posterior mean values of the fixed parameters and hyperparameters. We already presented and discussed these. We have not shown the following part of the output of the model yet.

```
Expected number of effective parameters(std dev):
6.004(4e-04)
Number of equivalent replicates: 40.47

Deviance Information Criterion (DIC) ...: 686.60
Effective number of parameters .........: 6.99
```

```
Marginal log-Likelihood:   -379.33
```

In the middle we have the DIC (686.60) and the effective number of parameters p_D (= 6.99). You can verify the value of p_D via

```
> pD <- I3$dic$mean.deviance -I3$dic$deviance.mean
```

It gives the 6.99 again. There is a second sentence in the output related to the effective number of parameters, but this time it is called 'Expected number of effective parameters'. That is an alternative approach to calculate the p_D; it goes via the trace of a matrix. One would hope that both approaches to calculate the p_D give similar values. In general the first approach (giving the 6.99 in this case) is more accurate. You can get the p_D directly by typing

```
> I3$dic$p.eff
```

The DIC can also be calculated manually.

```
> DIC <- I3$dic$mean.deviance + pD
```

It gives the same value as via `I3$dic$dic`. Since we are showing the numerical output, we might as well explain the two remaining pieces in the numerical output that we haven't mentioned yet.

The marginal likelihood is the likelihood in which all parameters have been integrated out. This likelihood can be used for model comparison. It can also be written as the probability of the data given the model: $p(Y \mid \text{Model})$.

Finally, there is the 'Number of equivalent replicates'. We have a sample size of 243 observations and the model has six regression parameters. That is about 40 observations per parameter. A general recommendation in statistics is to have at least 15 observations per parameter. In later chapters we will use mixed-effects models and models with spatial and temporal correlation. In those models it is more difficult to calculate the number of parameters and the 'Number of equivalent replicates' measures the number of observations per parameter. It is obtained by dividing the sample size by the 'Expected number of effective parameters'.

Model selection using DIC

Let us return to the DIC. In and of itself the DIC value does not tell us anything. We need to drop or add covariates and compare DIC values of different models. There is no `step` function that applies a backwards or forwards selection. If you want to carry out a backwards selection, then you have to do this manually. The code below drops 1 covariate in turn, and each time the new model is fitted in R-INLA.

```
> I3a <- inla(Latency ~  fSex + fColony + fRear,
             control.compute = list(dic = TRUE),
             data = Chimp)
```

```
> I3b <- inla(Latency ~ Age +  fColony + fRear,
              control.compute = list(dic = TRUE),
              data = Chimp)
> I3c <- inla(Latency ~ Age + fSex + fRear,
              control.compute = list(dic = TRUE),
              data = Chimp)
> I3d <- inla(Latency ~ Age + fSex + fColony,
              control.compute = list(dic = TRUE),
              data = Chimp)
> c(I3$dic$dic, I3a$dic$dic, I3b$dic$dic,
    I3c$dic$dic, I3d$dic$dic)

686.6012 692.3636 687.4633 685.0027 682.7205
```

The model with the lowest DIC is `I3d`, which is the model without fRear. Its DIC is lower than that of the full model; hence this covariate is not important. We can then proceed with model `I3d`, and drop each covariate in turn and obtain a DIC value. We continue this process until the DIC increases. Intermediate results of this backwards selection approach are not presented here, but the end result is a model with only age as the covariate. That is this model:

```
> I6 <- inla(Latency ~ Age,
             control.compute = list(dic = TRUE),
             data = Chimp)
```

If you did decide to remove the three monkeys with the highest latency values you end up with a model that contains age and fRear!

In case you set `control.compute = list(dic = TRUE)`, then `compute` is set by default to TRUE.

WAIC

Besides the DIC, it is also an option to use the widely applicable information criterion (WAIC), introduced by Watanabe (2010). It is also referred to as the Watanabe–Akaike information criteria.

Just like the AIC the WAIC consists of two terms representing the quality of fit and model complexity. Models with smaller WAIC values are preferred. WAICs are seen as an improvement on DICs; see Gelman et al. (2013). The WAIC is closely linked to Bayesian leave-one-out cross-validation. A technical explanation of WAICs is given in Watanabe (2010) and Gelman et al. (2013).

It should be noted that currently no selection criterion is deemed 'good' in the literature. AIC has problems with strong informative priors. DIC has problems if a posterior distribution is not well represented by its posterior mean. WAIC is based on data partition, which may cause trouble for spatial and temporal correlated data.

The WAIC value can be obtained by adding

```
control.compute = list(dic = TRUE, waic = TRUE)
```

to the inla command. The value itself is accessed via `I6$waic$waic`.

8.6.3 Out of sample prediction

In this subsection we discuss three tools that can also be used for model selection. However, somehow we always end up with examples in which these approaches don't seem to be useful. We decided to present the methods in this book so that you know that they exist, and perhaps they work for your data.

In Chapter 15 of Zuur et al. (2016a) we analysed egg-laying preferences of butterflies (*S. ilicis*). The data were originally analysed in Maes et al. (2014), who visually inspected 251 oaks in a nature reserve. Zuur et al. (2016a) modelled the number of *S. ilicis* eggs as a function of a set of covariates. Eventually we ended up with three potential models, namely a zero-inflated Poisson model, a zero-altered Poisson model, and a negative binomial GLMM. For the storyline of this section, it is not important that you are familiar with the technicalities of these models (we will encounter them later in this book). Instead of relying on an AIC or a DIC, a different approach was followed to select from the three candidate models. We randomly split the data into two exclusive sets, $Data_{Fit}$ and $Data_{pred}$. Each model was fitted using the $Data_{Fit}$ data set, and once we obtained the parameter estimates we used the covariates in $Data_{pred}$ to predict the number of eggs. Because we also had the observed number of eggs in $Data_{pred}$, we could quantify a prediction error for a particular model. The model with the lowest prediction error was deemed 'best'. This approach is also called 'cross-validation'.

R-INLA can also do cross-validation. The $Data_{Fit}$ contains all observed data except for one observation; that is the $Data_{pred}$. This is a 'leave one out' cross-validation. If you have 100 observations, R-INLA applies the 'leave one out' cross-validation on each observation (though luckily there is no need to run R-INLA 100 times).

R-INLA gives so-called CPO and PIT values. These are both 'leave one out' cross-validation scores. CPO stands for conditional predictive ordinate and PIT stands for probability integral transform.

CPO goes back to Geisser (1980). It is defined as the probability of an omitted observation given all the other data. For the monkey data this is

$$CPO_i = Pr(\text{omitted latency score } i \mid \text{all other observations})$$

A high CPO_i value means that an observed latency value for monkey i complies with (or supports) the model. A low probability means that the latency score is surprising in the light of the other observations, models, and prior knowledge. Another way of interpreting this value is that a high (close to 1) value of CPO_i implies a good model fit and a low CPO_i value means a poor fit of observation i (Lawson 2013).

PIT values are an alternative to CPO and are briefly discussed in Blangiardo and Cameletti (2015).

R-INLA can give us a CPO_i value for each observation, as we will illustrate next. We use all covariates again as the CPO is an alternative tool to the DIC.

```
> I7 <- inla(Latency ~ Age + fSex + fColony +
                        fRear, data = Chimp,
              control.compute = list(cpo = TRUE))
```

Note the `cpo = TRUE` argument. To get the CPO values we type

```
> I7$cpo
```

This list contains three objects. The first one is `I7$cpo$cpo` (these are the 243 CPO values we are after, one per observation), the second one is `I7$cpo$pit` (these are the 243 PIT values), and the third one is `I7$cpo$failure`. If an observation has a failure value equal to 1 then we cannot trust the corresponding CPO or PIT value. In this case all 243 failure values are equal to 0, so we can proceed with the interpretation of the CPO values.

The `I7$cpo$cpo` contains 243 probabilities and Figure 8.6 shows a Cleveland dotplot of all the CPO values. Note that all CPO values are on the small side, indicating a poor fit. This doesn't come as a surprise after seeing the poor fit in Figure 8.5.

If we ignore for the moment that we have a model with a poor fit, we proceed to the explanation of how to use CPO for model selection. We should now fit models with different sets of covariates and see ('eyeball') which one has better CPO values. Clearly, if you drop an important covariate then you may see differences between CPO values of different models. But in our experience it is quite a challenge to visually observe differences between CPO values of different models.

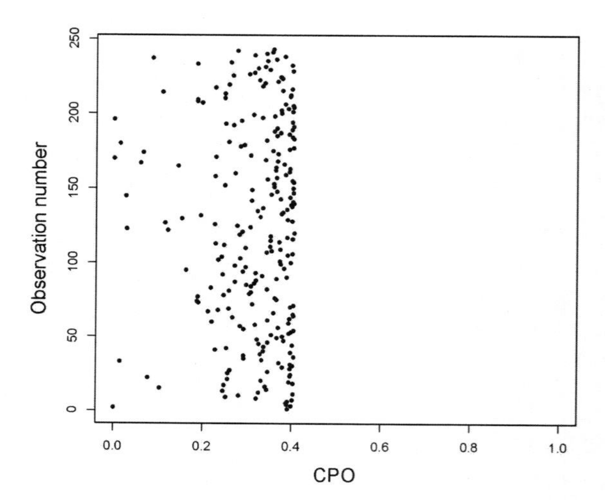

Figure 8.6. CPO values for the model with all four covariates.

It is also possible to summarise all 243 CPO values with one value. The sum of all the log CPO values can be used for model comparison; the larger the sum of all log CPO values, the better the model (Gneiting and Raftery 2007). To use this logarithmic score for model selection purposes we need to fit the model with all four covariates, get the CPOs, log transform them, and add them all up:

```
> sum(log(I7$cpo$cpo))
```

Then we have to remove each covariate in turn, refit the model, get the CPOs, and calculate the sum of the log CPO values again. That gives us four logarithmic scores. Pick the model with the larger score as it is the 'best' model. There is no function that implements a full forward or backwards selection using the sum of logarithmic scores, so you have to do it all yourself.

8.6.4 Posterior predictive check

One way to assess whether a model fits the data well is by inspecting residuals. Small residuals indicate a good fit, but when we work with scaled or normalised residuals (which is what we do with GLMs) then we can only look at (the absence) of patterns in residuals. Not so much in the absolute values of the residuals. Another way to assess the quality of a model is an R^2, but it is not always easy to calculate an R^2. An informal tool to assess the performance of a model is to plot the observed data versus the fitted data. If the scatter of points is roughly on a straight 45° line then the fit is likely to be good. A problem with this approach is that count data often have lots of zeros and some of the distributions (e.g. negative binomial and Poisson) allow for more variation for larger fitted values. This can make the interpretation of the scatterplot difficult.

Yet another way to assess the quality of the model is to perform posterior predictive checks. This can be done either by simulating data from the model or we can look at the probability of similar values of the response variable.

The first approach was used in Zuur et al. (2016a) in the context of MCMC, and it is also explained in Kéry and Royle (2016). It can also be done in a frequentist setting using bootstrapping. It works as follows. Suppose we have fitted a model (so we have the observed data and the parameters) and for a specific monkey (say monkey 1) we simulate 10,000 latency values *from* the model. For a good model we expect that these 10,000 simulated values are comparable to the observed latency value. To assess whether this is indeed the case we have multiple options. One option is to calculate residuals for the 10,000 simulated values, compare the 10,000 squared residuals with the original squared residual, and count how often the squared residuals for the simulated data are larger than for the observed data. If this is about fifty-fifty then the simulated data complies with the observed data. The process can then be repeated for all other data points.

Blangiardo and Cameletti (2015) looked at the probability of similar values of the response variable Pr(*Latency*$_i$* | all latency data). This is the probability for replicate latency values for monkey *i* (denoted by *Latency*$_i$*) given all latency data. This is called a posterior predictive distribution. We can easily get these for each observation in R-INLA:

```
> I7 <- inla(Latency ~ Age + fSex + fColony +
                         fRear,
             control.predictor = list(
                                   compute = TRUE),
             data = Chimp)
```

We end up with a posterior predictive distribution for each observation *Latency*$_i$, and they are all in `I7$marginals.fitted.values`. We can easily calculate the marginal fitted values for a specific observation. For example, for Monkey 1 we use:

```
> pmY1 <- I7$marginals.fitted.values[[1]]
```

Just as in Section 8.3 we can plot this distribution; see Figure 8.7. This graph shows the probability of other latency values for this specific observation. The key question is now: Where, in this distribution, is the observed latency value for Monkey 1? If it is in the tail then that is not good. If it is somewhere in the centre then that would be great. To quantify this the posterior predictive *p*-value is defined as Pr(*Latency*$_i$* ≤ *Latency*$_i$ | all latency data). If this is near 0 or 1 then the model does not fit the data well. If most values are close to 0.5 then the model fits the data well.

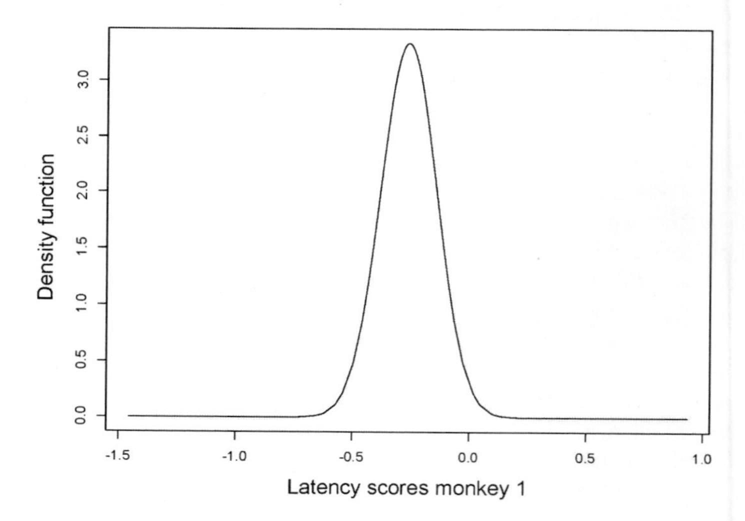

Figure 8.7. Posterior predictive distribution Pr(*Latency*$_1$* | all latency data) for Monkey 1. The *x*-axis shows potential other latency scores and the *y*-axis shows the corresponding distribution.

We calculate the Bayesian p-value for every observation using the following R code:

```
> pval <- rep(NA, nrow(Chimp))
> for (i in 1:nrow(Chimp)){
   pval[i] <- inla.pmarginal(q = Chimp$Latency[i],
      marginal = I7$marginals.fitted.values[[i]])}
```

Figure 8.8 shows a histogram of the 243 Bayesian p-values. Note that most Bayesian p-values are close to either 0 or 1, indicating a rather poor model fit!

These Bayesian p-values can be calculated for different models and the model with the best-looking histogram of Bayesian p-values can be selected.

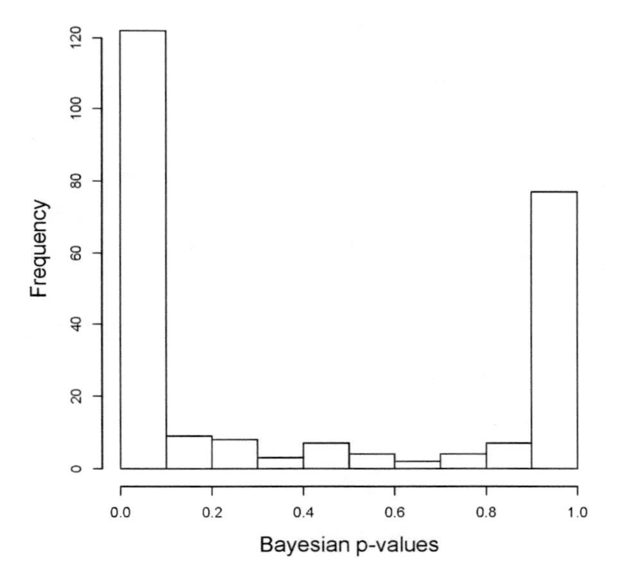

Figure 8.8. Bayesian *p*-values.

8.7 Visualising the model

In this section we visualise the optimal model, following the protocol presented in Zuur et al. (2016b). In Section 8.5 we used the DIC for model selection and we found that the model with age as the only covariate was the best model. On the one hand that is nice as it means that the model is rather simple to present. On the other hand it is unfortunate that there is only one covariate, as it doesn't allow us to fully explain the R-INLA code that is required to visualise fitted values of more complicated models. For this reason we present the results of the model with age and sex, and we will ignore the fact that the second covariate is not important in the model.

To explain what we want, we will first fit the model in a frequentist setting and visualise the results. We use the `lm` function to fit the model with age and sex as covariates.

```
> M8 <- lm(Latency ~ Age + fSex, data = Chimp)
```

To visualise the model we create a grid of artificial covariate values for which we will predict latency values. We will use 25 age values for the males and 25 age values for the females. The males and females may have different ranges for age, and because we don't want to extrapolate we use some `ddply` magic to avoid extrapolation of age.

```
> library(plyr)
> MyData <- ddply(Chimp,
                  .(fSex),
                  summarize,
                  Age = seq(from = min(Age),
                            to = max(Age),
                            length = 25))
```

We now have 25 age values for each sex. That is 50 rows in total. We can use the `predict` function to obtain the predicted latency values for our 50 covariate values. Before we do that we make a backup of `MyData` as we will need a clean version in a moment for R-INLA.

```
> MyData2 <- MyData  #Backup
> P8 <- predict(M8, newdata = MyData, se = TRUE)
```

If you don't want to use the `predict` function then we can also calculate the predicted values and standard errors ourselves.

```
> X <- model.matrix(~ Age + fSex, data = MyData)
> beta  <- coef(M8)
> MyFit <- X %*% beta
```

These fitted values are exactly the same as the ones in `P8`. We can even calculate the standard errors ourselves, though the underlying mathematical expression is not so easy on the eyes (but it can be found in any statistics textbook).

```
> MySE <- sqrt(diag(X %*% vcov(M8) %*% t(X)))
```

For `ggplot2` plotting purposes we add the fitted values and the 95% confidence intervals for the fitted values to `MyData`.

```
> MyData$mu    <- MyFit                     #Fitted values
> MyData$selow <- MyFit - 1.96 * MySE #Lower bound
> MyData$seup  <- MyFit + 1.96 * MySE #Upper bound
```

Now we are ready to plot the model fit with `ggplot2`; see the elementary R code below. The resulting graph is presented in Figure 8.9. This is a graph that should be added to a scientific paper; see the Zuur et al. (2016b) protocol.

```
> p <- ggplot()
> p <- p + geom_point(data = Chimp,
                      aes(y = Latency, x = Age),
                      shape = 16, size = 3)
> p <- p + xlab("Age") + ylab("Latency")
> p <- p + geom_line(data = MyData,
                     aes(x = Age, y = mu))
> p <- p + geom_ribbon(data = MyData, alpha = 0.5,
                       aes(x = Age,
                           ymax = seup,
                           ymin = selow ))
> p <- p + facet_grid(. ~ fSex, scales = "fixed")
> p
```

The question is now how to produce the same graph in R-INLA, as there is no `predict` function. There are two options to get the same graph; we present only one here. The second approach (via `inla.make.lincombs`) is illustrated in Chapter 9. We will also compare both approaches in Chapter 9.

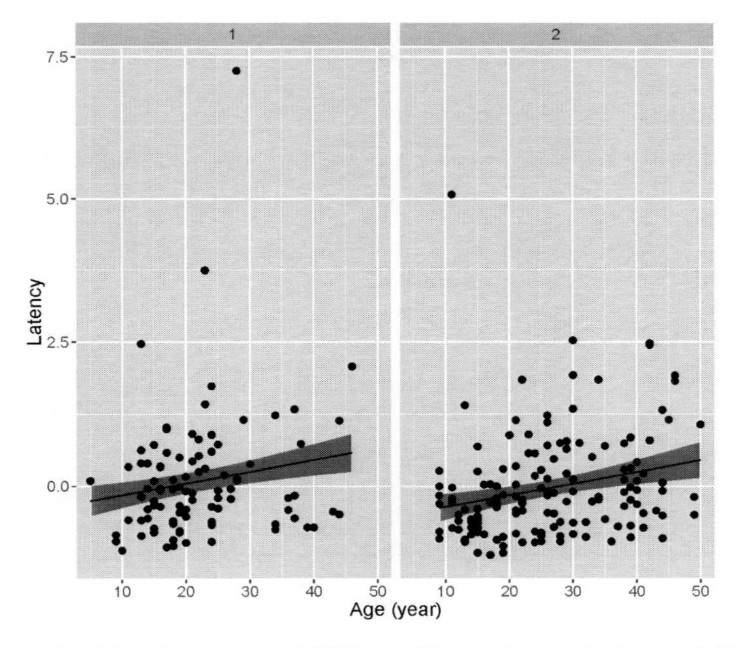

Figure 8.9. Fitted values and 95% confidence intervals for model M8 (obtained by a frequentist approach).

R-INLA does the same as JAGS; wherever it sees an NA for the response variable it will produce a posterior (marginal) distribution. What we have to do is add the 50 artificial covariate values from `MyData2` to the `Chimp` data object, and add 50 NAs to the response variable. But this will give an error message as the number of columns in MyData and Chimp don't match. We need a little bit of ugly R coding to combine the two data sets. First we prepare the artificial data in MyData2 by adding NAs. Recall that MyData2 is a copy of the original MyData (without any frequentist results). All that we do is add a column with NAs and call the new column 'Latency'.

```
> MyData3 <- MyData2
> MyData3$Latency <- NA
> head(MyData3, 4) #Show first 4 rows

  Latency fSex       Age
1      NA    1  5.000000
2      NA    1  6.708333
3      NA    1  8.416667
4      NA    1 10.125000
```

Next we extract the relevant columns of Chimp.

```
> Chimp2 <- Chimp[, c("Latency", "fSex", "Age")]
```

The crucial point is to ensure that the order of the columns in Chimp2 is the same as in MyData3 because we will combine them with the rbind function.

```
> Chimp.Comb <- rbind(Chimp2, MyData3)
```

This is a rather error-prone process, but once you have reached this stage without any warning or error messages the rest is easy. We rerun the model in R-INLA using the combined data set. Ensure that the compute = TRUE is selected in the control.predictor; otherwise R-INLA will not calculate the fitted values.

```
> I8 <- inla(Latency ~ Age + fSex,
           control.predictor = list(compute = TRUE,
                        quantiles = c(0.025, 0.975)),
           data = Chimp.Comb)
```

Rows 1 to 243 in I8$summary.fitted.values are for the observed data and rows 244 to 293 are for the added data with NA for latency. We extract the later ones and add the relevant pieces to MyData2. If you now rerun the ggplot2 code with MyData2 instead of MyData you will get exactly the same graph as in Figure 8.9.

```
> FIT <- I8$summary.fitted.values[244:293,]
> MyData2$mu    <- FIT[,"mean"]      #Fitted values
> MyData2$selow <- FIT[,"0.025quant"] #Lower bound
> MyData2$seup  <- FIT[,"0.975quant"] #Upper bound
```

If you don't run the frequentist code then you can just stick to MyData.

9 Mixed effects modelling in R-INLA to analyse otolith data

In this chapter we show how to fit linear mixed-effects models in R-INLA. We will use a fisheries data set to illustrate the important steps.

Prerequisite for this chapter: Knowledge of multiple linear regression and linear mixed effects models and how to fit these models in R-INLA is required (see Chapters 7 and 8).

9.1 Otoliths in plaice

Otoliths are earstones produced by 96% of fish species. Just like trees otoliths have rings, which can be used to determine the age of a fish. The otolith is made of calcium carbonate, which is primarily derived from water. Studying the trace elemental composition of the otolith, fisheries biologists hope to determine in which water bodies a fish has been (e.g. for stock identification). But if you want to say where a fish has been based on its trace elemental composition of the otolith, it is rather important to know what is more relevant for this composition: environmental or physiological factors.

The data used in this chapter were taken from Sturrock et al. (2015), who carried out an experiment to identify the main controls on otolith microchemistry in European plaice (*Pleuronectes platessa* L.). In the experiment, 25 fish were kept in near-natural conditions in a tank for 7–12 months. Physiological variables from each fish (total length, weight, Fulton's condition factor, growth rate, blood plasma protein, and elemental concentrations) and environmental variables (salinity, temperature, seawater elemental concentrations) were measured at least monthly. At the end of the experiment, otolith measurements were quantified retrospectively.

Concentrations like ^{7}Li, ^{26}Mg, ^{41}K, ^{48}Ca, ^{88}Sr, ^{138}Ba, and element / calcium ratios were determined for the seawater (environmental variables), the blood plasma (physiological variable), and from the otolith (response variable).

Other factors that may influence the response variable are sex of the fish (male versus female), origin of the fish (Irish Sea versus English Channel), and whether the fish received a certain hormone (GnRH) to encourage spawning.

9.2 Model formulation

Sturrock et al. (2015) applied a series of models using different response variables, for example the Li / Ca ratio in the otoliths, the K / Ca ratio in the otoliths, the Sr / Ca ratio in the otoliths, etc. In this chapter we will repeat one of their analyses, namely for the Sr / Ca ratio. It is one of the most used ratios in this field. Sr stands for strontium.

As covariates we will use sex of the fish, GnRH treatment, origin of the fish, the environmental variables salinity, temperature, Sr concentration in the water, the Sr / Ca ratio of the water, and the physiological variables age, total length, weight, Fulton's condition factor, growth rate, blood plasma protein, Sr concentration in the blood, and Sr / Ca ratio in the blood. This leads to a model of the form (in words):

$$\begin{aligned}\text{Sr / Ca ratio} = \;&\text{Intercept + Sex + GnRH treatment + Origin +}\\ &\text{Lots of environmental variables +}\\ &\text{Lots of physiological variables +}\\ &\text{Noise}\end{aligned} \tag{9.1}$$

Sturrock et al. (2015) also included all two-way interactions, but the data set is not large enough for our liking for interactions.

9.3 Dependency

The experiment consisted of 25 fish, but only 19 fish exhibited sufficient otolith growth suitable for the statistical analyses. We have multiple observations over time from the same fish, as can be seen from Figure 9.1. This means that we have dependency. We have time series that consist of only a few observations per fish, so perhaps we may be able to avoid a more complicated model with temporal dependency by applying a linear mixed-effects model with random intercept 'fish'. Such a mixed-effects model assumes that all Sr / Ca ratio values from the same fish are correlations with a value ϕ (the intraclass correlation) and Sr / Ca ratios from different fish are independent. This means that we need to implement a model of the form:

$$\begin{aligned}\text{Sr / Ca ratio} = \;&\text{Intercept + Sex + GnRH treatment + Origin +}\\ &\text{Lots of environmental variables +}\\ &\text{Lots of physiological variables +}\\ &\text{Random intercept Fish +}\\ &\text{Noise}\end{aligned} \tag{9.2}$$

The Sr / Ca ratio is continuous and strictly positive. A Gaussian distribution, or perhaps better a gamma distribution, is the obvious candidate for the distribution in the model. To keep the analysis simple we will use the Gaussian distribution, which means that the model in Equation (9.2) is a linear mixed-effects model. Once we have fitted this model we will need to assess the residuals for any temporal dependency.

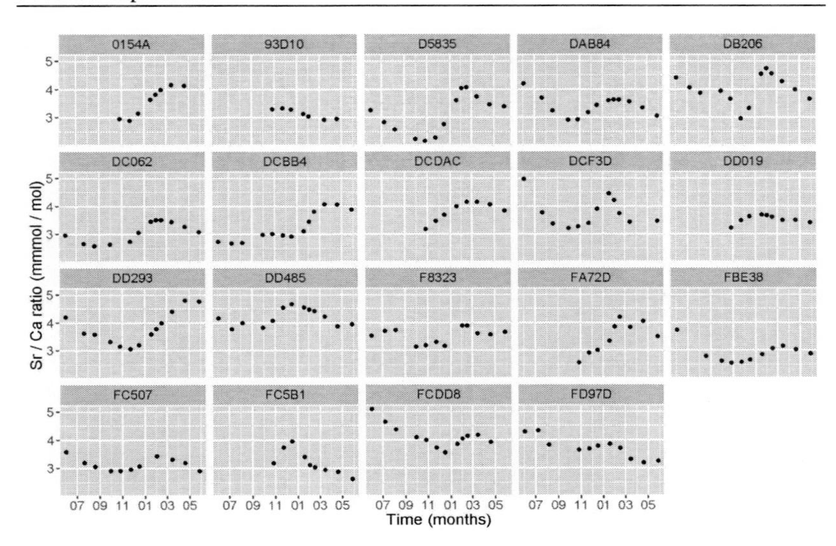

Figure 9.1. Plot of the Sr / Ca ratio versus time for each fish.

9.4 Data exploration

We carry out a short data exploration in which we focus on outliers and collinearity. Figure 9.2 shows Cleveland dotplots for all continuous variables. The response variable is O.Sr.Ca (O stands for otolith) and based on the Cleveland dotplot it seems that this variable has no extreme large or small values. The otolith growth rate has a couple of large values and the model may benefit from transforming this covariate, but we will not do this. The Cleveland dotplots of all other variables look ok (which means that nothing stands out from the rest).

We have around 17 covariates, which is a rather large number for the sample size (212 observations). A general rule in statistics is to have at least 15 observations per parameter. To assess the presence of collinearity we use variance inflation factors (VIF); see Montgomery and Peck (1992) or Zuur et al. (2010) for an explanation. Let's calculate these in R. We first import the data and load all required packages for the analysis and source our support function (which is available on the website for this book).

```
> OT <- read.csv("OTODATA.csv", header = TRUE)
> library(lattice)
> library(INLA)
> library(ggplot2)
> source("HighstatLibV10.R")
```

We use our support function `corvif` from the `HighstatLibV10.R` file to calculate the VIF values for selected variables.

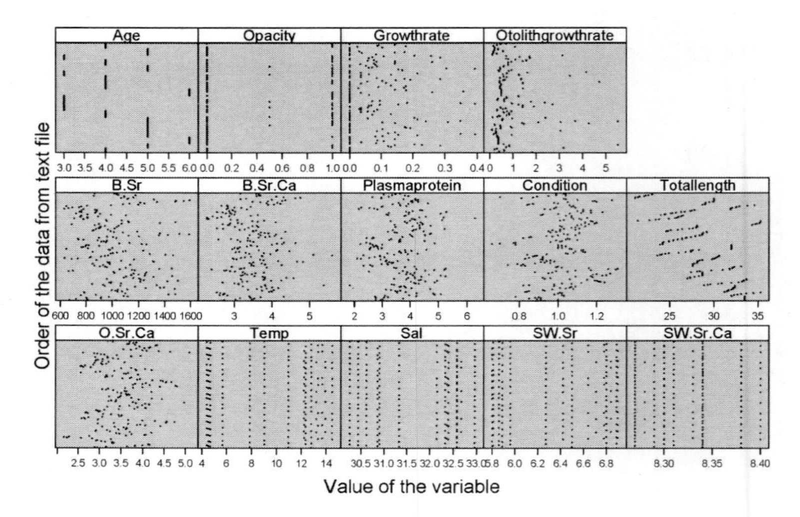

Figure 9.2. Cleveland dotplots. The response variable is O.Sr.Ca. SW.Sr is the Sr concentration in the water and B.Sr is the Sr in the blood plasma. Horizontal axes show the values of the variables and the vertical axes show the order of the data as imported from the data file.

```
> MyVar <- c("Sex", "Origin", "GnRH", "Temp",
             "Sal", "SW.Sr", "SW.Sr.Ca", "B.Sr",
             "B.Sr.Ca", "Plasmaprotein",
             "Condition", "Totallength", "Age",
             "Opacity", "Growthrate",
             "Otolithgrowthrate")
> corvif(OT[,MyVar])

                        VIF
Sex                3.718988
Origin             5.673576
GnRH               1.836395
Temp               6.142643
Sal                7.369388
SW.Sr             13.234485
SW.Sr.Ca           1.640356
B.Sr               3.313430
B.Sr.Ca            3.718258
Plasmaprotein      2.604822
Condition          2.877644
Totallength        7.419827
Age                1.954285
Opacity            1.363590
Growthrate         1.604538
Otolithgrowthrate  1.356995
```

High VIF values mean collinearity, where 'high' is defined subjectively. We prefer to drop as many collinear variables as possible, and therefore we tend to use a threshold of 3. But it is also an option to use 5 or 10.

The Sr concentration in the seawater (environmental variable) has the highest VIF value and it is therefore the most collinear covariate. It is collinear with one, or perhaps with multiple covariates; we don't know. To further investigate what is collinear with what, it may be an option to make a principal component analysis biplot (Jolliffe 2002; Zuur et al. 2007) or inspect Pearson correlation coefficients of the covariates. The Pearson correlation coefficient between Sr concentration in the seawater and salinity is 0.89, and between Sr concentration in the seawater and temperature is 0.82. Dropping either Sr concentration in the seawater or temperature and salinity from the analysis should be considered. We decided (rather arbitrarily) to omit Sr concentration in the seawater.

We then recalculated the VIF values, and there were still VIF values larger than 3 (e.g. total length, sex, origin). We removed total length, recalculated VIF values, and finally removed the Sr / Ca ratio in the blood to obtain a set of covariates in which all VIFs were smaller than 3. Our impression is that there was no strong collinearity between environmental and physiological variables, which is good from a biological point of view. If there would have been collinearity between these two groups then it would not have been possible to state which set of covariates, the environmental or physiological, is driving the Sr / Ca ratio.

Other data exploration steps were applied (e.g. plotting the response variable versus each covariate), but these did not reveal any major information and results are not presented here.

Note that the observed collinearity has implications for the conclusions of the model. If we find an important temperature effect then this may represent an Sr concentration in the seawater (or an unmeasured covariate may affect both temperature and Sr concentration).

Use VIFs, scatterplots, Pearson correlation coefficients, and principal component analysis biplots to detect collinearity. Collinearity has consequences for the interpretation of the results.

9.5 Running the model in R-INLA

Before executing the mixed-effects model, we drop a few rows with missing values. This makes model validation easier.

```
> MyVar <- c("O.Sr.Ca", "Fish", "Sex", "Origin",
             "GnRH", "Temp", "Sal", "SW.Sr.Ca",
             "B.Sr", "Plasmaprotein", "Condition",
             "Age", "Opacity", "Growthrate",
             "Otolithgrowthrate", "Time")
> OT2 <- na.exclude(OT[, MyVar])
```

We are now ready to fit the mixed-effects model in R. Obviously we can use the lme function from the nlme package (Pinheiro et al. 2016) or the lmer function from the lme4 package (Bates et al. 2015). But we are working towards mixed-effects models with spatial and temporal correlation in later chapters, and we therefore fit the model in R-INLA.

The mathematical notation for the mixed-effects model is as follows.

$$
\begin{aligned}
SrCa_{ij} &= Intercept + \text{Covariates} + a_i + \varepsilon_{ij} \\
a_i &\sim N\left(0, \sigma^2_{Fish}\right) \\
\varepsilon_{ij} &\sim N\left(0, \sigma^2\right)
\end{aligned}
\tag{9.3}
$$

The notation $SrCa_{ij}$ stands for the jth observation for fish i on the Sc / Ca ratio. The random effects a_i represent the fish effect. We will have 19 of these. We assume that the random effects $a_1, a_2, \ldots, a_{19}$ are identical and independent distributed with mean 0 and variance σ^2_{Fish}. To code this in R-INLA we use as covariate f(Fish, model = "iid"). The iid stands for independent and identical distributed.

Initially we ran the model using the original data but because the covariates have rather different ranges, we ended up with some posterior mean values that were rather small. A general recommendation is always to standardise continuous covariates because this avoids numerical estimation problems and improves interpretation of the regression parameters. Our support file has a function MyStd that can be used for standardisation. It is also an option to use the scale function from R.

```
> OT2$Temp.std            <- MyStd(OT2$Temp)
> OT2$Sal.std             <- MyStd(OT2$Sal)
> OT2$SW.Sr.Ca.std        <- MyStd(OT2$SW.Sr.Ca)
> OT2$B.Sr.std            <- MyStd(OT2$B.Sr)
> OT2$Plasmaprotein.std<- MyStd(OT2$Plasmaprotein)
> OT2$Condition.std       <- MyStd(OT2$Condition)
> OT2$Age.std             <- MyStd(OT2$Age)
> OT2$Growthrate.std      <- MyStd(OT2$Growthrate)
> OT2$Otolithgrowthrate.std <-
                          MyStd(OT2$Otolithgrowthrate)
```

We did not standardise Opacity as it only has the values 0, 0.5, and 1. It is an option to consider Opacity as a categorical covariate, but we will use it as a continuous covariate as it is measured on a continuous scale.

```
> I1 <- inla(O.Sr.Ca ~
             Sex + Origin + GnRH + Temp.std +
             Sal.std + SW.Sr.Ca.std + B.Sr.std +
             Plasmaprotein.std + Condition.std +
             Age.std + Opacity + Growthrate.std +
             Otolithgrowthrate.std +
             f(Fish, model = "iid"),
```

```
           control.predictor = list(
              compute = TRUE,
              quantiles = c(0.025, 0.975)),
           control.compute = list(dic = TRUE),
           data = OT2)
```

In Chapter 8 we spent some time on accessing the output from a linear regression model. The code to obtain the numerical output from a linear mixed-effects model is identical. For example, to obtain the posterior mean values and 95% credible intervals we copied the following two commands from Chapter 8.

```
> Beta1 <- I1$summary.fixed[, c("mean",
                  "0.025quant", "0.975quant")]
> Beta1
```

	mean	0.025quant	0.975quant
(Intercept)	3.374	3.072	3.668
SexM	-0.109	-0.483	0.257
OriginIS	0.453	-0.013	0.930
GnRHTreated	-0.263	-0.581	0.059
Temp.std	**-0.132**	**-0.226**	**-0.039**
Sal.std	**-0.111**	**-0.185**	**-0.036**
SW.Sr.Ca.std	0.047	-0.019	0.113
B.Sr.std	**0.154**	**0.078**	**0.231**
Plasmaprotein.std	**-0.117**	**-0.203**	**-0.031**
Condition.std	-0.069	-0.164	0.026
Age.std	-0.144	-0.315	0.026
Opacity	-0.097	-0.230	0.035
Growthrate.std	-0.046	-0.111	0.019
Otolithgrowthrate.std	-0.046	-0.111	0.019

The covariates temperature, salinity, Sr concentration in the blood, and plasma protein have 95% credible intervals that do not contain 0. Therefore we can state that these four covariates are important and the other covariates are less important. The effect of temperature, salinity, and plasma protein is negative and Sr in blood has a positive effect. So, we have two environmental variables and two physiological variables that are important. The results in Sturrock et al. (2015) are different, but that may be due to how they dealt with collinearity.

We can also access information for the hyperparameters. The code is identical to that in Chapter 8.

```
> I1$summary.hyperpar[, c("mean", "0.025quant",
                        "0.975quant")]
```

	mean	0.025quan	0.975quan
Precision Gauss. obs.	7.223	5.830	8.801
Precision for Fish	16.707	6.246	36.096

We have two variance terms in the model, but recall from Chapters 7 and 8 that R-INLA works with precision. We have the two precision parameters τ_{Fish} and τ. The posterior means and 95% credible intervals for τ_{Fish} and τ are given above. As explained in Chapter 8, R-INLA has support functions to convert the numerical information of τ_{Fish} and τ to σ_{Fish} and σ.

The first hyperparameter τ is equal to $1 / \sigma^2$ and is used for the residuals ε_{ij}. The numerical output above labels it as '`Precision Gauss. obs.`', but that is because the full name in the code didn't fit on the page. The full name in the code is '`Precision for the Gaussian observations`'. In Chapter 8 we used the R-INLA support function `inla.emarginal` to convert precision to the standard deviation parameter. Because we have two precision parameters in the mixed-effects model in Equation (9.3), we do this conversion twice:

```
> tau <- I1$marginals.hyperpar$`Precision for
                            the Gaussian observations`
> tau.Fish <- I1$marginals.hyperpar$`Precision
                            for Fish`
> MySqrt    <- function(x) { 1 / sqrt(x) }
> sigma     <- inla.emarginal(MySqrt, tau)
> sigmaFish <- inla.emarginal(MySqrt, tau.Fish)
```

We now have the posterior mean values of σ_{Fish} and σ.

```
> c(sigmaFish, sigma)

0.2635656 0.3736126
```

We can use these to calculate the intraclass correlation:

```
> sigmaFish^2 / (sigmaFish^2 + sigma^2)

0.3322926
```

This means that the correlation between any two Sr / Ca ratio observations from the same fish is 0.33. We compared the R-INLA results with those obtained by `lme4` and results are similar.

R-INLA works with precision instead of variance. Using the `$marginals.hyperpar` argument from an inla object and the `inla.emarginal` function we can convert the precision into variance or standard deviation.

9.6 Model validation

For model validation we need fitted values and residuals. The fitted values are available because we used `compute = TRUE` in the `control.predictor` argument in the `inla` function. This means that we can access the posterior mean fitted values via

```
> Fit.inla <- I1$summary.fitted.values[, "mean"]
```

These fitted values contain the effects of the covariates and also of the random intercepts (i.e. $\mathbf{X}_i \times \boldsymbol{\beta} + a_i$). If you want to have fitted values that only represent the effects of the covariates (i.e. $\mathbf{X}_i \times \boldsymbol{\beta}$, without the random effects a_i), then calculate such fitted values manually (though in a moment we will show that there is a second approach to get fitted values).

```
> X <- model.matrix(~ Sex + Origin + GnRH +
            Temp.std + Sal.std + SW.Sr.Ca.std +
            B.Sr + Plasmaprotein.std +
            Condition.std + Age.std + Opacity +
            Growthrate.std + Otolithgrowthrate.std,
            data = OT2)
> F1 <- X %*% Beta1[, "mean"]   #=X * beta
```

For model validation purposes, we will use the fitted values that contain the effects of the covariates and random intercepts (= $\mathbf{X}_i \times \boldsymbol{\beta} + a_i$); these are the ones in `Fit.inla`. Once we have the fitted values we can calculate residuals.

```
> E1 <- OT2$O.Sr.Ca - Fit.inla
```

From this point onwards we can apply standard model validation steps. We need to plot the residuals versus fitted values, versus each covariate in the model, and versus each covariate not in the model. We need to assess the residuals for spatial correlation (though that is not relevant here) and for temporal correlation.

Results are not presented here, but a graph of the residuals `E1` versus the fitted values `Fit.inla` shows homogeneity variance. We plotted the residuals `E1` versus each covariate and there are no clear non-linear patterns present, which means that there is no need to consider models with quadratic terms or smoothers. Normality of the residuals is also a valid assumption. So far so good. The R code for all these graphs is on the website for this book.

Unfortunately, the last model validation step spoils it all. Figure 9.3 shows the residuals plotted versus time for each fish. There are clear temporal patterns in the residuals. Before starting the analysis we expected that temperature or another seasonal covariate would capture the temporal patterns that we saw in Figure 9.1; our hopes haven't been realised. This means that the model is incorrect. Later in this book we will introduce regression models with temporal correlation. Such a model can be applied on this otolith data set. An alternative option is to lag some of the covariates with a certain time period.

Fitted values in R-INLA can be obtained via the `control.predictor(compute = TRUE)` argument; you can compute them manually. In Section 9.8 we will illustrate the use of `inla.make.lincombs`.

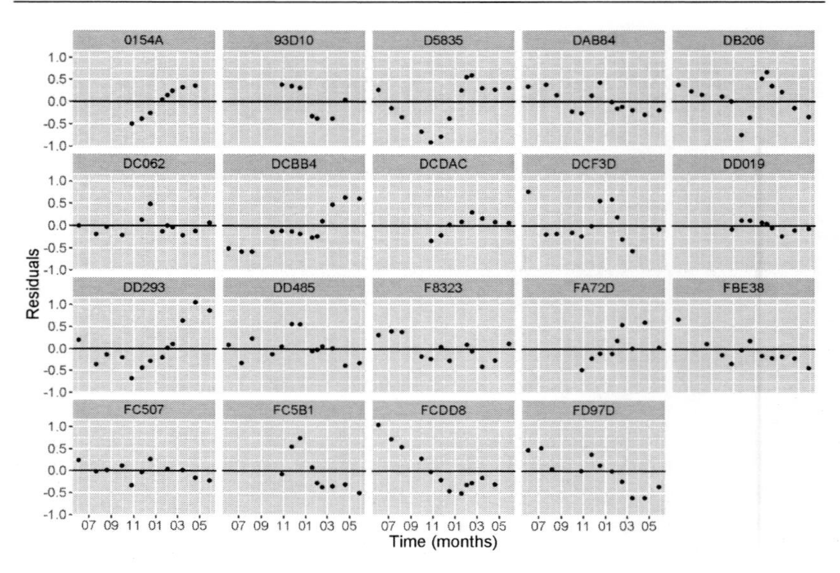

Figure 9.3. Plot of residuals versus time for each fish.

In principle we should deal with the temporal residual correlation before continuing with the analysis of this data set, but we haven't explained yet how to do this. We will pretend that there is no residual correlation and continue the analysis for teaching purposes. But obviously, continuing the analysis is not the correct thing to do.

We also need to check the assumptions for the random effects. We assume that these are independent and identically distributed. However, we only have 19 random effects. There is no point in making a histogram of the 19 posterior mean values of a_1, a_2, ..., a_{19} because we will not be able to assess normality using only 19 points. So, we will pass on that one.

In case you are wondering how to access the posterior mean values of the random effects in R-INLA, here is how you do it. Recall that the posterior mean and 95% credible interval for the fixed parameters and hyperparameters are obtained with

```
> I1$summary.fixed
> I1$summary.hyperpar
```

To obtain the posterior mean and 95% credible intervals for the random effects we use

```
> I1$summary.random
```

To be more precise, use

```
> a <- I1$summary.random$Fish[,"mean"]
```

Now you have 19 values. If we had more random intercepts then we could make a histogram to check normality.

Use `$summary.fixed`, `$summary.hyperpar`, and `$summary.random` obtained from an R-INLA object to access the posterior mean values and 95% credible intervals for the fixed parameters, hyperparameters, and random effects.

9.7 Model selection

If you want to do model selection, use the DIC. Have fun with the backwards selection; it will take a while to do this because of the large number of covariates. We will not carry out model selection here.

9.8 Model interpretation

To understand what the modelling results tell us we can either look at the numerical output (posterior mean values and 95% credible intervals) or we can try to visualise the results.

The numerical output presented in Section 9.5 showed that we have four important covariates in the model and nine covariates that are not important.

In this section we will focus on a visual presentation of the model. We have 14 covariates so we need to do some jiggling to make a graph. One option is to choose average values for 13 covariates and show the effect of the 14th. We will show the effect of plasma protein while keeping all other covariates constant. A sensible constant value seems to be the average of continuous covariates, and for the categorical covariates sex, origin, and treatment we just pick a level as none of them was important. Because the continuous covariates are standardised the average equals 0. This means that we need to carry out the following steps in R:

1. Create a grid of artificial covariate values by:
    a. Choosing a range of sensible plasma protein values, say 25 values between the smallest and largest observed values.
    b. Setting all other continuous covariates to their mean values.
    c. Selecting arbitrary levels for sex, origin, and hormone treatment.
2. Calculate predicted values for these covariate values using R-INLA.
3. Obtain 95% credible intervals for these predictions in R-INLA.
4. Plot the results with standard plotting tools or `ggplot2`.

In Chapter 8 we mentioned that there are two options to do this in R-INLA and we illustrated one approach. The approach that we demonstrated requires the extension of the data set `OT2` with our grid of artificial covariate values and setting the corresponding entries for the response variable to NA. The approach that we have not shown yet uses the argument `inla.make.lincombs` in R-INLA.

We will first show the option in which we extend the data set with the grid of artificial covariate values.

9.8.1 Option 1 for prediction: Adding extra data

We first make a grid of artificial covariate values, which we call `MyData`. The crucial point is that the response variable O.Sr.Ca only contains NAs. R-INLA will then automatically estimate such values.

One issue is the random effects. We want to obtain predicted values due to the covariates only; we do not want to add random effects. But we have to fill in something from `Fish` otherwise we cannot combine our artificial data with `OT2`; filling in NA for the variable `Fish` in `MyData` will do the job!

We make the predictions for female fish from the English Channel that are not treated with a hormone (these levels were chosen arbitrarily). All continuous covariates are set to their average values (which is 0 because the covariates are standardised), except for plasma protein; we choose 25 values from the smallest to the largest observed (standardised) value.

```
> MyData <- data.frame(
   O.Sr.Ca  = rep(NA, 25),#<== CRUCIAL
   Fish     = factor(NA, levels =levels(OT2$Fish)),
   Sex      = factor("F", levels = c("F", "M")),
   Origin   = factor("EC", levels = c("EC", "IS")),
   GnRH     = factor("Non-treated",
              levels = c("Non-treated","Treated")),
   Temp.std                = 0,
   Sal.std                 = 0,
   SW.Sr.Ca.std            = 0,
   B.Sr.std                = 0,
   Plasmaprotein.std       = seq(from = -2.44,
                                 to = 3.36,
                                 length = 25),
   Condition.std           = 0,
   Age.std                 = 0,
   Opacity                 = mean(OT2$Opacity),
   Growthrate.std          = 0,
   Otolithgrowthrate.std = 0
  )
```

Because we are going to use some clumsy R code to combine `OT2` with `MyData` using the `rbind` function, it is essential that the order of the variables in both data sets match. And both data sets also need to have the same number of variables. You may want to investigate whether there are any packages in R that do this more efficiently (e.g. `dplyr`).

```
> OT3 <- rbind(OT2[, c("O.Sr.Ca", "Fish", "Sex",
                       "Origin", "GnRH", "Temp.std",
                       "Sal.std ", "SW.Sr.Ca.std ",
```

```
                        "B.Sr.std ",
                        "Plasmaprotein.std ",
                        "Condition.std ", "Age.std ",
                        "Opacity", "Growthrate.std ",
                        "Otolithgrowthrate.std ")],
              MyData)
```

The new data set OT3 contains the original OT2 data that will be used to fit the model and also the 25 artificial values from MyData. Because the response variable has NAs for these 25 observations, these 25 rows do not affect the results. R-INLA will automatically provide posterior mean values and 95% credible intervals for the 25 O.Sr.Ca observations with NAs.

We rerun R-INLA; we only changed the data argument.

```
> I1 <- inla(O.Sr.Ca ~ Sex + Origin + GnRH +
                       Temp.std + Sal.std +
                       SW.Sr.Ca.std + B.Sr.std +
                       Plasmaprotein.std +
                       Condition.std + Age.std +
                       Opacity + Growthrate.std +
                       Otolithgrowthrate.std +
                       f(Fish, model = "iid"),
             control.predictor = list(
                    compute = TRUE,
                    quantiles = c(0.025, 0.975)),
             data = OT3)
```

In Chapter 8 we showed how to access the 25 predicted values. Rows 1 to 209 in OT3 are for the original data and rows 210 to 234 are for the 25 artificial covariate values. We need the latter ones.

```
> FIT <- I1$summary.fitted.values[210:234,]
> MyData$mu    <- FIT[,"mean"]        #Fitted values
> MyData$selow <- FIT[,"0.025quant"] #Lower bound
> MyData$seup  <- FIT[,"0.975quant"] #Upper bound
```

We now have the predicted values and 95% credible intervals. From here onwards it is standard ggplot2 coding to visualise the results. The only tricky thing is to back-standardise plasma protein.

```
> MyData$Plasmaprotein<-MyData$Plasmaprotein.std *
                        sd(OT2$Plasmaprotein) +
                        mean(OT2$Plasmaprotein)
> p <- ggplot()
> p <- p + geom_point(data = OT2,
                      aes(y = O.Sr.Ca,
                          x = Plasmaprotein),
                      shape = 1,
                      size = 1)
```

```
> p <- p + xlab("Plasma protein")
> p <- p + ylab("O.Sr.Ca")
> p <- p + theme(text = element_text(size = 15))
> p <- p + geom_line(data = MyData,
                     aes(x = Plasmaprotein,
                         y = mu),
                     colour = "black")
> p <- p + geom_ribbon(data = MyData,
                       aes(x = Plasmaprotein,
                           ymax = seup,
                           ymin = selow ),
                       alpha = 0.3)
> p
```

The resulting graph is presented in Figure 9.4. It shows the posterior mean predicted values and 95% credible intervals. We also added the observed data, but that is misleading. Not all data points have average covariate values.

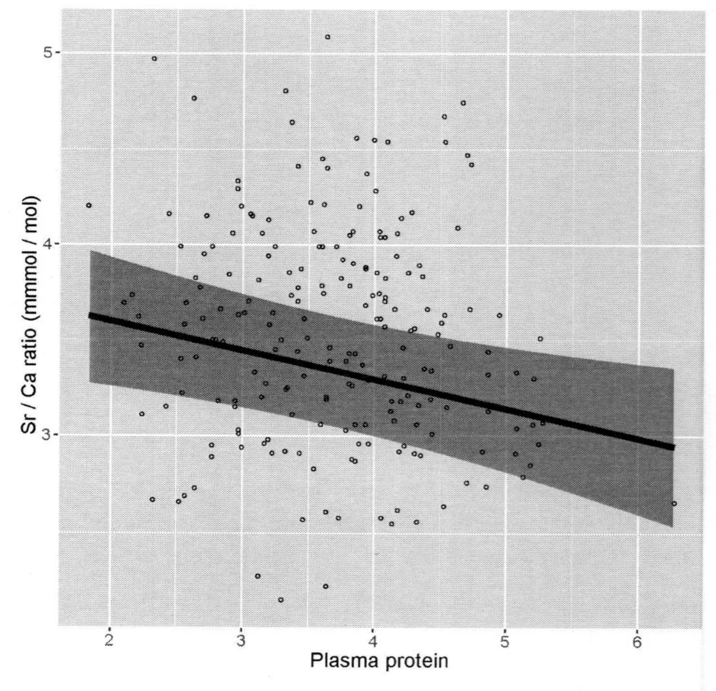

Figure 9.4. Partial fit of plasma protein for average values of all other continues covariates and for the females from the English Channel who did not receive a hormone treatment.

When discussing the results, we must take into account that the temperature effect may represent an 'Sr concentration in the water' effect due to collinearity. The latter covariate was actually one of the significant variables in Sturrock et al. (2015). Hence, any differences between our set of important variables and the significant covariates in Sturrock et al. (2015) are most likely due to collinearity.

To create Figure 9.4 we created a grid of artificial covariate values, which we added to the data set. The crucial step is to add NAs for the response variable for this grid. R-INLA will then estimate the posterior means for such values. Creating the actual graph requires some extensive R coding, but it is not overly complicated.

9.8.2 Option 2 for prediction: Using the `inla.make.lincombs`

In the previous subsection we added extra data in order to do the predictions. In this subsection we will use the `inla.make.lincombs` option in the `inla` function. It will produce similar (though not identical) results. The steps are now as follows.

1. Create a grid of covariate values.
2. Make a design matrix using the `model.matrix` function.
3. Run the model in R-INLA with the `inla.make.lincombs` option selected.
4. Visualise the results using for example `ggplot2`.

We will start from scratch. First we define the grid of covariate values. The code below defines a data frame that contains average values for all continuous covariates and a range of plasma protein values. Compared to the `MyData` object that was formulated in the previous subsection, there are two small changes: we did not define values for the response variable or for the fish variable.

```
> MyData <- data.frame(
   Sex      = factor("F", levels = c("F", "M")),
   Origin   = factor("EC", levels = c("EC", "IS")),
   GnRH     = factor("Non-treated",
              levels = c("Non-treated","Treated")),
   Temp.std              = 0,
   Sal.std               = 0,
   SW.Sr.Ca.std          = 0,
   B.Sr.std              = 0,
   Plasmaprotein.std     = seq(from = -2.44,
                               to = 3.36,
                               length = 25),
   Condition.std         = 0,
   Age.std               = 0,
   Opacity               = mean(OT2$Opacity),
   Growthrate.std        = 0,
```

```
   Otolithgrowthrate.std = 0)
```

Next we create a design matrix using the `model.matrix` function.

```
> Xmat <- model.matrix(~Sex + Origin + GnRH +
                       Temp.std + Sal.std +
                       SW.Sr.Ca.std + B.Sr.std +
                       Plasmaprotein.std +
                       Condition.std + Age.std +
                       Opacity + Growthrate.std +
                       Otolithgrowthrate.std,
                       data = MyData)
```

Actually, we could have used the `MyData` object from the previous subsection because the `model.matrix` function only uses the relevant pieces from `MyData`.

The `model.matrix` function adds some text strings to the `MyData` object, and to avoid error messages in R-INLA it may be wise to remove this via the `as.data.frame` function:

```
> Xmat <- as.data.frame(Xmat)
```

We are now ready to run R-INLA again. There are two small changes in the `inla` function as compared to the previous subsection; we specify the `lincomb` argument (which stands for 'linear combinations') and we use the `OT2` data set again.

```
> lcb <- inla.make.lincombs(Xmat)
> I1 <- inla(O.Sr.Ca ~ Sex + Origin + GnRH +
                       Temp + Sal + SW.Sr.Ca +
                       B.Sr + Plasmaprotein +
                       Condition + Age + Opacity +
                       Growthrate +
                       Otolithgrowthrate +
                       f(Fish, model = "iid"),
            lincomb = lcb,
            control.inla = list(
                  lincomb.derived.only = FALSE),
            control.predictor = list(
                  compute = TRUE,
                  quantiles = c(0.025, 0.975)),
            data = OT2)
```

The object `I1$summary.lincomb` contains the predicted values and 95% credible intervals using the covariate values specified in `MyData`, and we can add the relevant pieces of data (posterior mean and 95% credible intervals) to `MyData`:

```
> MyData <- cbind(MyData,
            I1$summary.lincomb[,c("mean",
                  "0.025quant", "0.975quant")])
```

In order to use exactly the same ggplot2 code as in the previous subsection we need to rename the last three columns. This can be done with the rename function in the reshape package (Wickham 2007); alternatively, have a look at the function with the same name in the dplyr package (Wickham and Francois 2016).

```
> library(reshape)
> MyData <- rename(MyData,
                   c("mean" = "mu",
                     "0.025quant" = "selow",
                     "0.975quant" = "seup"))
```

We can now rerun the ggplot2 code that was presented in Subsection 9.8.1. It will produce exactly the same graph as in Figure 9.4 (results are not presented here).

The inla.make.lincombs option in the inla function allows the user to specify linear combinations of certain covariates for which R-INLA will calculate the posterior mean values and credible intervals.

9.8.3 Adding extra data or inla.make.lincombs?

You may now have the question which approach is better to get fitted values: adding extra data or using the inla.make.lincombs option? For sketching fitted values it does not matter at all. The R code for the inla.make.lincombs option seems slightly less error prone. Another advantage of the inla.make.lincombs option is that we can also use it for post-hoc tests on factors with more than two levels.

9.9 Multiple random effects

Suppose we want to fit a model with multiple random intercepts. As an example, suppose that the otolith in the fisheries data set were sampled by 10 different observers. Readings from the same observer may be more similar than readings from different observers. Ignoring an observer effect introduces pseudo-replication and a sensible solution is to add a random intercept 'Observer' to the model, resulting in

$$
\begin{aligned}
\text{Sr / Ca ratio} = &\ \text{Intercept} + \text{Covariates} + \\
&\ \text{Random intercept Fish} + \\
&\ \text{Random intercept Observer} + \\
&\ \text{Noise}
\end{aligned}
\tag{9.4}
$$

To fit such a model in R-INLA we need to include the term f(Fish, model = "iid") as before, and also the term f(Observer, model = "iid") in the inla function.

For two-way nested random effects the R-INLA code is identical but you have to ensure that the levels of each random intercept are uniquely coded.

9.10 Changing priors of fixed parameters

In all the R-INLA analyses presented so far we used default prior distributions for the fixed regression parameters and the hyperparameters. As long as we work with linear regression models or linear mixed-effects models with only one random intercept there is probably not much reason to change the default priors that are used by R-INLA (though the results presented in Section 9.11 counter-argue with this). However, when we reach the GLMMs with multiple random effects and the GLMs with spatial correlation in later chapters it is advisable to check the effect of the prior distributions, especially for the sigmas. That raises the question which default prior distributions R-INLA is using for the fixed parameters and for the hyperparameters, how to change them, and what to change them to.

In this section we look at priors for regression parameters for the covariates and in the next section we will focus on priors for the hyperparameters.

The default prior distribution for a fixed parameter is a normal distribution with mean 0 and precision $\tau = 0.001$. Because $\tau = 1 / \sigma^2$, this translates as $\sigma = 31.62$. Hence, the default prior distribution for a fixed parameter in R-INLA is of the form

$$\beta_i \sim N\left(0, 31.6^2\right)$$

For a normal distribution, three times the standard deviation covers around 99% of the possible values. If we take this as guidance then a precision of $\tau = 0.001$ implies that we expect that the regression parameter β_i is most likely somewhere around 0, potentially between –94.8 and +94.8, but it can even be smaller or larger than this! The 94.8 comes from 3×31.6.

This diffuse prior is being used for all fixed regression parameters, except for the intercept. The precision for the intercept is 0, which means that the corresponding sigma is large.

We will present an example in which we first use diffuse priors and then informative priors. We start with the first one. In order not to get dizzy due to the standardisation of covariates we will work with the unstandardised variables. To keep the R code and numerical output compact, we only present a mixed-effects model containing the four important covariates.

The mixed-effects model with diffuse priors is executed in R-INLA with the following code:

```
> I3 <- inla(O.Sr.Ca ~ Temp + Sal + B.Sr +
                       Plasmaprotein  +
                       f(Fish, model = "iid"),
             data = OT2)
```

The numerical output for the fixed parameters is as follows.

```
> Beta3 <- I3$summary.fixed[, c("mean", "sd",
                  "0.025quant", "0.975quant")]
> Beta3

                mean     sd 0.025qua 0.975quan
(Intercept)    7.475  1.069    5.371     9.574
Temp          -0.021  0.009   -0.040    -0.001
Sal           -0.118  0.035   -0.187    -0.049
B.Sr           0.000  0.000    0.000     0.001
Plasmaprotein -0.218  0.048   -0.313    -0.122
```

Maybe now you realise why we standardised the covariates in the previous sections; note the very small posterior mean value for strontium in the blood.

Now let us use informative priors. Suppose that a similar otolith study was carried out by colleagues, and their results (which can be frequentist and Bayesian) showed that for a one-unit change of the plasma protein, the Sr / Ca ratio reduced with –0.22, and the standard error of this slope was 0.01. This can be written as

$$\beta_{Plasma} \sim N\left(-0.22, 0.01^2\right)$$

This is all before standardising the covariate! We want to use their results for plasma protein as prior information in our model. First we need to convert the variance into precision: $\tau_{Plasma} = 1 / \sigma^2_{\text{Plasma}} = 1 / 0.01^2 =$ 10,000. This means that that the prior distribution $\beta_{Plasma} \sim N(-0.22, 0.01^2)$ translates into $\beta_{Plasma} \sim N(-0.22, 10{,}000)$ inside R-INLA. Such a prior distribution can be specified in R-INLA as follows.

The `control.fixed` option in the `inla` function is used to specify diffuse normal priors for the slopes of temperature, salinity, B.St, and the intercept. The mean of the prior distribution for the slope of plasma protein is set to –0.22 and its precision to 10,000.

```
> I3a <- inla(O.Sr.Ca ~ Temp + Sal + B.Sr +
                        Plasmaprotein  +
                        f(Fish, model = "iid"),
              control.fixed = list(
                mean = list(Temp = 0,
                            Sal  = 0,
                            B.Sr = 0,
                            Plasmaprotein = -0.22),
                prec = list(Temp = 0.001,
                            Sal  = 0.001,
                            B.Sr = 0.001,
                            Plasmaprotein = 10000),
                mean.intercept = 0,
                prec.intercept = 0),
              data = OT2)
```

The results of this model are as follows.

```
> print(Beta3a, digits = 2)

                 mean     sd  0.025q  0.975q
(Intercept)     7.475  1.067   5.375   9.569
Temp           -0.021  0.009  -0.039  -0.002
Sal            -0.118  0.034  -0.186  -0.049
B.Sr            0.000  0.000   0.000   0.001
Plasmaprotein -0.219  0.018  -0.256  -0.183
```

Note that the posterior mean value for the plasma protein parameter is much smaller as compared to the results from the model in which it has a diffuse prior distribution. An informative prior influences the corresponding posterior distribution, but we already knew this from Chapter 7.

Changing priors for fixed parameters is done via the `control.fixed` option in R-INLA. Either specify one value for all covariates or provide a list with covariate-specific priors.

9.11 Changing priors of hyperparameters

The model that we fitted in the previous section has two hyperparameters: σ_{Fish} for the random intercepts and the σ for the residuals. In Chapter 7 we explained that prior distributions for hyperparameters require special attention. One reason is that the standard deviation parameters σ_{Fish} and σ are strictly positive and therefore we cannot use a normal distribution for the prior distribution of such parameters. In Chapter 7 we used the uniform distribution for the prior of a standard deviation parameter, and we also mentioned that the gamma distribution can be used.

Another reason for paying extra attention when choosing a prior distribution for a hyperparameter is that these are not your friends. If the software gives numerical warnings or worse, error messages, changing the priors is one of the first things to try (besides standardising the covariates and simplifying the model) to solve the problem(s).

Once we have chosen a prior distribution for the hyperparameters then the next problem is to figure out where, and how, to implement it in R-INLA. The 'where' part is not too difficult. The prior for the σ for the normal distribution of the response variable is set via `control.family` argument in the `inla` function. The prior for σ_{Fish} is specified within the `f(..., model = "iid", ...)` function in `inla`.

Now we know where to specify a prior distribution. Next we focus on the 'how' part. In Chapter 7 we specified a uniform prior distribution (between 0 and 20) for the standard deviation parameters, and inside the JAGS code we converted the sigmas into precision via the equation precision = 1 / sigma^2. In R-INLA we need to specify the prior distribution for the log of the precision. Let us clarify this. We have σ_{Fish} and σ. These are converted in R-INLA into precision via

$$\tau_{Fish} = \frac{1}{\sigma^2_{Fish}}$$

$$\tau = \frac{1}{\sigma^2}$$

R-INLA is using a log gamma distribution for the priors of $\log(\tau_{Fish})$ and $\log(\tau)$ with shape parameter a = 1 and inverse scale parameter b = 0.00005.

$$\log(\tau) \sim LogGamma(1, 0.00005)$$

$$\log(\tau_{Fish}) \sim LogGamma(1, 0.00005)$$

Let us first clarify the log gamma distribution. We say that a variable $\log(\tau)$ follows a log gamma distribution if τ follows a gamma distribution. So, we can keep our explanation simple and focus on the gamma distribution (though the `VGAM` package has tools to visualise a log gamma distribution).

In our experience many scientists are not familiar with a gamma distribution. A gamma distribution can be used for a variable that is strictly positive and continuous, and it has two parameters: the shape parameter and a scale parameter. The shape parameter is typically denoted by a. The scale parameter can be a source of confusion. The R-INLA manual uses b for the 'inverse' scale, whereas the `dgamma` function in R uses b for the scale parameter. So the b in R-INLA is the $1 / b$ in the `dgamma` function in R. If you ever decide to delve into this, just keep an eye on the mathematical expression of the density function.

With the definition used by R-INLA, the mean of a variable that follows a gamma distribution with shape parameter a and inverse scale parameter b (also called rate parameter) is a / b and its variance is given by a / b^2. R-INLA uses a gamma(a = 1, b = 0.00001) distribution for the prior distribution of τ_{Fish} and τ.

Carroll et al. (2015) carried out an extensive simulation study in which the performances of a range of gamma priors were compared for Poisson GLMMs in R-INLA, WinBUGS, and OpenBUGS. They found that a gamma(1, 0.5) performed considerably better than the default gamma(1, 0.00005) prior used by R-INLA. Figure 9.5 shows the gamma(1, 0.5) density function; small values of the precision parameter are most likely. And the larger the tau, the smaller the value of the density function.

At this point you may start to see stars before your eyes. What does a small or large tau mean in terms of sigma? Figure 9.6 shows a histogram of 1,000 simulated τ values from a gamma(1, 0.5) distribution, except that we transformed them to sigmas. The shape of the histogram for the sigmas indicates that when we use a gamma(1, 0.5) prior for a hyperparameter we would expect the sigma to be somewhere between 0 (though not 0 itself) and 5.

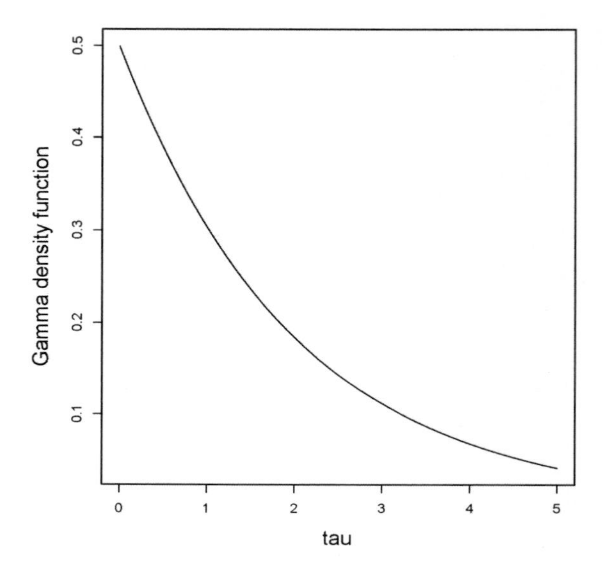

Figure 9.5. Density function of gamma(1, 0.5).

R-INLA uses a log gamma(1, 0.00005) prior for the log of a precision parameter. This is equivalent to assuming that a precision parameter is gamma(1, 0.00005). A simulation study carried out by Carroll et al. (2015) showed that for Poisson GLMMs a gamma(1, 0.5) is better. This roughly corresponds to stating that σ is most likely somewhere between 0 and 5.

As a critical note to the Carroll et al. (2015) paper, a σ of 5 is very large for a Poisson GLMM due to the log link. Imagine we have a Poisson GLMM with only an intercept. The upper bound of the 95% confidence interval for the fitted values is then given by exp(Intercept + $1.96 \times \sigma$) = exp(Intercept) $\times$ exp($1.96 \times \sigma$). If $\sigma = 5$, then exp($1.96 \times \sigma$) equals 18,033.74, which is a rather large multiplication value. As a result the confidence intervals will be much larger.

Figure 9.5 was created with the following R code. First we set the shape and inverse scale parameters a and b respectively. Then we specify a range of possible values for the precision and calculate the corresponding gamma density values.

```
> a   <- 1
> b   <- 0.5
> X   <- seq(0.001, 5, length = 100)
```

```
> tau <- dgamma(x = X, shape = a, scale = 1/b)
> plot(x = X,
       y = tau,
       type ="l",
       xlab = "tau",
       ylab = "Gamma density function")
```

The histogram in Figure 9.6 was created as follows.

```
> hist(1 / sqrt(rgamma(1000,
                       shape = a,
                       scale = 1/b)),
       xlab = "sigma values", main = "",
       xlim = c(0, 15))
```

We certainly recommend changing the default values of the priors for the hyperparameters and seeing how this affects the final results. Let us implement our own advice. The model with the default gamma(1, 0.00005) prior distributions for the two hyperparameters is given by

```
> I4a <- inla(O.Sr.Ca ~ Temp + Sal + B.Sr +
                        Plasmaprotein  +
                        f(Fish, model = "iid"),
              data = OT2)
```

Implementation of the two gamma(1, 0.5) priors is done as follows. One of them (τ) is part of the distribution and needs to be specified via the `control.family` argument.

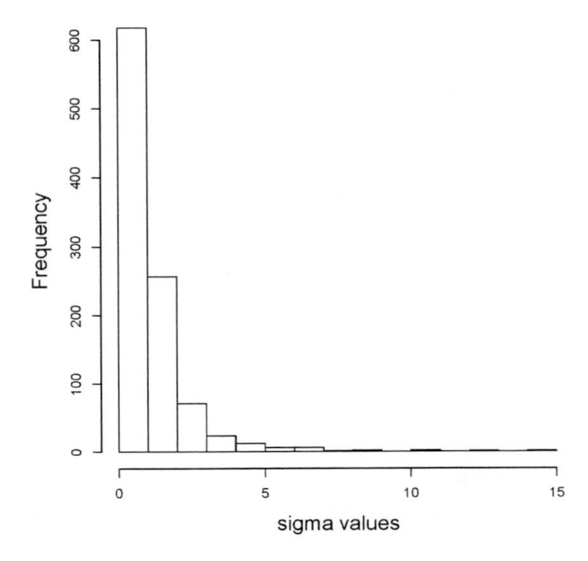

Figure 9.6. 1,000 simulated values from a gamma(1, 0.5) distribution converted into sigmas.

```
> I4b <- inla(O.Sr.Ca ~
              Temp + Sal + B.Sr + Plasmaprotein +
              f(Fish, model = "iid", hyper = list(
                        prec = list(
                            prior = "loggamma",
                            param = c(1, 0.5)))),
              control.family = list(
                        hyper = list(
                          prec = list(
                            prior = "loggamma",
                            param = c(1, 0.5)))),
              data = OT2)
```

The `f()` function has a `hyper` argument and the `control.family` has also a `hyper` argument. Both arguments are lists with further lists within them.

The posterior mean values, 95% credible intervals, and posterior marginal distribution of the fixed parameters obtained by the models with the different priors for the hyperparameters are nearly identical and are not presented here. The posterior marginal distributions for σ_{Fish} are different; see Figure 9.7A. The gamma(1, 0.5) gives a posterior mean that is nearly twice the one obtained by the gamma(1, 0.00005) prior. The results for σ are comparable; see Figure 9.7B.

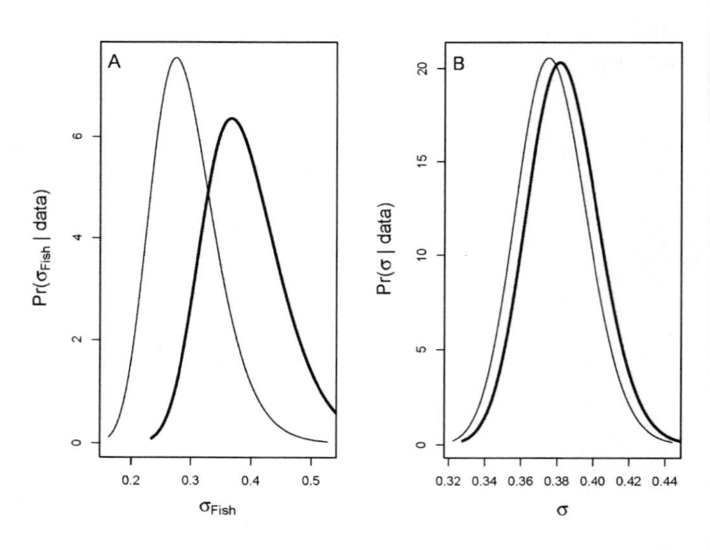

Figure 9.7. A: Marginal posterior distributions for σ_{Fish}. The thick lines are obtained with the gamma(1, 0.5) priors and the thinner lines with the default gamma(1, 0.00005) prior. B: Same as panel A but now for σ.

Prior distributions for hyperparameters linked to the distribution of the response variable are set via the `control.family` option. Priors for random effects and other hyperparameters are set inside the `f()` function.

Later in this book we will show how prior knowledge of a sigma (e.g. uniform between 0 and 10) can be converted into the parameters of a gamma distribution.

To make Figure 9.7 we first need the marginal distributions for τ and τ_{Fish}. We can calculate these for model `I4a` (default gamma priors) and also for model `I4b` (modified priors).

```
> taua <- I4a$marginals.hyperpar$`Precision for
                  the Gaussian observations`
> tauFish4a <- I4a$marginals.hyperpar$`Precision
                  for Fish`
> tau4b <- I4b$marginals.hyperpar$`Precision for
                  the Gaussian observations`
> tauFish4b <- I4b$marginals.hyperpar$`Precision
                 for Fish`
```

These are then converted in posterior distributions of σ_{Fish} and σ using the `MySqrt` function that we defined earlier in this chapter.

```
> Posterior.SigmaFish4a <- inla.tmarginal(MySqrt,
                                     tauFish4a)
> Posterior.SigmaFish4b <- inla.tmarginal(MySqrt,
                                     tauFish4b)
> Posterior.Sigma4a <- inla.tmarginal(MySqrt,
                                     tau4a)
> Posterior.Sigma4b <- inla.tmarginal(MySqrt,
                                     tau4b)
```

And the rest is simple plotting code.

```
> plot(Posterior.SigmaFish4a, type = "l",
       xlab = expression(paste(sigma [Fish])),
       ylab = expression(paste("Pr(",
                                sigma [Fish],
                                " | data)")))
> lines(Posterior.SigmaFish4b, lwd = 3)
> plot(Posterior.Sigma4a, type = "l",
       xlab = expression(paste(sigma)),
       ylab = expression(paste("Pr(", sigma,
                                " | data)")))
> lines(Posterior.Sigma4b, lwd = 3)
```

9.12 Should we change priors?

A question we often encounter is: When would you change the prior distribution of a regression parameter? The question translates as: When would you use an informative prior and what happens if you use the wrong prior information?

If your main motivation for using Bayesian techniques is lack of good frequentist software for fitting GLMs with spatial and temporal correlation or advanced GLMMs, then you are not a true Bayesian. There is nothing wrong with such an attitude! Just keep all priors diffuse (assuming this does not cause any major estimation problems).

On the other hand, perhaps you are willing to use results of your colleagues or you have a fair degree of common sense. An example that cries for informative priors is the osprey example (Chapter 7) in which DDD is used to model eggshell thickness. We probably all agree that DDD can't have a positive effect on eggshell thickness. Why not use that information?

What happens if we use the wrong prior information? You can easily answer this question yourself. Rerun model `M3a` but instead of using –0.22 for the posterior mean of plasma protein use 0.22. You will notice that the posterior mean is close to 0.22. Extract the residuals of this model and apply a detailed model validation. You will detect serious problems with the residuals.

In Chapter 14 we will discuss Penalised Complexity (PC) priors as an alternative to the priors discussed in this chapter.

10 Poisson, negative binomial, binomial and gamma GLMs in R-INLA

In this chapter we explain how to apply Poisson, negative binomial (NB), Bernoulli, binomial, and gamma generalised linear models (GLM) in R-INLA. Using R-INLA for such models is certainly overkill as it is more convenient to use the `glm` function, but it prepares us for things that come later in this book.

Prerequisite for this chapter: You need to be familiar with GLMs; see for example Zuur et al. (2007; 2009a; 2013) or Hilbe (2014).

10.1 Poisson and negative binomial GLMs in R-INLA

10.1.1 Introduction

In this section we use data from Timi et al. (2008), who analysed abundances of parasites infecting the body surface and grills of the Brazilian sand perch *Pinguipes brasilianus*. Fish samples were collected in three main areas of the Argentine Sea. We also have information on sex, length, and weight of the fish, and we would like to know which of these variables are useful for explaining the numbers of parasites in a fish. The data were also used in Zuur et al. (2016a) to compare models with a Poisson, negative binomial, and generalised Poisson distribution.

We import the data from the text file Turcoparasitos.txt.

```
> TP <- read.table(file = "Turcoparasitos.txt",
                   header = TRUE,
                   dec = ".")
```

A detailed data exploration was carried out following the protocol described in Zuur et al. (2010). We made Cleveland dotplots to investigate the presence of potential outliers. None were present. We also made multi-panel scatterplots and calculated variance inflation factors for the covariates. Not surprisingly, there was high collinearity between fish length, fish weight, and sagittal length. We selected fish length (abbreviated as LT in the data file) for the analysis. Figure 10.1 shows a scatterplot of the number of fish versus length for each of the three locations; we have enough observations per location to fit a model with an interaction between length and location (location represents temperature and we expect that the length effect on the number of parasites changes with temperature). Although the three lines in the graph clearly show a difference, we need to apply a model with an interaction term to prove that there is an important difference between the locations.

The R code for these steps of the data exploration are given on the website for this book.

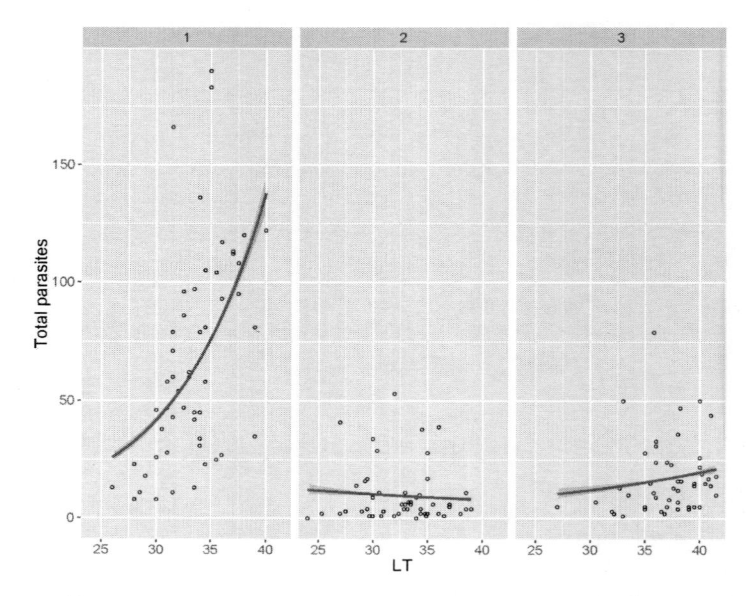

Figure 10.1. Scatterplot of total number of parasites per fish plotted versus length of the fish. Each panel corresponds to a location. In each panel a curve obtained from a Poisson GLM was added.

10.1.2 Poisson GLM in R-INLA

To investigate whether the relationship between total numbers of parasites and length differs per location, while taking into account a sex effect, we apply the following Poisson GLM.

$$
\begin{aligned}
&TP_i \sim Poisson(\mu_i) \\
&E(TP_i) = \mu_i \quad \text{and} \quad \text{var}(TP_i) = \mu_i \qquad (10.1)\\
&\log(\mu_i) = Intercept + Sex_i + LT_i + Location_i + LT_i \times Location_i
\end{aligned}
$$

TP_i is the total number of parasites in fish i, where $i = 1, \ldots, 155$. LT_i is the length of fish i. The other variables are self-explanatory. Before we can execute the model we define the two categorical covariates as factors in R.

```
> TP$fSex <- factor(TP$SEX)
> TP$fLoc <- factor(TP$Location)
```

We are now ready to execute the model in R-INLA. The `dic = TRUE` argument in the `control.compute` option ensures that we obtain a DIC, which we can use for model selection.

```
> library(INLA)
> M1 <- inla(Totalparasites ~ fSex + LT * fLoc,
             control.compute = list(dic = TRUE),
             family = "poisson",
             data = TP)
> summary(M1)
```

```
Fixed effects:
             mean    sd 0.025q   0.5q 0.975q   mode kld
Intercept   0.171 0.201 -0.225  0.172  0.567  0.172   0
fSex2       0.008 0.036 -0.062  0.008  0.079  0.008   0
LT          0.118 0.006  0.106  0.118  0.130  0.118   0
fLoc2       2.982 0.462  2.069  2.983  3.885  2.987   0
fLoc3       0.899 0.483 -0.058  0.901  1.842  0.907   0
LT:fLoc2   -0.145 0.014 -0.173 -0.145 -0.118 -0.145   0
LT:fLoc3   -0.071 0.013 -0.096 -0.071 -0.045 -0.071   0

The model has no random effects
The model has no hyperparameters
Expected number of effective parameters (std dev):
7.028 (0.00)
Number of equivalent replicates: 22.05

Deviance Information Criterion (DIC): 3021.75
Effective number of parameters: 7.028

Marginal log-Likelihood:  -1551.76
Posterior marginals for linear predictor and
fitted values computed
```

These results show that we can write the fitted model as follows.

$$\log(\mu_i) = \begin{cases} 0.171 + 0.118 \times LT_i & \text{for location 1 \& Sex 1} \\ 3.154 + 0.118 \times LT_i & \text{for location 2 \& Sex 1} \\ 1.071 + 0.118 \times LT_i & \text{for location 3 \& Sex 1} \\ 0.180 + 0.118 \times LT_i & \text{for location 1 \& Sex 2} \\ 3.162 - 0.027 \times LT_i & \text{for location 2 \& Sex 2} \\ 1.079 - 0.047 \times LT_i & \text{for location 3 \& Sex 2} \end{cases}$$

Before delving into model selection and model interpretation, we investigate whether the Poisson distribution and the model itself are appropriate for these data. To check the validity of the Poisson distribution we need to assess whether the model is over- or underdispersed. To see whether the model itself is adequate, we need to inspect the residuals for patterns and we also need to assess whether there is potentially a pseudoreplication problem (we always need to do this).

Figuring out whether a Bayesian Poisson GLM is under- or overdispersed is slightly more elaborate than in a frequentist setting.

Obviously, we can follow the same steps that we do in a frequentist analysis (see Zuur et al. 2007), namely calculate the dispersion statistic by obtaining the Pearson residuals, squaring these, dividing them by the sample size minus the number of parameters, and comparing this to 1.

```
> mu1 <- M1$summary.fitted.values[, "mean"]
> E1  <- (TP$Totalparasites - Fit1) / sqrt(mu1)
> N   <- nrow(TP)
> p   <- nrow(M1$summary.fixed)
> Dispersion <- sum(E1^2) / (N - p)
> Dispersion
```

```
18.17258
```

In a frequentist analysis this would indicate severe overdispersion. The problem is that this approach comes from a chi-square test, which has the label 'frequentist' written all over it. Obviously, the value of 18.172 does indicate overdispersion in this case as the results from a frequentist model and a Bayesian model with diffuse priors are similar. So the diagnostics and conclusions from both models should be similar too. But it would be nice to come to the conclusion that the model is overdispersed using Bayesian tools.

Using Markov chain Monte Carlo (MCMC) techniques we execute an algorithm that generates thousands upon thousands of realisations of the regression parameters. Using these realisations we can easily simulate count data from the model, and for each MCMC iteration we can compare a summary statistic (e.g. the sum of squared Pearson residuals) obtained from the simulated data with that of the observed data. If the model fits the data reasonably well, then for the majority of the MCMC iterations the summary statistics (plural!) for the simulated data and the summary statistic (not plural!) for the original data should be similar. See Zuur et al. (2016a) for extensive examples.

But we don't have MCMC iterations when we run R-INLA. Instead we can tell R-INLA to simulate from the posterior distributions of the regression parameters and use these to generate simulated count data. This idea leads to the implementation of the following simulation study to assess whether the Poisson GLM is under- or overdispersed.

1. Apply the Poisson GLM in R-INLA.
2. Simulate one set of regression parameters from the posterior distributions; $\beta_1,\ldots, \beta_7$.
3. Calculate expected values using this set of regression parameters: $\boldsymbol{\mu} = \exp(\boldsymbol{X} \times \boldsymbol{\beta})$.
4. Simulate count data using the `rpois` function. This function requires the specification of a mean value; use the ones that were calculated in step 3.
5. Calculate the Pearson residuals for the simulated data set; square them and add them up.
6. Repeat steps 2–5 a thousand times.

7. Compare the thousand sums of squared Pearson residuals with the sum of squared Pearson residuals for the observed data.

There is no under- or overdispersion if about half of the sums of squared Pearson residuals for the simulated data are larger than that of the observed data. We would also be happy if this is 60% of the time, or 40%, or 70%, or 30%, but certainly not if it is 90% or 10% of the time (or more extreme).

Step 1: Apply the model in R-INLA

This may sound obvious as we already executed the Poisson GLM in R-INLA, but we need to do it again, with a small modification this time. The `config = TRUE` option allows us to simulate regression parameters in the next step.

```
> M2 <- inla(Totalparasites ~ fSex + LT * fLoc,
             control.compute = list(
                 config = TRUE,
                 dic = TRUE),
             family = "poisson",
             data = TP)
```

Step 2: Simulate regression parameters

We use the function `inla.posterior.sample` to simulate from the model. The output is stored in the `Sim` object.

```
> set.seed(12345)
> Sim <- inla.posterior.sample(n = 1, result = M2)
```

Typing

```
> Sim[[1]]$latent
```

gives an object with 162 rows for this specific data set and model. The first 155 rows are simulated values for $\boldsymbol{\eta} = \boldsymbol{X} \times \boldsymbol{\beta}$, where $\boldsymbol{X}$ is the matrix with covariates, and the last 7 rows are simulated regression parameters. This is just one set of simulated values (due to `n = 1` in the function above).

Step 3: Calculate predicted values

We have multiple options to do this step. We can either take the first 155 rows and exponent them to get $\boldsymbol{\mu} = \exp(\boldsymbol{\eta}) = \exp(\boldsymbol{X} \times \boldsymbol{\beta})$, or we access the last 7 rows and calculate the fitted values via

```
> RowNum <- 156:162
> X      <- model.matrix(~ fSex + LT * fLoc,
                         data = TP)
> Betas  <- Sim[[1]]$latent[RowNum]
> mu     <- exp(X %*% Betas)
```

Instead of typing that we want rows 156 to 162 we can also do this semi-automatically, but that requires more advanced R code.

Step 4: Simulate count data

We use the `rpois` function for this.

```
> Ysim <- rpois(n = nrow(TP), lambda = mu)
```

We now have 155 simulated values that correspond to the observed covariate values. We can calculate how many zeros this simulated data set has, determine the maximum value, the minimum value, the range, etc.

Step 5: Calculate summary statistic

Instead of the numbers of zeros, or the maximum value, we can also calculate the Pearson residuals for the simulated data, and square and sum them.

```
> Es <- (Ysim - mu1) /sqrt(mu1)
> sum(Es^2)
```

```
185.458
```

We used the fitted values `mu1` from the original model as this is the one from which we simulated the data. By the way, the sum of squared Pearson residuals for the original data and model is 2689.54, which is considerably larger!

Step 6: Repeat steps 2–5 a thousand times

Repeating the simulation step a large number of times requires only minor modification of the code presented in step 2.

```
> NSim    <- 1000
> SimData <- inla.posterior.sample(n = NSim,
                                   result = M2)
```

Now we have a thousand simulated mean values from the model. Processing the information requires more coding. We first determine the number of rows in the data (`N`), and then create a matrix `Ysim` that can hold a thousand simulated data sets (one in each column of `Ysim`). Then we start a loop and in each iteration we extract the simulated betas and predict the response variable.

```
> N    <- nrow(TP)
> Ysim <- matrix(nrow = N, ncol = NSim)
> for (i in 1:NSim){
    Betas <- SimData[[i]]$latent[RowNum]
    Ysim[,i] <- rpois(n = N,
                      lambda = exp(X %*% Betas))
  }
```

We now have a thousand simulated data sets with numbers of parasites.

Step 7: Compare simulation results and observed data

In this step we compare the simulation results with the observed data. There are many things that we can do with the simulated data sets. For example, for each simulated data set we can count the number of zeros, and see whether the percentages of zeros are comparable to the percentage of zeros in the observed data. If the model does not produce enough zeros then we can start thinking about zero-inflated models. However, zero inflation is not an issue for this data set.

The sum of squared Pearson residuals for the observed data is calculated as follows.

```
> E1 <- (TP$Totalparasites - mu1) /sqrt(mu1)
> SS <- sum(E1^2)
```

As mentioned above, this value is 2689.54. We calculate the same statistic for each of the simulated data sets.

```
> SS.sim <- NULL
> for(i in 1:NSim){
      e2 <- (Ysim[,i] - mu1) /sqrt(mu1)
      SS.sim[i] <- sum(e2^2)
  }
> mean(SS > SS.sim)

1
```

The sum of squared Pearson residuals for the observed data is always larger than that of the simulated data sets, and this indicates that the variation in the observed data is larger than is allowed for by a Poisson distribution. And that is called overdispersion.

Overdispersion can be caused by an outlier, a missing covariate, missing interaction, wrong link function, non-linear patterns that are modelled as linear, zero inflation, dependency, or a large variance (Hilbe 2014). It is the task of the researcher to identify the cause of the overdispersion; selecting the wrong solution may result in biased parameters.

Prime tools to figure out what is causing the problem are common sense, biological knowledge, data exploration, and model validation. In model validation we take the Pearson residuals and plot them against fitted values, and also against each covariate in the model and not in the model. If the data have a spatial element then variograms need to be made, and the same holds for time-series and auto-correlation functions. Only if no clear reason can be found for the overdispersion should the negative binomial distribution be considered.

We do not show all the model validation steps for fish data, but the conclusion is that a negative binomial GLM should be applied.

10.1.3 Negative binomial GLM in R-INLA

The NB GLM version of the model in Equation (10.1) is defined as follows.

$$TP_i \sim NB(\mu_i)$$

$$E(TP_i) = \mu_i \quad \text{and} \quad \text{var}(TP_i) = \mu_i + \frac{\mu_i^2}{k} \tag{10.2}$$

$$\log(\mu_i) = Intercept + Sex_i + LT_i + Location_i + LT_i \times Location_i$$

The parameter k allows for extra variation. The confusing thing is that we now have a dispersion parameter k, an overdispersion statistic (the sum of squared Pearson residuals divided by the sample size minus the number of parameters), and overdispersion if the overdispersion statistic is larger than 1. And to add more confusion, some packages in R use $\mu_i + \mu_i^2 / k$, whereas others use $\mu_i + \mu_i^2 \times k$. And a final source of confusion is that in the R-INLA help files k is called 'size' but also 'n'.

The NB model is executed with the following code.

```
> M3 <- inla(Totalparasites ~ fSex + LT * fLoc,
             control.compute = list(
                  config = TRUE,
                  dic = TRUE),
             family = "nbinomial",
             data = TP)
```

The only thing we had to change was the `family` argument. The numerical output of this model is as follows.

```
Fixed effects:
              mean    sd  0.025q  0.975q   mode kld
Intercept   -1.021 1.471  -3.897   1.885 -1.040   0
fSex2        0.011 0.161  -0.306   0.327  0.012   0
LT           0.154 0.044   0.065   0.242  0.154   0
fLoc2        4.383 1.934   0.592   8.192  4.372   0
fLoc3        1.784 2.153  -2.419   6.043  1.753   0
LT:fLoc2    -0.187 0.058  -0.302  -0.073 -0.187   0
LT:fLoc3    -0.098 0.060  -0.218   0.021 -0.097   0

The model has no random effects

Model hyperparameters:
        mean    sd  0.025q  0.975q   mode
size    1.51 0.170   1.212   1.872  1.487

Expected number of effective parameters (std dev):
6.993(0.0011)
Number of equivalent replicates: 22.17
Deviance Information Criterion (DIC): 1276.47
```

```
Effective number of parameters: 7.956

Marginal log-Likelihood:  -671.17
Posterior marginals for linear predictor and
fitted values computed
```

The main difference between the Poisson and NB GLMs is that the 95% credible intervals of the latter model are all wider. The results of the NB GLM fitted in R-INLA closely match the frequentist results fitted with the `glm.nb` function from the `MASS` (Venables and Ripley 2002) package (see the online R code).

The 'size' is the parameter k and is a hyperparameter, which means that we may have to think about its prior distribution. The default settings in R-INLA specify a (log-)gamma prior with parameters 1 and 0.1 on *theta* = log(k). But the output is defined in terms of the size.

In the numerical output presented above we cheated a little bit; the output does not mention 'size' but 'size for the nbinomial observations (1/overdispersion)'. But that was way too long to get the output in one sentence.

In Chapter 9 we introduced functions to calculate the posterior mean values and plot the marginal distribution of hyperparameters. The same functions can be applied here, except that we do not need the 1 / sqrt(x) as the output is already in terms of the size parameter. The posterior mean value of k is obtained as follows.

```
> k.pd <- M3$marginals.hyperpar$`size for the
              nbinomial observations
             (1/overdispersion)`
> k.pm <- inla.emarginal(function(x) x, k.pd)
> k.pm

1.517348
```

The long name for the size parameter in the R-INLA output makes the code rather cumbersome.

In Chapter 9 the `$marginals.hyperpar` part gave us the posterior distribution of a precision parameter, which we then needed to convert into the posterior distribution of a standard deviation parameter with the `inla.tmarginal` function. There is no need to do that type of conversion here because the numerical output is already specified in terms of the size parameter (and not in terms of the precision parameter). However, the `inla.tmarginal` function gives a slightly smoother curve than the one in `$marginals.hyperpar`, as can be seen in Figure 10.2.

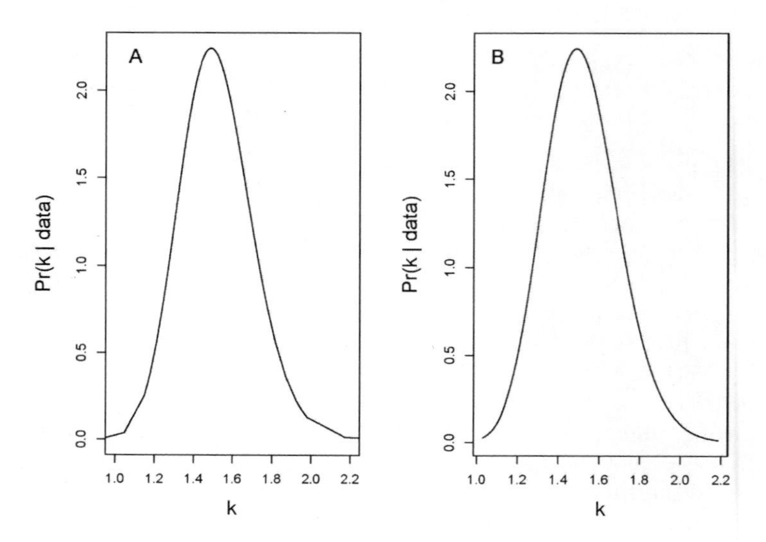

Figure 10.2. A: Posterior distribution of *k* (size) using the information in `$marginals.hyperpar`. B: Posterior distribution of *k* (size) using the conversion function `inla.tmarginal`.

Figure 10.2 was created with the following R code. First we apply the conversion function, and the rest is a matter of plotting the information in `$marginals.hyperpar` and in `k.marg`.

```
> k.marg <- inla.tmarginal(function(x) x, k.pd)
> plot(k.pd,
       type = "l",
       xlim = c(1,2.2),
       xlab = expression(paste(k)),
       ylab = expression(paste("Pr(", k," |
                                    data)")))
> plot(k.marg,
       type = "l",
       xlab = expression(paste(k)),
       ylab = expression(paste("Pr(", k," |
                                    data)")))
```

If you want to set the *k* to a specific value, say 1, and keep it fixed, then use the following code. An NB GLM in which $k = 1$ is also called a geometric distribution.

```
> HypNB <- list(size = list(initial = 1,
                                    fixed = TRUE))
> M4 <- inla(Totalparasites ~ fSex + LT * fLoc,
             control.compute = list(config = TRUE,
                                         dic = TRUE),
```

```
                family = "nbinomial",
                control.family = list(hyper = HypNB),
                quantiles = c(0.025, 0.975),
                data = TP)
```

When doing model selection it may be wise to keep the *k* fixed to the value of the full model; see Zuur et al. (2012a) for a justification. The `step` and `drop1` functions do this as well for frequentist NB GLMs. However, for this data set there is no need to specify $k = 1$, so we will focus on the results of `M3`.

We can calculate the fitted values and Pearson residuals as follows. The fitted values are trivial, but the variance of the Pearson residuals requires a small modification; see also the expression for the variance in Equation (10.2).

```
> mu3 <- M3$summary.fitted.values[,"mean"]
> E3 <- (TP$Totalparasites - mu3) /
                   sqrt(mu3 + mu3^2 / k.pm)
```

Using both the frequentist way of assessing for overdispersion and the simulation study, we found that there is no serious indication of overdispersion. The full R code for the simulation study follows that of the Poisson example and is given on the website for this book. The only difference in the code is that we now use the `rnegbin` function from the `MASS` package to simulate the count data.

10.1.4 Model selection for the NB GLM

Model selection is a controversial field. Many scientists oppose it, and other scientists just apply classical model selection using tools like backward selection. We briefly discussed the pros and cons of model selection in previous chapters. Here, we just show the mechanism for how to do it. It is up to you whether you want to simplify the model or present the full model and state what is and what is not important.

The following code applies the full model, the model without sex, and the model without the interaction term, and compares the DIC and WAIC values. In all models we use the posterior mean of *k* from the full model.

```
> Hyper.NB <- list(size = list(initial = k.pm,
                                fixed = TRUE))
> M5 <- inla(Totalparasites ~ fSex + LT * fLoc,
                control.compute = list(dic = TRUE,
                                       waic = TRUE),
                family = "nbinomial",
                control.family = list(
                                hyper = Hyper.NB),
                data = TP)
# Drop fSex
> M5a <- inla(Totalparasites ~  LT * fLoc,
                 control.compute = list(dic = TRUE,
```

```
                                        waic = TRUE),
             family = "nbinomial",
             control.family = list(
                                    hyper = Hyper.NB),
             data = TP)
# Drop LT + fLoc interaction
> M5b <- inla(Totalparasites ~ fSex + LT + fLoc,
             control.compute = list(dic = TRUE,
                                        waic = TRUE),
             family = "nbinomial",
             control.family = list(
                                  hyper = Hyper.NB),
             data = TP)
```

Next we extract the DICs and WAICs and add some row labels.

```
> dic5  <- c(M5$dic$dic, M5a$dic$dic, M5b$dic$dic)
> waic5 <- c(M5$waic$waic, M5a$waic$waic,
             M5b$waic$waic)
> Z     <- cbind(dic5,waic5)
> rownames(Z) <- c("Full model", "Sex",
                   "LT x Loc")
> Z

               dic5    waic5
Full model 1362.371 1377.916
Sex        1360.374 1373.805
LT x Loc   1385.809 1398.695
```

The results suggest that Sex is not an important covariate. Dropping this covariate and continuing the model selection in the same fashion shows that the model with the main terms LT and Loc and the interaction between them is preferred. That is this model:

```
> M6 <- inla(Totalparasites ~  LT * fLoc,
             family = "nbinomial",
             control.family = list(
                                  hyper = Hyper.NB),
             data = TP)
```

We extracted the fitted values and residuals of this model via

```
> mu6 <- M6$summary.fitted.values[,"mean"]
> E6  <- (TP$Totalparasites - mu6) /
                             sqrt(mu6 + mu6^2 / k.pm)
```

and plotted them against each other. We also plotted the residuals versus each covariate in the model. There were no clear patterns. We don't have spatial coordinates of the sampling positions of the fish or any other information on the sampling (e.g. multiple fish from the same catch). Absence of this knowledge is of course is not an excuse to disregard any potential pseudoreplication issues. But for the moment we assume that there are no such issues.

10.1.5 Visualisation of the NB GLM

The last thing we do in this analysis is to create a visualisation of the optimal model. We discussed in detail how to do this in Section 9.8. We will use the `inla.make.lincombs` approach. Recall that these were the steps that we have to execute.

1. Create a grid of covariate values.
2. Make a design matrix using the `model.matrix` function.
3. Run the model in R-INLA with the `inla.make.lincombs` option.
4. Visualise the results using `ggplot2`.

We follow the four steps. We first make a grid of covariate values.

```
> library(plyr)
> MyData <- ddply(TP,
                  .(fLoc),
                  summarize,
                  LT = seq(min(LT),
                           max(LT),
                           length = 25))
> head(MyData)

  fLoc       LT
1    1 26.00000
2    1 26.58333
3    1 27.16667
4    1 27.75000
5    1 28.33333
6    1 28.91667
```

We now have 25 length values for each area that go from the smallest to the largest observed values so that we are not extrapolating. That is 75 in total. We convert this into a covariate matrix with `model.matrix`.

```
> Xmat <- model.matrix(~ LT * fLoc, data = MyData)

  (Intercept)     LT fLoc2 fLoc3 LT:fLoc2 LT:fLoc3
1           1 26.000     0     0        0        0
2           1 26.583     0     0        0        0
3           1 27.166     0     0        0        0
4           1 27.750     0     0        0        0
5           1 28.333     0     0        0        0
6           1 28.916     0     0        0        0
```

And we run the model in R-INLA with the `lincomb` option.

```
> Xmat <- as.data.frame(Xmat)
```

```
> lcb  <- inla.make.lincombs(Xmat)
> M7 <- inla(Totalparasites ~  LT * fLoc,
             lincomb = lcb,
             control.inla = list(
                 control.predictor = list(
                     compute = TRUE,
                     quantiles = c(0.025, 0.975)),
             family = "nbinomial",
             control.family = list(
                 hyper = Hyper.NB),
             data = TP)
```

As explained in Chapter 9, the relevant results are in the object `$summary.lincomb`. There is one small catch. In Chapter 9 we used a Gaussian distribution with the identity link. Here we are using a Poisson distribution with the log link, and it is the log link that makes things slightly complicated. Obviously, we could take the information in `$summary.lincomb`, which looks like this:

```
> head(M7$summary.lincomb.derived)

     ID mean    sd 0.025q  0.5q 0.975q  mode kld
lc01  1 3.00 0.196   2.61 3.00   3.38  3.00   0
lc02  2 3.09 0.183   2.73 3.09   3.44  3.09   0
lc03  3 3.17 0.169   2.84 3.17   3.51  3.17   0
lc04  4 3.26 0.156   2.96 3.26   3.57  3.26   0
lc05  5 3.35 0.143   3.07 3.35   3.63  3.35   0
lc06  6 3.44 0.130   3.19 3.44   3.70  3.44   0
```

We have 75 rows in this object, one row for each artificial covariate value in `MyData`. These are summary statistics of the posterior distribution for each predicted value, hence the phrase 'summary' in the name `summary.lincomb.derived`. The output above is on the log-scale; the exponent has not been taken yet. We could do this ourselves via

```
> MyData$mu.wrong   <- exp(Out1$'mean')
> MyData$selo.wrong <- exp(Out1$'0.025quant')
> MyData$seup.wrong <- exp(Out1$'0.975quant')
```

and now we could run the `ggplot2` code from Section 9.8 again. But strictly speaking this is not correct because $E(\exp(x)) \neq \exp(E(x))$ if x is a stochastic variable. And we are performing a Bayesian analysis, so x is stochastic. The 'E()' stands for the expected value and exp stands for the exponential function. This was not a problem in Chapter 9 with the identity link function, but with the log link function it is.

The solution is to use support functions from R-INLA that convert the distribution of x into the distribution of $\exp(x)$. The starting point is the `$marginals.lincomb.derived`, which contains the marginal posterior distribution for each predicted value, and we have 75 of them.

```
> Pred.marg <- M7$marginals.lincomb.derived
```

It is slightly confusing to access the 75 marginal posterior distributions. The marginal posterior distribution for the first predicted value is obtained via

```
> Pred.marg[[1]]
```

This contains an entire posterior distribution, which can be plotted if you want to. What we really want is its posterior mean (or median) and a 95% credible interval. That is obtained with

```
> inla.qmarginal(c(0.025, 0.5, 0.975),
                 inla.tmarginal(exp,
                                Pred.marg[[1]]))

 13.64528 19.99457 29.29126
```

These are the 2.5%, 50%, and 97.5% quartiles for the first predicted value. For the posterior mean use

```
> inla.emarginal(exp, Pred.marg[[1]])

20.39682
```

This is only for the first predicted value; we need to do this for all 75 values and add the results to `MyData`. We can either write a loop and do the calculations above for *i* in 1 to 75, or use some fancy `unlist` and `lapply` code that does it all at once. The fancy code goes like this

```
> MyData$mu <- unlist(
              lapply(
               Pred.marg,
               function(x) inla.emarginal(exp,x)))
> MyData$selo <- unlist(
              lapply(
               Pred.marg,
               function(x)
                 inla.qmarginal(c(0.025),
                    inla.tmarginal(exp, x))))
> MyData$seup <- unlist(
              lapply(
               Pred.marg,
               function(x)
                 inla.qmarginal(c(0.975),
                    inla.tmarginal(exp, x))))
```

If you don't like fancy coding (like us), just run the following loop. It repeats the steps that we just explained for each predicted value. Using a loop means that we have to think less. The results are exactly the same.

```
> for (i in 1:75){
      MyData$mu2[i] <- inla.emarginal(exp,
```

```
                                   Pred.marg[[i]])
     lo.up <- inla.qmarginal(
                  c(0.025, 0.975),
                  inla.tmarginal(exp,
                                 Pred.marg[[i]]))
     MyData$selo2[i] <- lo.up[1]
     MyData$seup2[i] <- lo.up[2]
}
```

The results are presented in Figure 10.3 and show a strong length effect in location 1. We suggest that you include these types of graphs routinely in your scientific papers and reports. The R code that we used to create Figure 10.3 is exactly the same as in Section 9.8, and is not reproduced here.

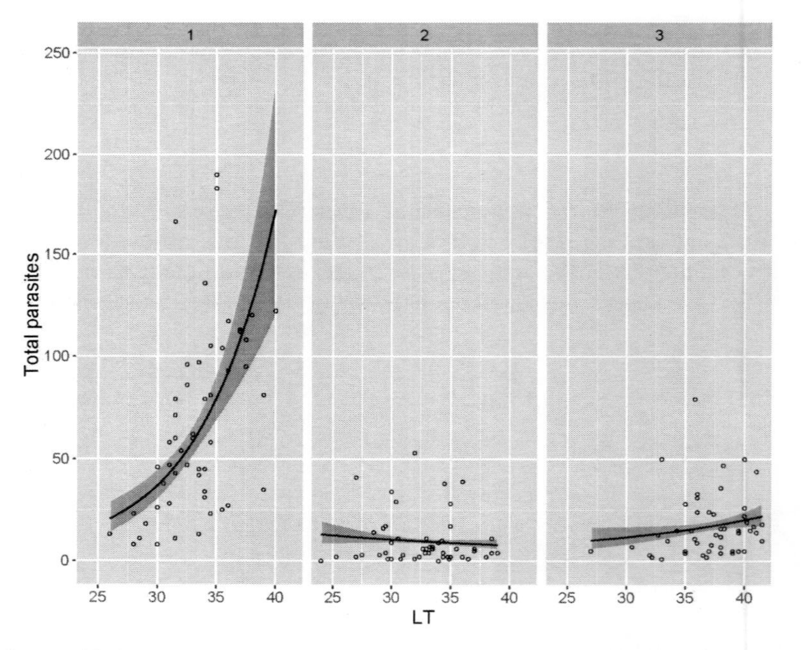

Figure 10.3. Posterior mean fitted values and 95% credible intervals. The R code to create this graph was taken from Chapter 9 and is also available on the website for this book.

10.2 Bernoulli and binomial GLM

In this section we will apply a Bernoulli GLM and also a binomial GLM. The R-INLA code for these models is nearly identical to the code used in Section 10.1. In the remainder of this chapter we will only show R code that shows something new. All the R code can be found on the website for this book.

10.2.1 Bernoulli GLM

Fukuda et al. (2015) analysed survival/nonsurvival of humans who were attacked by a crocodile in Australia. A wide variety of covariates were used, for example, alcohol consumption of the victim, location of the attack (on land, in water, on water), and the difference in body weight between the crocodile and the victim (called DeltaWeight in the data file). We will use the crocodile data to show how to execute a Bernoulli GLM in R-INLA. To keep things simple we only use the covariate DeltaWeight.

The response variable $Survival_i$ is either 1 (person i survived) or 0 (person i did not survive). Because of the binary nature of the response variable a Bernoulli GLM is needed. Such a model is formulated as follows.

$$\begin{aligned}
&Survival_i \sim Bernoulli(\pi_i) \\
&E(Y_i) = \pi_i \quad \text{and} \quad \text{var}(Y_i) = \pi_i \times (1 - \pi_i) \\
&\pi_i = \frac{\exp(\beta_1 + \beta_2 \times DeltaWeight_i)}{1 + \exp(\beta_1 + \beta_2 \times DeltaWeight_i)}
\end{aligned} \tag{10.3}$$

We have 85 observations. We assume that $Survival_i$ is Bernoulli distributed with probability π_i. The mean of a variable that is Bernoulli distributed is π_i, and its variance is $\pi_i \times (1 - \pi_i)$. We model π_i as a function of covariates. To ensure that the fitted probabilities are always between 0 and 1 we use the logistic link function. The last expression in Equation (10.3) can also be written as $\text{logit}(\pi_i) = \beta_1 + \beta_2 \times DeltaWeight_i$.

The Bernoulli GLM is executed in R-INLA with the following code.

```
> Crocs <- read.table("Crocodiles.txt",
                      header = TRUE,
                      dec = ".")
> M1 <- inla(Survived01 ~ DeltaWeight,
             family = "binomial",
             Ntrials = 1,
             data = Crocs)
```

The `Ntrials = 1` is not needed, but we added it as a bridge to the next subsection in which we apply a binomial GLM. The variable Survived01 contains only zeros and ones. Let's have a look at the numerical output.

```
> summary(M1)

                mean    sd 0.025q 0.975q   mode kld
(Intercept)    2.659 0.516  1.734  3.763  2.557   0
DeltaWeight   -0.016 0.003 -0.023 -0.010 -0.015   0

The model has no random effects
The model has no hyperparameters
```

```
Expected number of effective parameters(std dev):
2.00(0.00)
Number of equivalent replicates: 43.50
Marginal log-Likelihood:  -37.99
```

These results show that the fitted model is equal to

$$\pi_i = \frac{\exp(2.659 - 0.016 \times DeltaWeight_i)}{1 + \exp(2.659 - 0.016 \times DeltaWeight_i)} \tag{10.4}$$

The larger the weight difference, the smaller the probability of survival. We visualised the model in Figure 10.4. To survive an attack you better not be a skinny person and encounter a large crocodile!

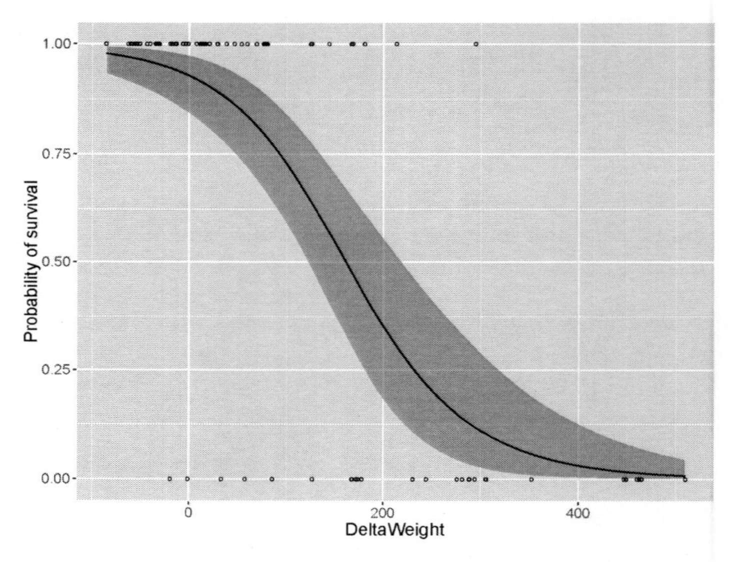

Figure 10.4. Fitted values of the Bernoulli model applied on the crocodile attack data.

The code to create Figure 10.4 is slightly different from the code presented in the previous section, and we therefore present and discuss it here. First we make a data frame `MyData` that contains 100 values for delta weight.

```
> MyData <- data.frame(
              DeltaWeight = seq(
                  from = min(Crocs$DeltaWeight),
                  to   = max(Crocs$DeltaWeight),
                  length = 100))
```

We will generate the posterior mean values and 95% credible intervals that correspond to these 100 covariate values; these are the ones that are used in Figure 10.4. We create a covariate matrix that contains the intercept and delta weight using the `model.matrix` function.

```
> Xmat <- model.matrix(~ DeltaWeight,
                        data = MyData)
> Xmat <- as.data.frame(Xmat)
```

Just as in the previous section we use the `lincomb` option in R-INLA to calculate the fitted values at the 100 covariate values.

```
> lcb <- inla.make.lincombs(Xmat)
> M2  <- inla(Survived01 ~ DeltaWeight,
              lincomb = lcb,
              control.inla = list(
                  lincomb.derived.only = FALSE),
              control.predictor = list(
                  compute = TRUE),
              family = "binomial",
              Ntrials = 1,
              data = Crocs)
```

We extract the posterior distributions for each of the 100 values in `MyData`.

```
> Pred.marg <- M2$marginals.lincomb.derived
```

We now face the task of converting the posterior distribution from the logistic scale to the scale of the data, and we need to do this for each of the 100 predicted values. We are using a logistic link function, which means that we cannot use the exp function that we used in the previous section. Instead we create a function that defines the predicted values on the scale of the data.

```
> MyFun <- function(x) {exp(x) / (1 + exp(x))}
```

The code below was taken from the previous section; we only changed the function exp by `MyFun`.

```
> MyData$mu <- unlist(
                 lapply(
                  Pred.marg,
                  function(x) inla.emarginal(MyFun,
                                             x)))
> MyData$selo <- unlist(
                   lapply(
                    Pred.marg,
                    function(x)
                     inla.qmarginal(c(0.025),
                     inla.tmarginal(MyFun, x))))
> MyData$seup <- unlist(
                   lapply(
                    Pred.marg,
```

```
                       function(x)
                        inla.qmarginal(c(0.975),
                        inla.tmarginal(MyFun, x))))
```

Recall from the previous section that these three commands take the posterior mean and 95% credible intervals for each of the 100 posterior distributions. The rest is a matter of elementary `ggplot2` coding which is presented on the website for this book. The results are presented in Figure 10.4.

10.2.2 Model selection with the marginal likelihood

If there are more covariates, and you want to apply model selection with the DIC or WAIC, then just follow exactly the same steps as in the previous section.

Because this is a short subsection, we use the opportunity to introduce something new. To assess whether delta weight is important we can inspect its posterior distribution and see whether 0 is in the 95% credible interval. Or we apply a model with and without delta weight and compare DICs or WAICs. We have seen examples of this. As a fourth option, we can compare the marginal log likelihoods of both models. We will discuss the motivation behind this.

Suppose we have two models; model 1 contains the intercept and delta weight, and model 2 contains only the intercept. The Bayes factor is defined as

$$\text{Bayes factor} = \frac{\text{Prob}\left(\textit{Model } 1 \,|\, \textit{Data}\right)}{\text{Prob}\left(\textit{Model } 2 \,|\, \textit{Data}\right)}$$

and represents the change in odds of a model after seeing the data. In simple language, if the Bayes factor is larger than 1 then model 1 is preferred. Using Bayes' theorem (see Chapter 7), we can rewrite this expression as

$$\begin{aligned}\text{Bayes factor} &= \frac{\text{Prob}\left(\textit{Model } 1 \,|\, \textit{Data}\right)}{\text{Prob}\left(\textit{Model } 2 \,|\, \textit{Data}\right)} = \\ &= \frac{\text{Prob}\left(\textit{Data} \,|\, \textit{Model } 1\right)}{\text{Prob}\left(\textit{Data} \,|\, \textit{Model } 2\right)} \times \frac{\text{Prob}\left(\textit{Model } 1\right)}{\text{Prob}\left(\textit{Model } 2\right)}\end{aligned}$$

We just applied the rule P(A | B) = P(A | B) × P(B). The term Prob(*Data* | *Model* 1) is the marginal likelihood of the data for model 1. It is the probability of the data given the model. The marginal log likelihood value is presented in the summary output. For model 1 it is –37.99, and for the model with only an intercept it is –54.429. Before using these two values we get rid of the second ratio. The Prob(*Model* 1) and Prob(*Model* 2) are priors for the models. If our expectations for both models are the same, then this ratio is 0.5 / 0.5 = 1, in which case we end up with

$$\text{Bayes factor} = \frac{\text{Prob}(Data \mid Model\ 1)}{\text{Prob}(Data \mid Model\ 2)}$$

Taking the log on both sides and substituting the estimated values gives

$$\log(Bayes\ \text{factor}) = \log\left(\frac{\text{Prob}(Data \mid Model\ 1)}{\text{Prob}(Data \mid Model\ 2)}\right)$$
$$= -37.989 + 54.429 = 16.439$$

This means that the probability of model 1 (intercept + delta weight) is much larger than the probability of model 2 (intercept only). In simple language, the model with the intercept and delta weight is much better.

The marginal log likelihood can only be used for model selection if the priors are defined properly. For ordinary GLMs and GLMMs that should be ok, but the rw1 and rw2 models that are used in Chapter 15 require some modifications of the marginal log likelihood.

10.2.3 Binomial GLM

Maggi (unpublished data) tested the effect of four commercial acaricides (a pesticide) on *Varoa* sp. mites infesting honeybees. Untreated infestations of these mites that are allowed to increase will lead to the death of a honeybee colony, causing damage in the apiary.

The file DrugsMites.txt contains the number of dead mites per batch, collected 24 hours after the four treatments were applied. The explanatory variables are the type of acaricide and the concentration. We would like to know which acaricide is the most efficient and at which concentration. We import the data and define the toxic treatment level as a categorical covariate.

```
> Mites <- read.table(file = "DrugsMites.txt",
                      header = TRUE,
                      dec = ".")
> Mites$fToxic <- factor(Mites$Toxic)
```

To get an impression of the setup of the data we show the first six lines.

```
> head(Mites)

ID Concentration Toxic Dead_mites Total Proportion fToxic
1           0.00     1          0     4       0.00      1
2           0.00     1          0     4       0.00      1
3           0.00     1          1     4       0.25      1
4           0.00     1          0     4       0.00      1
5           0.00     1          0     4       0.00      1
6           0.25     1          2     4       0.50      1
```

The response variable is the number of dead mites and per batch we have 3–5 mites. The response variable is a count, but we cannot have more dead mites than the number of mites in a batch. That is a count with a lower limit at 0 and an upper limit. Assuming that each mite in a batch has the same probability of dying and that the mites in a batch are independent, we can use a binomial GLM to analyse the data. Such a model is given by

$$DeadMites_i \sim Binomial(\pi_i, N_i)$$
$$E(DeadMites_i) = \pi_i \times N_i \quad \text{and} \quad \text{var}(DeadMites_{ii}) = N_i \times \pi_i \times (1 - \pi_i)$$
$$\text{logit}(\pi_i) = Concentration_i + Toxic_i + Concentration_i \times Toxic_i$$

DeadMites$_i$ is the number of dead mites in each batch, and N_i is the total number of mites per batch (which is the column 'Total' in the data file). We have 115 observations. To fit this model in R-INLA we use the following code.

```
> M1 <- inla(Dead_mites ~ Concentration  * fToxic,
             family = "binomial",
             Ntrials = Total,
             control.compute = list(dic = TRUE,
                                    waic = TRUE),
             data = Mites)
```

The response variable is the number of dead mites and *Ntrials* contains the number of mites per batch.

A binomial model can be under- or overdispersed; hence we need to calculate the dispersion statistic again. Alternatively, we can simulate data from the model. The first approach gives

```
> Pi   <- M1$summary.fitted.values[, "mean"]
> ExpY <- Pi * Mites$Total
> VarY <- Mites$Total * Pi * (1 - Pi)
> E1   <- (Mites$Dead_mites - ExpY) / sqrt(VarY)
> N    <- nrow(Mites)
> p    <- nrow(M1$summary.fixed)
> Dispersion <- sum(E1^2) / (N - p)
> Dispersion

1.12427
```

A dispersion statistic of 1.12 is small enough to ignore. The R code above was copied from the Poisson GLM example that was presented earlier in this chapter. We only changed the definition of the mean and the variance.

To see whether the interaction term is important we also executed the code for the DIC and WAIC. This is also a copy-paste exercise from the Poisson GLM. Results are as follows.

```
                   dic     waic
Full model    230.7484 233.5206
Conc x Toxic  253.2437 256.1772
```

On the first line we have the DIC and WAIC for the model with concentration, toxic level, and the interaction term, and on the second line we dropped the interaction. Both the DIC and WAIC indicate that the model with the two main terms and the interaction should be used.

The Pearson residuals can be used for model validation purposes and modelling results can be visualised in the same way as for the negative binomial and Bernoulli GLMs.

10.3 Gamma GLM

In the last section of this chapter we present an example of a gamma GLM. A gamma GLM can be used if the response variable is continuous and positive. The data that we use in this section are taken from Ligas (2008). The goal of his work was to gather and interpret existing information on red swamp crayfish (*Procambarus clarkii*) populations with emphasis on the dynamic features (size-frequency distributions, sex-ratio, growth of juveniles, etc.) in the Salica River in the southern part of Tuscany (Italy). Six morphometric variables were taken of 746 crayfish specimens. In this section we will model weight as a function of length and sex. We also include an interaction term.

We import the data into R.

```
> Crayfish <- read.table(file = "Procambarus.txt",
                         header = TRUE,
                         dec = ".")
```

Because weight is strictly positive, we use a gamma distribution. Don't confuse this with the gamma priors that we have been using for hyperparameters.

$$
\begin{aligned}
&Weight_i \sim Gamma(\mu_i, \phi) \\
&E(Weight_i) = \mu_i \quad \text{and} \quad \text{var}(Weight_i) = \frac{\mu_i^2}{\phi} \qquad (10.5) \\
&\log(\mu_i) = Length_i + Sex + Length_i \times Sex_i
\end{aligned}
$$

The gamma distribution can take all kinds of shapes; see Zuur et al. (2016a) for examples. To fit this model in R-INLA we use the following code. First we define sex as a categorical covariate.

```
> Crayfish2$fSex <- factor(Crayfish2$Sex)
> I1 <- inla(Weight ~ CTL * fSex,
             family = "gamma",
             control.compute = list(
                                dic = TRUE,
                                waic = TRUE),
             control.family = list(
                     link = "log",
                     hyper = list(
```

```
                          prec = list(
                            prior = "loggamma",
                            param = c(1, 0.5)))),
                 data = Crayfish2)
```

The `family` argument is used to specify that we want to use the gamma distribution for the response variable. To see which components are important we calculate the DIC and WAIC. The parameter ϕ is like the k from the negative binomial distribution and is a hyperparameter. The online manual of the gamma distribution in R-INLA is slightly confusing with respect to the variance term of the distribution. It talks about a scale (which is set to 1) and precision τ, but at the end of the day it comes down to

$$\text{var}\left(Weight_i\right) = \frac{\mu_i^2}{\phi}$$

where a gamma prior is defined on theta = log(ϕ). So, we have a gamma distribution for the response variable $Weight_i$, and a gamma distribution is used for the prior of theta = log(ϕ). Are you still following us? And to add a bit more confusion, the ϕ is called the precision parameter in the help file, and the output contains information on ϕ directly (and not for theta).

To see whether we need the interaction term in the model, we also fitted the model without the interaction and extracted the DICs and WAICs of both models.

```
                 dic     waic
Full model  1518.942 1519.527
CTL x fSex  1517.319 1517.370
```

Both the DIC and WAIC indicate that the model without the interaction term is slightly better. The marginal log likelihood values for the model with the interaction term and without the interaction term are –800.75 and –91.40 respectively. The second value is less negative (i.e. better) and that confirms the conclusions from the DICs and WAICs.

The numerical output of the model without the interaction term is as follows.

```
            mean     sd  0.025q  0.975q    mode kld
Intercept -0.412  0.052  -0.515  -0.309  -0.412   0
CTL        0.073  0.001   0.071   0.075   0.073   0
fSex2      0.106  0.020   0.066   0.145   0.106   0

The model has no random effects

Model hyperparameters:
                  mean     sd  0.025q  0.975q  mode
Gamma precision  33.30  2.498   28.69   38.36 33.05
...
```

In the numerical output the 'Gamma precision' is actually called 'Precision parameter for the Gamma observations' but such a long line does not fit on one line together with the numbers. It represents the ϕ.

The fitted values and Pearson residuals are obtained via

```
> mu   <- I2$summary.fitted.values[,"mean"]
> phi  <- I2$summary.hyperpar[,"mean"]
> VarY <- mu^2 / phi
> E1   <- (Crayfish2$Weight - mu) / sqrt(VarY)
```

Plotting the Pearson residuals versus the covariate length shows a clear non-linear pattern. One solution is to log-transform the covariate length.

11 Matérn correlation and SPDE

In this chapter we will explain how spatial dependency can be added to regression models that are applied on geostatistical data. We will use simulated data, and in Chapter 12 a detailed example using real data (Irish pH) is presented.

If you are not interested in a conceptual explanation of the theory of regression models with spatial correlation, then you can skip this chapter and continue on to Chapter 12 in which we show how to run the models in R-INLA.

Prerequisite for this chapter: Knowledge of R, multiple linear regression, R-INLA (Chapter 7), and spatial correlation (Chapter 4) is required.

11.1 Continuous Gaussian field

Suppose that we have N spatial locations with coordinates $s_1, \ldots, s_N$ at which we sample variables, for example *pH*. This means that we have a sample $y(s_1), \ldots, y(s_N)$ of a stochastic process $Y(s)$ defined on the continuous domain D. If we were to repeat the experiment under exactly the same conditions we will end up with a different set of realisations $y(s_1)$ to $y(s_N)$ of the same stochastic process $Y(s)$. If we assume that the $y(s_i)$ are normal distributed then we have a continuous Gaussian field (GF). As part of this normal distribution we need a mean and a covariance matrix.

In a moment we will specify the covariance but we will not do this for the response variable. Instead we will do it for a residual term $u(s_i)$. We can follow the same story: We have realisations $u(s_1)$ to $u(s_N)$ of a stochastic process $U(s)$. We assume that these $u(s_i)$s are normal distributed, and we need their covariance. $U(s)$ is a continuous Gaussian random field.

11.2 Models that we have in mind

An ordinary linear regression model applied on spatial data can be written as

$$\begin{aligned} &y(s_i) \sim N\left(\mu(s_i), \sigma^2\right) \\ &\mu(s_i) = \text{Covariates}(s_i) \end{aligned} \tag{11.1}$$

and its Poisson GLM equivalent is

$$\begin{aligned} &y(s_i) \sim Poisson\left(\mu(s_i)\right) \\ &\log\left(\mu(s_i)\right) = \text{Covariates}(s_i) \end{aligned} \tag{11.2}$$

These models ignore spatial dependency. To take this into account we use the following models:

$$\begin{aligned} y(s_i) &\sim N\left(\mu(s_i),\sigma^2\right) \\ \mu(s_i) &= \text{Covariates}(s_i)+u(s_i) \end{aligned} \tag{11.3}$$

and the Poisson GLM equivalent

$$\begin{aligned} y(s_i) &\sim Poisson\left(\mu(s_i)\right) \\ \log\left(\mu(s_i)\right) &= \text{Covariates}(s_i)+u(s_i) \end{aligned} \tag{11.4}$$

The distribution for the response variable is irrelevant for our discussion on spatial correlation. The crucial point is the presence of the residual $u(s_i)$ terms in Equations (11.3) or (11.4). We will use this term to incorporate dependency.

11.3 Matérn correlation

To obtain a covariance matrix for the $u(s_i)$ components we will use Euclidean distances[1] between the sampling locations and the Matérn correlation function. The Matérn correlation function was introduced in Chapter 4, but in this section we will delve into it a bit more deeply. We reproduce its mathematical specification from Chapter 4.

$$cor_{Matern}\left(U(s_i),U(s_j)\right)=\frac{2^{1-\nu}}{\Gamma(\nu)}\times\left(\kappa\times\left\|s_i-s_j\right\|\right)^{\nu}\times K_{\nu}\left(\kappa\times\left\|s_i-s_j\right\|\right)$$

We explained in Chapter 4 that s_i and s_j define the spatial positions for observations i and j, and the notation || || stands for Euclidean distance. This defines the actual distance (in centimetres, metres, kilometres, miles, or whatever unit you use) between two spatial locations with coordinates s_i and s_j. Let's do a small simulation study to explain this. Suppose that we have 100 sampling locations (or sites) with coordinates s_1 to s_{100}. We simulate the spatial positions of these sites by drawing 100 values from a uniform distribution and the sampling locations are plotted in Figure 11.1.

```
> x1 <- runif(100, min = 0, max = 100)
> y1 <- runif(100, min = 0, max = 100)
```

We can combine these coordinates via the `cbind` function:

```
> S <- cbind(x1, y1)
```

Each row of `S` represents an s_i, where i = 1, ..., 100. Using `dist(S)` we can calculate the distance between each of the 100 sampling locations and it can be useful to make a histogram of them to see the range of the distances (Figure 11.2A) as it gives an indication of how many sites are

[1] These are obtained via the Pythagorean theorem.

separated by certain distances. In our simulation example the maximum distance between any two sites is 127.68, and around 50% of the combinations of sites are separated by less than 50.

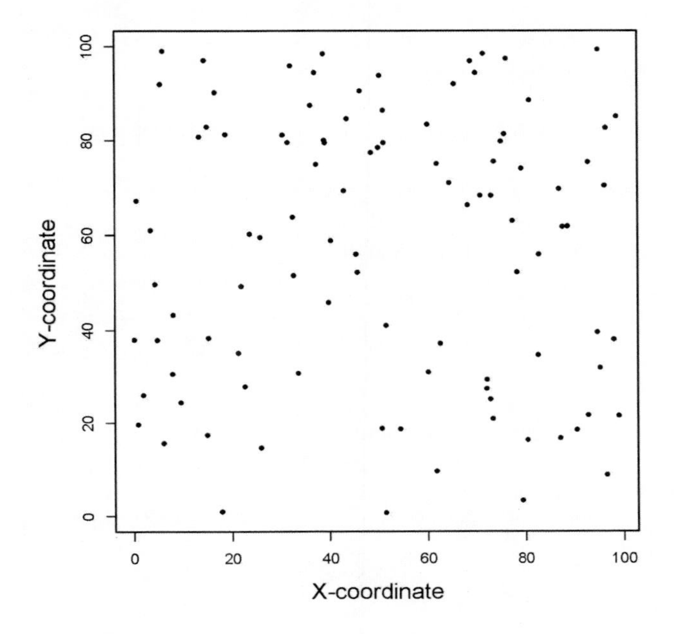

Figure 11.1. Simulated spatial positions of 100 sites.

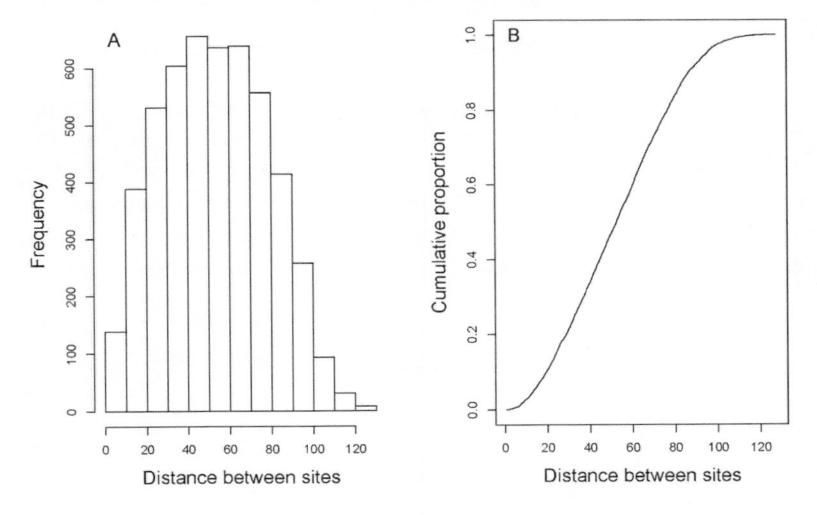

Figure 11.2. A: Histogram of distances between sites in the simulation study. B: Cumulative proportion versus distance between sites.

Let us return to the Matérn correlation function. Now we know that calculating $||s_i - s_j||$ is just a matter of typing `dist(S)`, and this clarifies one of the components of the Matérn correlation function. The K_ν is a mathematical function that is best considered as a black box. If we set ν = 1, which is also the default value in R-INLA, then that makes the formula for the Matérn correlation a little easier on the eye.

$$cor_{Matern}\left(U(s_i),U(s_j)\right)=\kappa\times\left\|s_i-s_j\right\|\times K_1\left(\kappa\times\left\|s_i-s_j\right\|\right)$$

This reads as: The correlation between $U(s_i)$ and $U(s_j)$ equals an unknown parameter κ multiplied by the distance between sites s_i and s_j multiplied with a mathematical function that depends on the same unknown parameter and distance. It is relatively easy to calculate this correlation in R, *provided* that we know the parameter κ. Suppose that we use $\kappa = 0.1$.

```
> kappa <- 0.1
```

We define a vector `d` of length 100 that contains a range of relevant distance values (e.g. 100 values between 0 and the largest observed distance in Figure 11.1). We can now use the `besselK` function from the `fields` package (Nychka et al. 2015) to calculate the Matérn correlation for each of these 100 distances.

```
> library(fields)
> d         <- seq(0, 127.68, length = 100)
> CorMatern <- kappa * d * besselK(kappa * d, 1)
> CorMatern[1] <- 1
```

We now have the Matérn correlations for 100 distance values. But why did we use $\kappa = 0.1$ and what happens if we use a smaller or larger value? The answer to the first question is that we just wanted to show how to calculate the Matérn correlation in R. As to the second question, Figure 11.3 shows the Matérn correlation for nine different κ values. The larger the value of κ, the smaller the distance at which we have dependency. For example, if $\kappa = 2$, then only sites that are separated by less than 1½ units are dependent. If $\kappa = 0.03$ then sites separated up to a distance of 100 units are still dependent.

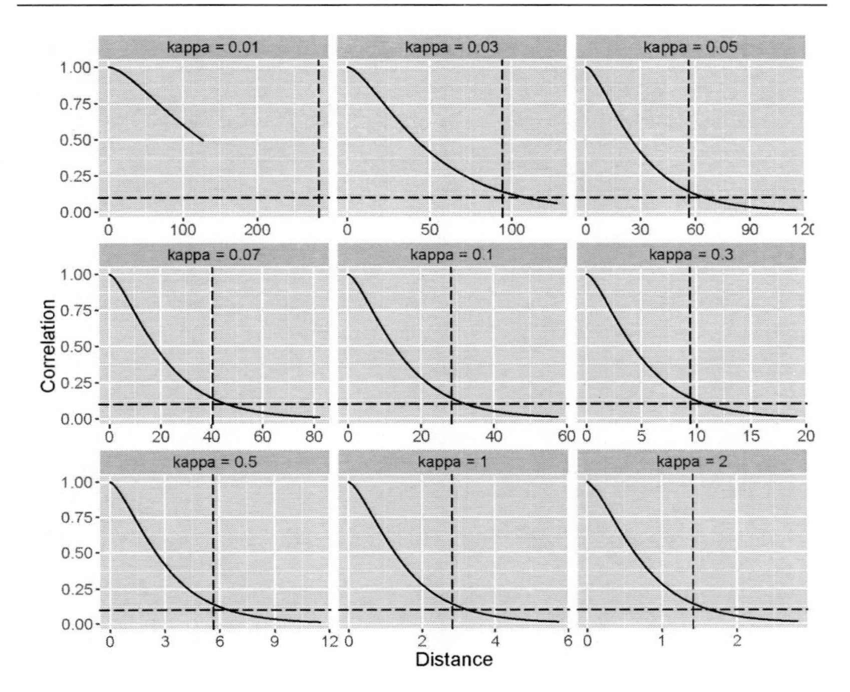

Figure 11.3. Matérn correlation plotted versus distance for the simulated data set in Figure 11.1. Each panel corresponds to a different κ value. Note that the range of the *x*-axis differs per panel. The two dotted lines in each panel are explained in the text.

Recall from Chapter 4 that the range r in a variogram is defined as the distance at which the spatial dependency diminishes. Based on empirical results the following expression seems to hold between the range and κ.

$$r = \frac{\sqrt{8 \times \nu}}{\kappa} = \frac{\sqrt{8 \times 1}}{\kappa} \tag{11.5}$$

In line with R-INLA, we used $\nu = 1$. Let's see whether this relationship holds for our simulation study and what it means. We quasi-arbitrarily define 'diminishing dependency' as correlation below 0.1, and this is visualised as the horizontal dotted line in each panel in Figure 11.3. The vertical dotted line is drawn at the value of r. So, when $\kappa = 2$, then based on Equation (11.5) we have $r = 1.41$. In all panels the expression in Equation (11.5) gives indeed the (approximate) distance at which the Matérn correlation becomes smaller than 0.1.

The Matérn correlation function has an unknown parameter κ that is indirectly linked to the distance at which the correlation diminishes. The κ of the Matérn correlation function and the range r of a variogram are brother and sister.

Now that we are all experts in the Matérn correlation function, let us discuss how we can use it for the covariance of the Gaussian random field. First of all, for a normal distribution we need covariance and not correlation. To go from the correlation between $U(s_i)$ and $U(s_j)$ to covariance between $U(s_i)$ and $U(s_j)$ is a matter of multiplying the Matérn correlation function with a variance parameter:

$$\mathbf{\Sigma} = \text{cov}_{Matern}\left(U(s_i), U(s_i)\right) = \sigma_u^2 \times cor_{Matern}\left(U(s_i), U(s_i)\right)$$

We can now write the models that we have in mind as

$$\begin{aligned} & y(s_i) \sim N\left(\mu(s_i), \sigma^2\right) \\ & \mu(s_i) = \text{Covariates}(s_i) + u(s_i) \\ & u \sim GF\left(\mathbf{0}, \mathbf{\Sigma}\right) \end{aligned} \tag{11.6}$$

or the GLM equivalent as

$$\begin{aligned} & y(s_i) \sim Poisson\left(\mu(s_i)\right) \\ & \log\left(\mu(s_i)\right) = \text{Covariates}(s_i) + u(s_i) \\ & u \sim GF\left(\mathbf{0}, \mathbf{\Sigma}\right) \end{aligned} \tag{11.7}$$

where $\boldsymbol{u} = (\text{u}(s_1), .., u(s_N))$ and $\mathbf{\Sigma}$ is the covariance matrix of dimension $N \times N$ in which the i, jth element is defined by the Matérn covariance function. This means that we need to estimate σ_u and κ in order to quantify the covariance of the spatial Gaussian field [denoted by GF in Equations (11.6) and (11.7)]. And herein lies a problem: how can we do this in a computationally convenient and fast way? And don't forget that we also need to calculate the inverse of this matrix.

It is possible to use maximum likelihood estimation to obtain the regression parameters for the covariates and the hyperparameters κ and σ_u. This would allow us to calculate the covariance matrix $\mathbf{\Sigma}$. However, such an approach may not work well for large data sets and non-Gaussian distributions.

Using the Matérn correlation function to define covariance for the continuous Gaussian field means that we end up with two unknown parameters, κ and σ_u. This sounds simple, but $\mathbf{\Sigma}$ tends to be large (which means long computing time).

11.4 SPDE approach

Using a full covariance matrix $\mathbf{\Sigma}$ for the Gaussian field is going to cause trouble when the sample size *N* increases. One approach to simplify the computations is to introduce a Markovian element again (see also Chapter 7). This means that only local information will be used and consequently the covariance matrix becomes dense (i.e. it contains lots of zeros). This is called a Gaussian Markovian Random Field (GMRF). Hopefully, the covariance matrix of the GMRF approximates the covariance matrix of the GF.

To simplify numerical calculations we will use a GMRF. This is a Gaussian random field that has a covariance matrix in which only neighbouring sites have non-zero covariance values. This is shortcut number 1.

Instead of working with a GMRF for the residual term $U(s)$, we make two more shortcuts. We are now going to show you an equation. You may want to take a deep breath before looking at the next equation.

$$\left(\kappa^2 - \Delta\right)^{\alpha/2} \tau U(\boldsymbol{s}) = W\left(\boldsymbol{s}\right) \tag{11.8}$$

That is an impressive equation. It is the same kappa κ as from the previous section. The upwards-pointing triangle stands for the Laplace operator. That is something you are exposed to when studying for a degree in mathematics or engineering. It is a mathematical notation for taking the sum of all second-order derivatives of the $U(\mathbf{s})$ elements. The $W(\mathbf{s})$ is another brain teaser; it is the Gaussian spatial white noise process (which means that there is no spatial correlation in the $W(\mathbf{s})$ terms). So be it. The expression in Equation (11.8) is called an SPDE, which stands for continuous domain stochastic partial differential equations. Turns out that the SPDE is a standard expression in physics and engineering.

This is all beautiful mathematics, but why in heaven's name would anyone try to understand the expression in Equation (11.8)? Luckily, there is no need to understand where it comes from. What is important is that the components in Equation (11.8) are linked to the components of the Matérn correlation function. So, if we have software that can solve Equation (11.8) then we have all the hyperparameters that we need for the covariance matrix of the GMRF. Lindgren and Rue (2015) show that

$$\nu = \alpha - \frac{d}{2} \tag{11.9}$$

where d is the dimension of the space. For spatial data we work in a two-dimensional space, hence $d = 2$. For time series we have one-dimensional data and $d = 1$. If we set $\alpha = 2$ (the default setting in R-INLA), then $\nu = 1$ and that was the setting we used to simulate data in the previous section. Another result of solving Equation (11.8) is

$$\sigma_u^2 = \frac{\Gamma(\nu)}{\Gamma(\alpha) \times (4\pi)^{d/2} \times \kappa^{2\nu} \times \tau^2} \tag{11.10}$$

The first time we saw Equations (11.8)–(11.10) we had the temptation to close the computer and go to the beach. If you have the same sensation at this moment, then have a look at the comment in the owl. That is all you have to remember.

Solve Equation (11.8) and you have all the hyperparameters that are needed for the covariance of the GMRF. This is the second shortcut.

So, how can we solve the expression in Equation (11.8)? The short answer is to run R-INLA. The slightly longer answer may prompt some beach daydreams again. Lindgren et al. (2011) present a series of theorems. They start by showing that the elements of the covariance matrix of a GMRF that is defined on a regularly spaced two-dimensional lattice can be obtained by solving Equation (11.8). The problem with geostatistical data is that such data are not necessarily on a regular grid. Therefore Lindgren et al. (2011) went on with their theorems and used an irregular grid, also called a mesh.

Mesh

When making a mesh, we divide the study area into a large number of non-overlapping triangles. These triangles only have one common edge with each neighbour and two common corners with each neighbor. The three corners are called vertices. The software will start to place the initial vertices of the mesh at the sampling locations and additional vertices are added following certain rules that can be specified by the user (e.g. maximum length of the edges, minimal allowed angles). This brings us to triangularisation. Delaunay triangularisation, for example, divides the study area in lots of small, non-overlapping triangles that cover the entire area, and it does this by maximising the minimum interior angle of the triangles. As an example of a mesh we present the triangularisation for the 100 simulated sites from the previous section; see Figure 11.4.

R-INLA has a function to create a mesh. It has various options that we will discuss in Chapter 12.

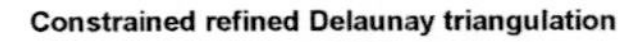

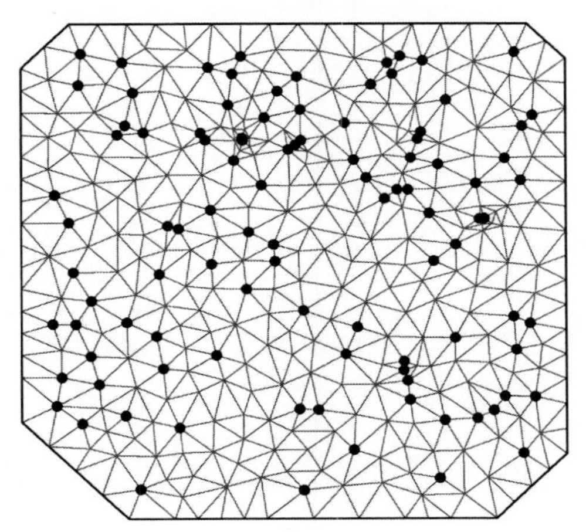

Figure 11.4. Triangularisation for simulated data. Black dots are sampling locations. Where triangles come together is called a vertex.

Finite element approach

Once Lindgren et al. (2011) moved from a regular lattice to an irregular mesh, they came with a third shortcut, the so-called finite element approach. The key equation for this is

$$u(s_i) = \sum_{k=1}^{G} a_k(s_i) \times w_k \qquad (11.11)$$

This expression states that the value of the error term u at site s_i, which is what we have been after all the time, is equal to the sum of a large number of $a_k(s_i)$s multiplied by w_ks. The simulation study with 100 sites is too large to conveniently explain the expression in Equation (11.11).

Simulation study with five points

Let's simulate a data set with only five spatial points.

```
> set.seed(123)
> S <- cbind(x1 = runif(5), y1 = runif(5))
```

We now have the spatial coordinates of five sampling locations and the `inla.mesh.2d` function is used to calculate the mesh. We will explain the `max.edge` option in Chapter 12.

```
> mesh <- inla.mesh.2d(S, max.edge = 1)
> plot(mesh)
> points(S, pch = 16, cex = 2)
```

The resulting mesh and the sampling locations are plotted in Figure 11.5.

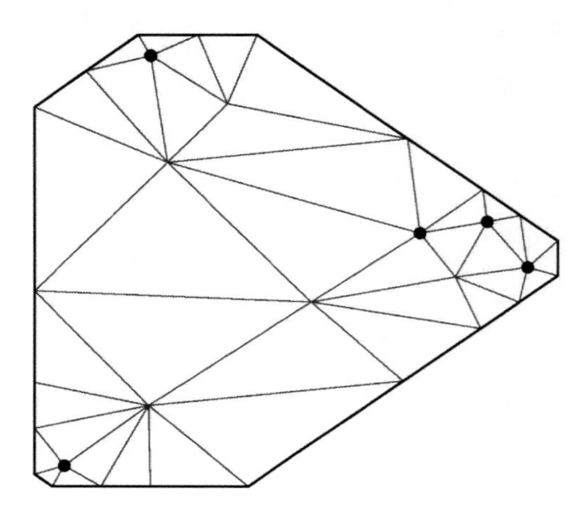

Figure 11.5. Mesh for five simulated points. The black dots are the sampling locations.

Let us now focus on Equation (11.11). The G is the number of vertices in the mesh. You can either count them in Figure 11.5 or you can type

```
> mesh$n
31
```

We have 31 vertices in the mesh and at 5 of them we have the sampling locations (these are the black dots in Figure 11.5). This means that for the first sampling location, which is the black dot in the lower-left corner in Figure 11.5, we can write Equation (11.11) as

$$u(s_1) = \sum_{k=1}^{31} a_k(s_1) \times w_k$$

The $a_k(s_i)$ is a known value. In this simple example these are either 1 (sampling location s_i is on vertex k) or 0 (sampling location s_i is not on vertex k). We need to inform R-INLA about the configuration of the random field and tell it where the sampling locations are with respect to the mesh. This is done via a projector matrix A.

```
> A <- inla.spde.make.A(mesh, loc = S)
> A
```

```
5 x 31 sparse Matrix of class "dgCMatrix"

. . . . . . . 0 1 . . . . . . . . . . . . . 0 . . . . . . . .

. . . . . . . . . 1 . . . 0 . . . . . . . . . . . . 0 . . . .

. . . . . . . . . . 1 . . . . . . . . . . 0 . . . . 0 . . . .

. . . . . . . . . . . 1 0 . . 0 . . . . . . . . . . . . . . .

. . 0 . . . . . . . . . 1 . . . . . . . . . . 0 . . . . . .
```

There are 5 rows and 31 columns in this matrix. The first row is for site 1 and it has 28 dots, 2 zeros, and 1 one. A dot means a 0. If you type `mesh$loc[, 1:2]` in R then you see the coordinates of all the 31 vertices in Figure 11.5. Site 1 (the lower-left point) is vertex number 9 on the mesh. This explains why the 9th column on the first row of A equals 1.

The first row of A tells us: $a_1(s_1) = 0$, $a_2(s_1) = 0$, $a_3(s_1) = 0$, $a_4(s_1) = 0$, ..., $a_8(s_1) = 0$, $a_9(s_1) = 1$, $a_{10}(s_1) = 0$, ..., $a_{31}(s_1) = 0$. The second sampling location is on the 10th point of the mesh. The second row in A tells us $a_1(s_2) = 0$, $a_2(s_2) = 0$, $a_3(s_2) = 0$, $a_4(s_2) = 0$, ..., $a_9(s_2) = 0$, $a_{10}(s_2) = 1$, $a_{11}(s_2) = 0$, ..., $a_{31}(s_2) = 0$. And the same can be done for sampling sites 3, 4, and 5.

R-INLA will use a slightly modified version of Equation (11.8) to obtain the w_ks and the corresponding covariance matrix. As a result of this we get the w_1 to w_{31}, and the corresponding covariance matrix. Because we know the $a_k(s_i)$ values, we can simply calculate the five $u(s_i)$ values via Equation (11.11). The covariance of the w_ks find its way into the covariance of the $u(s_i)$s.

A bonus of the finite element approach is that we get the 31 w_1 to w_{31} values. This is one value for each vertex in the mesh. Using spatial interpolation for the areas inside the triangles and graphical tools we can present this component as a nice picture representing the residual spatial pattern.

The owl below summarises the storyline of this section.

1. We have N sampling locations s_1 to s_N.
2. We have a random effect $u(s_i)$ at each sampling location.
3. We assume that the GF $u(s_1)$,..., $u(s_N)$ are normal distributed with mean 0 and covariance $\mathbf{\Sigma}_{GF}$.
4. Problem: How do we get $\mathbf{\Sigma}_{GF}$?
5. Shortcut 1: Assume a Markovian (= local) behavior. This gives a GMRF with a sparse (lots of zeros) $\mathbf{\Sigma}_{GMRF}$.
6. Shortcut 2: To quantify the covariance matrix $\mathbf{\Sigma}_{GMRF}$ we use the Matérn correlation function. It has a couple of parameters that we need to estimate. And we also need the inverse of $\mathbf{\Sigma}_{GMRF}$.
7. Shortcut 3: The parameters in the SPDE are related to

the parameters that we need.

8. Shortcut 4: Using the finite element approach (defined on an irregularly spaced mesh) in combination with a modified SPDE gives us all the parameters that we need [i.e. the $u(s_i)$s and all covariance terms].
9. Bonus: Step 8 provides a spatial component at each vertex of the mesh. Using linear interpolation and graphical software we can visualise it.
10. R-INLA does it all.

Weighting factors ... again

In the mesh in Figure 11.5 R-INLA puts each sampling location on a vertex, and as a consequence only one $a_k(s_i)$ was non-zero per site. Formulated differently, each row in A had exactly one element equal to 1, and the rest were 0. It is also possible to put the sampling locations inside a triangle; see Figure 11.6. It shows the same five sampling locations again but this time a different mesh is used. Note that none of the sampling locations are on a vertex of the mesh. Instead, each sampling location is inside a triangle. We will show later how to do this.

This mesh has 27 vertices and the first row of A is given by

```
............. 0.338 .... 0.1 ................ 0.562.
```

We have $a_{14}(s_1) = 0.338$, $a_{19}(s_1) = 0.1$ and $a_{36}(s_1) = 0.562$. All other $a_k(s_1)$ for this site are 0. Vertices 14, 19, and 36 are the three vertices of the triangle in which sampling location 1 is placed. Following the expression in Equation (11.11) we can calculate $u(s_1)$ as a weighted average:

$$u(s_1) = 0.338 \times w_{14} + 0.1 \times w_{19} + 0.562 \times w_{36}$$

The finite element approach will give the ws and the covariance matrix of the us.

When making a mesh we will hopefully not have big and small triangles next to each other as it will give poor interpolation results. And we also want to avoid triangles in which one side is long and the other two sides small. The reason why we don't want these can be seen from our example in Figure 11.6. The triangle used for site 1 does have one edge that is considerably longer than the other two. Although vertex 19 is relatively close to site 1, its weight is the lowest among the three due to the poor configuration of the triangle. When we calculate $u(s_1)$ as the weighted average of vertices 14, 19, and 36, the influence of vertex 19 may be low.

When making a mesh try to have triangles that have similar edge lengths and no sharp angles.

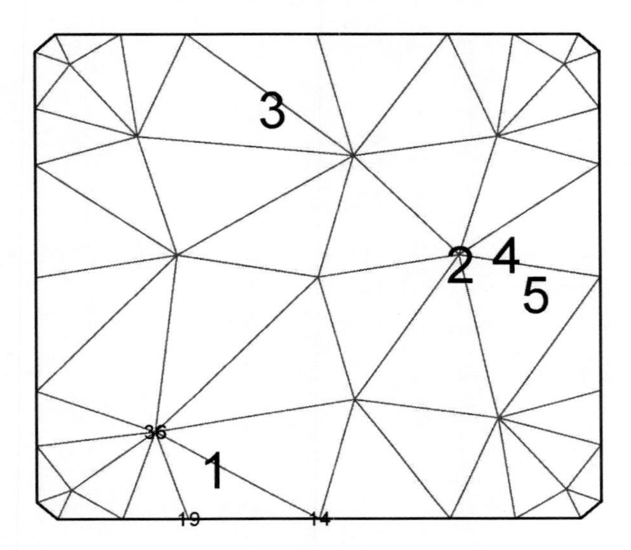

Figure 11.6. Different mesh for the five sampling locations. In this case all sampling locations are inside triangles and not on the vertices. Sampling location 1 is in a triangle with vertices 14 (bottom right vertex), 19 (bottom left vertex), and 36 (top vertex).

12 Linear regression model with spatial dependency for the Irish pH data

In this chapter we will apply a multiple linear regression model with a spatial dependency component using the SPDE approach on the Irish pH data. Preliminary analyses of this data set were presented in Chapters 2 and 4.

Prerequisite for this chapter: It is recommended that you have a conceptual understanding of the SPDE approach that was discussed in Chapter 11. However, if you skipped Chapter 11 you can still catch up because this chapter contains conceptual revisions.

12.1 Introduction

The Irish pH data were discussed at various points in previous chapters, and therefore we will keep this introduction short. Sampling took place at 257 locations (see Figure 2.1) along rivers in Ireland in 2003. We will use data from 210 locations in the analysis. The aim of the study is to model pH of the water as a function of SDI (sodium dominance index), altitude (log transformed; see Section 4.2), and whether a site is forested (categorical variable with the values yes and no).

We argued in Chapter 2 that spatial dependency between pH values at sites sampled close to one another is likely. To avoid pseudoreplication we need a model that allows for the spatial dependency.

Data exploration was (partly) carried out in Section 2.2 and is not repeated here.

12.2 Model formulation

In Chapter 2 we kept the models for this data set simple but here we apply the model with all main terms, two-way interactions, and the three-way interaction; see Equation (12.1).

$$\begin{aligned}
& pH_i \sim N(\mu_i, \sigma^2) \\
& E(pH_i) = \mu_i \quad \text{and} \quad \text{var}(pH_i) = \sigma^2 \\
& \mu_i = \alpha + \beta_1 \times SDI_i + \beta_2 \times LogAltitude_i + \beta_3 \times Forested_i \\
& \qquad \beta_4 \times SDI_i \times LogAltitude_i + \beta_5 \times SDI_i \times Forested_i + \\
& \qquad \beta_6 \times LogAltitude_i \times Forested_i + \\
& \qquad \beta_7 \times SDI_i \times LogAltitude_i \times Forested_i
\end{aligned} \tag{12.1}$$

In the case of numerical estimations problems in R-INLA, the first thing to try for solving the problem is to standardise the continuous covariates. We did not do that here because working with a Gaussian distribution and a sample size of 210 observations makes us reasonably optimistic that we will not encounter numerical problems.

12.3 Linear regression results

The model in Equation (12.1) can be fitted in R-INLA with the following code. We import the data with the `read.table` function, define a categorical covariate fForested, log transform altitude, and then apply the linear regression model in R-INLA.

```
> iph <- read.table(file = "IrishPh.txt",
                    header = TRUE,
                    dec = ".")
> iph$fForested <- factor(iph$Forested,
                          levels = c(1, 2),
                          labels = c("Yes", "No"))
> iph$LogAlt <- log10(iph$Altitude)
> library(INLA)
> I1 <- inla(pH ~ LogAlt * SDI * fForested,
          family = "gaussian",
          control.predictor = list(compute=TRUE),
          data = iph)
```

The `compute = TRUE` for the `control.predictor` option ensures that R-INLA calculates the fitted values. The numerical output of the linear regression model is stored in the list `I1`, which has various objects that we can inspect. The first two objects are `I1$summary.fixed` and `I1$summary.hyperpar`.

```
> Beta1 <- I1$summary.fixed[, c("mean", "sd",
                    "0.025quant", "0.975quant")]
> print(Beta1, digits = 3)

                       mean    sd  0.025q 0.975q
(Intercept)          10.035 1.918   6.267 13.801
LogAlt               -0.774 0.931 -2.604   1.053
SDI                  -0.036 0.033 -0.100   0.028
fForestedNo          -1.783 2.063 -5.839   2.268
LogAlt:SDI            0.005 0.016 -0.026   0.036
LogAlt:fForestedNo    0.880 1.009 -1.102   2.860
SDI:fForestedNo       0.008 0.037 -0.065   0.081
LogAlt:SDI:fForestedNo -0.003 0.018 -0.039  0.032
```

The first column shows the posterior mean and the third and fourth columns the 95% credible intervals. The 95% credible interval for β_8 goes from –0.039 to 0.032. This means that there is a 95% probability that β_8 is in this interval. Because 0 is also in this interval we state that β_8 is 'not important' as compared to the frequentist phrase 'not significant'.

To get the posterior mean of σ we need the `$summary.hyperpar` part of the output. As explained in Chapter 8, R-INLA works with precision, which is defined as $\tau = 1 / \sigma^2$. The `$summary.hyperpar` output contains information on the precision parameter τ, which is probably not that interesting. Using support functions from R-INLA we can convert the information from τ into the posterior mean of σ.

```
> tau <- I1$marginals.hyperpar$`Precision for
              the Gaussian observations`
> sigma <- inla.emarginal(function(x) (1/sqrt(x)),
                              tau)
> sigma
0.374952
```

Now that we have the most important numerical output, let us try to understand what it tells us. We have a model with a three-way interaction term, one of which is a categorical variable with two levels. The concept of a categorical variable adjusting the intercept and an interaction adjusting the slope is the same as in frequentist analysis. This means that we can write the fitted model as

$$
\begin{aligned}
& pH_i \sim N\left(\mu_i, 0.374^2\right) \\
& E\left(pH_i\right) = \mu_i \quad \text{and} \quad \text{var}\left(pH_i\right) = 0.374^2
\end{aligned}
$$

If a site is forested we have

$$
\begin{aligned}
\mu_i = & 10.035 - 0.036 \times SDI_i - 0.774 \times LogAltitude_i + \\
& 0.005 \times SDI_i \times LogAltitude_i
\end{aligned}
$$

and if it is not forested the mean value takes the form

$$
\begin{aligned}
\mu_i = & 8.251 - 0.028 \times SDI_i + 0.105 \times LogAltitude_i \\
& 0.001 \times SDI_i \times LogAltitude_i
\end{aligned}
$$

Before expending effort making a visual representation of this model we perform model validation. However, visualising the model fit can also be useful for model validation!

12.4 Model validation

Now that we have the numerical results we can proceed to the next step and apply model validation. For this we need the fitted values and residuals. In Chapter 8 we explained how to get the fitted values. Recall that there are two options. We can access the fitted values from R-INLA because we used `control.predictor = list(compute = TRUE))`. Alternatively, because we have a linear regression model with a Gaussian link we can also calculate the fitted values ourselves using the posterior mean values of the regression parameters (i.e. calculate $\mathbf{X} \times \boldsymbol{\beta}$). We will opt for the first approach.

```
> Fit1 <- I1$summary.fitted.values[, "mean"]
> E1   <- iph$pH - Fit1
```

We can use the residuals `E1` for model validation. This means that we have to plot them against fitted values to check for homogeneity, plot the residuals versus each covariate in the model and each covariate not in the model to check for patterns, and we also need to inspect the residuals for spatial dependency.

We already know from the frequentist analysis of this data set that there is spatial dependency in the residuals. Because the frequentist and Bayesian results for this relatively simple model are nearly identical, all problems we discovered for the frequentist analysis are also present for the Bayesian analysis. We therefore proceed to the next section and fit a model with spatial dependency included.

12.5 Adding spatial correlation to the model

The spatial model that we have in mind is given in Equation (12.2).

$$\begin{aligned}
&pH_i \sim N(\mu_i, \sigma^2) \\
&E(pH_i) = \mu_i \quad \text{and} \quad \text{var}(pH_i) = \sigma^2 \\
&\mu_i = \alpha + \beta_1 \times SDI_i + \beta_2 \times LogAltitude_i + \beta_3 \times Forested_i \\
&\qquad \beta_4 \times SDI_i \times LogAltitude_i + \beta_5 \times SDI_i \times Forested_i + \\
&\qquad \beta_6 \times LogAltitude_i \times Forested_i + \\
&\qquad \beta_7 \times SDI_i \times LogAltitude_i \times Forested_i + u_i \\
&u_i \sim GMRF(0, \Sigma)
\end{aligned} \tag{12.2}$$

The only difference between Equations (12.1) and (12.2) is the u_i term. This is a random intercept that is assumed to be spatially correlated with mean 0 and covariance matrix $\boldsymbol{\Sigma}$. As explained in Chapter 11 there is a whole drama involved in getting this covariance matrix. This drama has the following storyline. First we assume that the u_is are normal distributed, which makes it a Gaussian field. Then we assume that the covariance is Markovian in nature, which means that only observations from neighbouring sites are correlated. This makes the u_is a Gaussian Markov random field (GMRF). Because the covariance matrix is large (it has as many rows and columns as the sample size), computing time to calculate it can still be large, and therefore a mathematical function is imposed on its structure (see also Chapter 4). This is done with the Matérn correlation function. This function has a couple of parameters that we need to estimate. Once we know these parameters we know the covariance matrix $\boldsymbol{\Sigma}$. And we also need the inverse of $\boldsymbol{\Sigma}$. That was part 1 of the drama.

It turns out that there is this fancy mathematical expression called the SPDE (which stands for 'continuous domain stochastic partial differential equation'). The beauty of the SPDE is that when it is solved, we can use simple expressions to calculate the Matérn correlation parameters as a function of the SPDE parameters. So the problem shifts from estimating the parameters of the Matérn correlation function to solving the SPDE. To do this, we need to put a dense triangular grid on the sampling area. This grid, also called a mesh, consists of a large number of triangles. These triangles are all connected via their edges and the end points of the triangles are called vertices. If we have 253 sampling locations then we can easily end up with thousands of small triangles and vertices. This was part 2 of the drama.

In the final part of the drama the so-called finite element approach is applied. This means that for each vertex we get a value w_k, where k is an index for the vertices. So for each vertex we get a w_k. These w_ks also form a GMRF. The SPDE is used to get the w_ks and its covariance matrix. Once we have the w_ks we can calculate the u_is as a weighted sum of known factors a_{ik}s times the w_ks.

This means that we end up with the posterior distribution of the u_is (the spatially correlated random effects at the sampling locations) and also of the w_ks. The latter ones can be plotted so that we get an impression of the spatial patterns. That was the final part of the drama. Figure 12.1 shows a sketch of all the steps.

In order to implement this whole process in R-INLA we need to carry out the following steps.

1. Make a mesh.
2. Define the weighting factors a_{ik} (also called the projector matrix).
3. Define the SPDE.
4. Define the spatial field.
5. Make a stack. In this process we tell R-INLA at which points on the mesh we sampled the response variable and the covariates. We also need to inform R-INLA (via the stack) at which points of the mesh we have any other terms, e.g. the random effects as we know them from Chapter 8.
6. Specify the model formula in terms of the response variable, covariates, and the spatial correlated term.
7. Run the spatial model in R-INLA.
8. Inspect the results.

We will carry out each of these steps in the remaining sections of this chapter.

Figure 12.2 shows the steps in a trapezoid ring together with the corresponding R code.

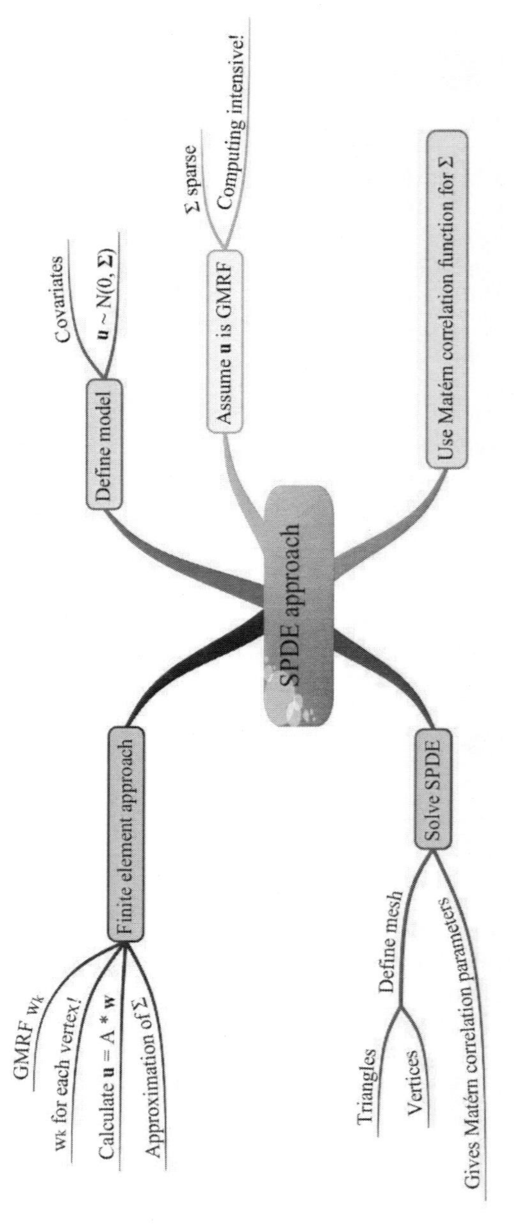

Figure 12.1. Summary of the SPDE approach. We specify a model with a spatial correlated random effect u. After making a series of assumptions (u is Markovian, its covariance matrix Σ is modelled with the Matérn correlation function) and numerical approximations (use SPDE for a GMRF defined on each vertex of a mesh) we end up with an approximation of u and its covariance matrix.

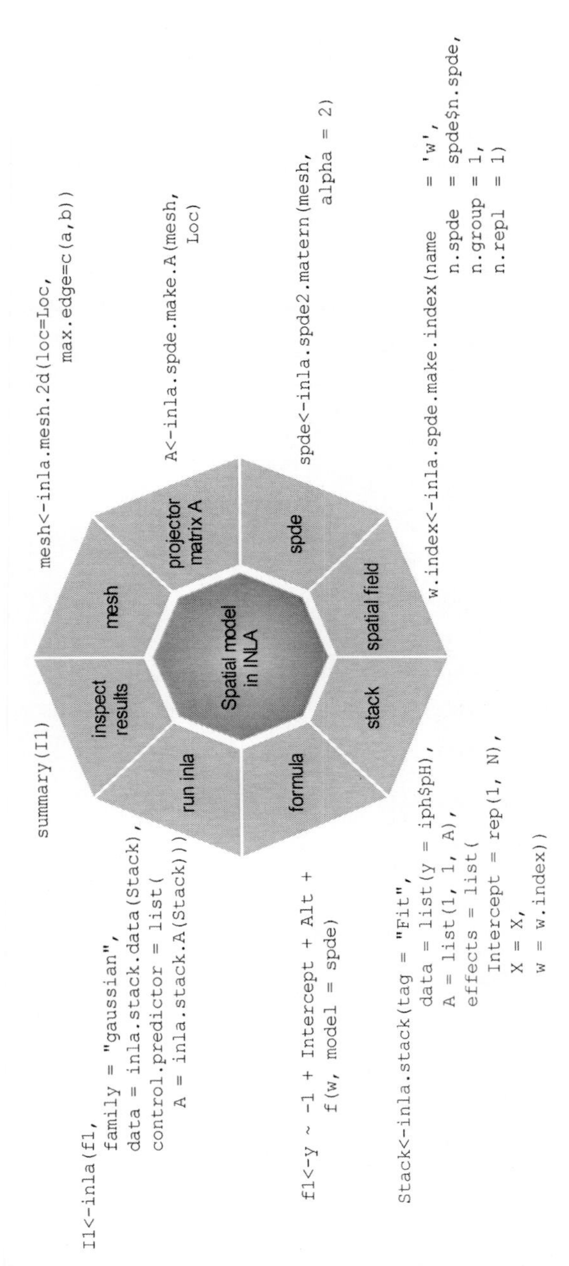

Figure 12.2. Outline of the required steps to apply a model with a spatial correlated random intercept in R-INLA. Starting point is the mesh. The R code, which is at the outside of the trapezoid ring, is discussed in the next sections.

12.6 Defining the mesh for the Irish pH data

In this section we choose a mesh for the Irish pH data. When confronted with a spatial data set the first task is to figure out the scale of the spatial coordinates. Are sites separated by metres, kilometres, or on a much larger scale? And what are the units of the variables in the data set that define the spatial positions? To avoid problems with extreme large or small parameters for the Matérn correlation function we express the coordinates in kilometres.

```
> iph$EastingKM  <- iph$Easting / 1000
> iph$NorthingKM <- iph$Northing / 1000
```

Figure 1.1 showed the spatial locations of the sites; we certainly have a large study area! To gain further insight into the distribution of the distances between sites we make a histogram; see Figure 12.3A. Figure 12.3B indicates that 50% of the distances between sites is less than 200 kilometres (and 50% is larger than this).

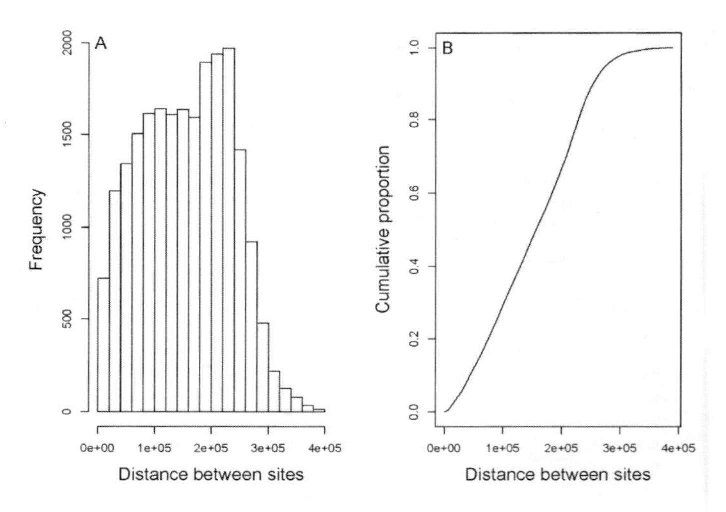

Figure 12.3. A: Histogram of distances between the 210 sites in Ireland. B: Cumulative proportion of distances versus distance.

Step 1 of making a mesh: Get a sense for the distribution of distances between sampling locations.

Now that we know that the distances differ on a scale of thousands of metres we proceed to the next step, making a mesh. In reality we make a large number of meshes and pick one. The second question that we should ask ourselves is whether the mesh should be build on the sampling locations, or whether a boundary for the study area should be supplied. The boundary can be a simple square or rectangular, but it can also be a shapefile. To make a mesh based on the sampling locations we use the following code.

```
> mesh1 <- inla.mesh.2d(loc = Loc,
                        max.edge = c(10, 10),
                        cutoff = 0)
```

The `max.edge` argument specifies the largest allowable edge length for the inner part and for the outer part of the mesh. It is the prime method to control the number of triangles in a mesh. The smaller these two values are, the more triangles there will be in the mesh. The selected values for `max.edge` ensure that the edges of the triangles must be smaller than 10 kilometres. The resulting mesh is presented in the upper-left graph in Figure 12.4.

Note that the mesh has a thick line that divides the study area into an inner part and an outer part. The mesh is used to calculate the w_ks and the algorithm uses neighbouring information in this process. If a triangle is at the edge of the study area, then it has fewer neighbours and the estimated w_k values for the three corresponding vertices may have a large variance. To avoid a boundary effect an outer area is used. This is the area between the two black lines without any sampling locations. Consider the outer area as a buffer zone that is being used to ensure that there is no edge effect for the vertices on the inner side. It is possible to control the extent of the inner and outer parts of the mesh with the `offset` argument. It is recommended that the outer area is at least as large as the range (the distance at which the dependency diminishes). We will show examples of the use of the `offset` in `inla.mesh.2d` in Chapter 13.

The `cutoff` argument ensures that sampling locations that are within a distance that is smaller than the cut-off value are replaced by a single vertex. This is useful if there are sampling locations with the same spatial locations, or if some are very close to one another.

The R code below creates eight additional meshes, and they are all plotted in Figure 12.4.

```
> mesh2 <- inla.mesh.2d(Loc,
                        max.edge = c(10, 10),
                        cutoff = 10)
> mesh3 <- inla.mesh.2d(Loc,
                        max.edge = c(50, 50))
> mesh4 <- inla.mesh.2d(Loc,
                        max.edge = c(75, 75),
                        cutoff = 1)
> mesh5 <- inla.mesh.2d(Loc,
                        max.edge = c(25, 50),
                        cutoff = 1)
> mesh6 <- inla.mesh.2d(Loc,
```

```
                              max.edge = c(50, 80),
                              cutoff = 1)
> mesh7 <- inla.mesh.2d(Loc,
                              max.edge = c(100, 120),
                              cutoff = 1)
> mesh8 <- inla.mesh.2d(Loc,
                              max.edge = c(150, 150),
                              cutoff = 1)
```

In mesh 3 the edges of the triangles within the inner area must be smaller than 25 kilometres, but the ones in the outer area can be 50 kilometres. In general it is wise to use an outer area to avoid a boundary effect, but there is no need to use a fine grid in the outer area. If the second argument for `max.edge` is omitted then there is no outer area.

Step 2 of making a mesh: The primary tool to control the shape of the triangles is `max.edge`. Other useful arguments are the `cutoff` and `offset`. Use an outer area to avoid a boundary effect. The outer area can be less fine than the inner area to reduce computing time.

In all eight meshes we used the `max.edge` argument to control the shape and number of triangles. An alternative approach is to specify a boundary area so that all sampling locations are within this boundary area. An example follows.

```
> Bound <- inla.nonconvex.hull(loc)
> mesh9 <- inla.mesh.2d(boundary = Bound,
                              max.edge = 50,
                              cutoff = 5)
```

The resulting mesh is presented the lower-right graph in Figure 12.4. In this case the sampling locations are not necessary on a vertex of the mesh, and R-INLA will use three non-zero weighting factors w_k per triangle (see the previous section and also the next section). In Chapter 13 we will use a shapefile with the contour lines of an island as the boundary.

To plot a mesh, simply type `plot(mesh1, asp = 1)`, where the `asp = 1` argument ensures that the aspect ratio is fixed. Before deciding which mesh to use it is also useful to look at the number of vertices per mesh.

```
> c(mesh1$n, mesh2$n, mesh3$n, mesh4$n, mesh5$n,
    mesh6$n, mesh7$n, mesh8$n, mesh9$n)

4943 4921 656 526 731 527 518 518 175
```

Mesh 1 has 4943 vertices. This means that the model will estimate 4943 of the w_ks. As a result the solution will be precise but long computing time may be an issue. On the other hand, mesh 9 has 175 vertices and computing time will be short. But when we plot the spatial component it may show a non-smooth pattern.

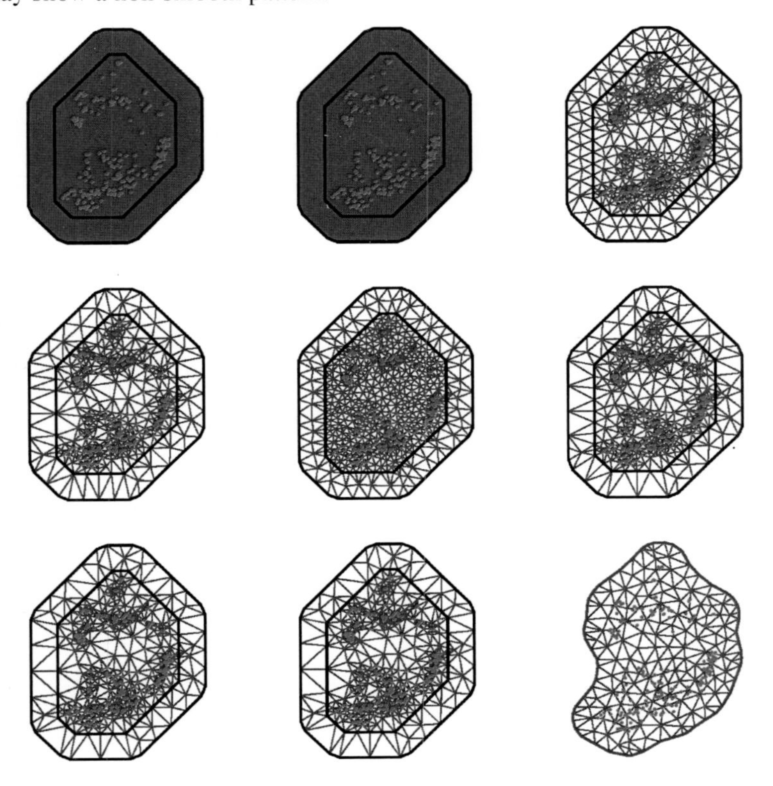

Figure 12.4. Various meshes. Top row from left to right: meshes 1 to 3. Middle row from left to right: meshes 4 to 6. Bottom row from left to right: meshes 7 to 9.

Ultimately the choice of the mesh is a trade-off between the quality of the approximation of the SPDE and computing speed. A mesh with 700–800 vertices takes a few seconds, whereas a mesh with 4000–5000 vertices takes a few minutes. Our general strategy is to work with a mesh of size 700–800 vertices during the initial analysis and for final presentation use a finer mesh. For the moment we will use mesh 5; see Figure 12.5. It has 731 vertices.

In Chapter 13 we will discuss how to adjust the mesh in the event there are boundaries inside the mesh, e.g. due to an island or fjord for fisheries data.

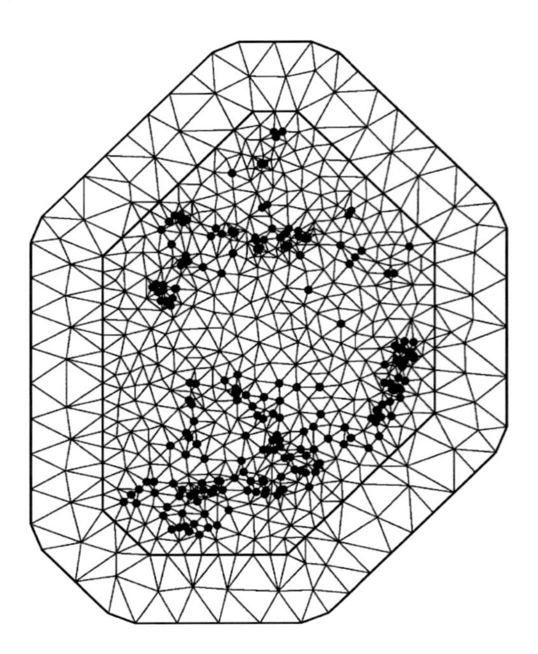

Figure 12.5. Selected mesh 5. The black dots represent the sampling locations.

12.7 Define the weight factors a_{ik}

In the previous section we selected the mesh in Figure 12.5 for the Irish pH data. It has 731 vertices. For each of these 731 vertices R-INLA will estimate a w_k value. Hence, we will end up with posterior mean values (and credible intervals) for w_1, w_2, w_3, ..., w_{731}. The model in Equation (12.2) contains a spatial random effect u_i. There are 210 sampling locations. Hence we have 210 of these spatial random effects: u_1, u_2, ..., u_{210}. Depending on the type of mesh that is selected, the sampling locations (which are the dots in Figure 12.5) are either on a vertex or they are inside a triangle. In the mesh in Figure 12.5 every sampling location is on a vertex. This means that if sampling location i is on vertex k then the value of u_i is equal to the w_k. If sampling location i is inside a triangle, then u_i is a weighted average of the three w_ks that correspond to the vertices that make up the corresponding triangle. In both cases we can calculate u_is as follows (see Chapter 11 for details).

$$u_i = \sum_{k=1}^{731} a_{ik} \times w_k \tag{12.3}$$

The a_{ik}s are the known weight functions. To review the idea behind this equation (see Chapter 11 for more details) we have sketched two scenarios in Figure 12.6. Let us focus on panel A. It shows one triangle from a mesh. Suppose that the big dot is the sampling location of site $i = 15$. It is positioned on vertex 502. As a result all $a_{15,1}$, …, $a_{15,731}$ values are equal to 0, except for $a_{15,502}$ which is 1. The software will estimate w_1, .., w_{731}, but only w_{502} affects u_{15}. To be more precise, following Equation (12.3) we have $u_{15} = 1 \times w_{502} = w_{502}$.

In panel B the sampling location is inside the triangle. In this case all $a_{15,1}$, …, $a_{15,731}$ are 0, except for $a_{15,501}$, $a_{15,502}$, and $a_{15,503}$. Their values depend on the position of the sampling location relative to each vertex. For example $a_{15,501} = 0.3$, $a_{15,502} = 0.1$, and $a_{15,503} = 0.6$. The sampling location is closest to vertex 503 so it has the highest value. The weights add up to 1. In this case we have $u_{15} = 0.3 \times w_{501} + 0.1 \times w_{502} + 0.6 \times w_{503}$. And the w_ks are obtained from the SPDE.

In Figure 12.5 the majority of the sampling locations are on a vertex, which simplifies the construction of the a_{ik}s. If sampling location i is on vertex k then $a_{ik} = 1$, and $a_{ik} = 0$ otherwise. In this case the a_{ik}s are just simple indices that match the sampling locations with the w_ks. If a sampling location is inside a triangle we end up with the situation described in Figure 12.6B.

The a_{ik}s are obtained in R via

```
> A5 <- inla.spde.make.A(mesh5, loc = Loc)
> dim(A5)

210 731
```

We have 210 sampling locations and the mesh has 731 vertices. We can type `A5[1,]` to see the values of $a_{1,1}$ to $a_{1,731}$. It shows a large number of dots (representing zeros) and somewhere a '1'. The '1' tells us that at that specific vertex we have the sampling location.

Figure 12.6. A: Sampling location is on one of the vertices. B: Sampling location is inside the triangle.

12.8 Define the SPDE

For a model with a random intercept with spatial correlation we type the following R code.

```
> spde <- inla.spde2.matern(mesh5, alpha = 2)
```

As long as spatial data are analysed, there is no need to make any changes to this code. It informs R-INLA about the details of the SPDE approach. This means that it now knows the mesh and what type of Matérn correlation function we want to use. As explained in Chapter 11, `alpha = 2` is the default value. This means that the ν in Chapter 11 equals 1, which is fine for two-dimensional spatial data. We will need to adjust it for time-series data.

12.9 Define the spatial field

Next we set up a list for the spatial random intercept **u**. As explained in Section 12.6, **u** is rewritten internally as $\mathbf{A} \times \mathbf{w}$. We need to specify the **w** in R-INLA. This is done with the `inla.spde.make.index` function. We will call the output of this function `w.index`.

```
> w.index <- inla.spde.make.index(
                        name    = 'w',
                        n.spde  = spde$n.spde,
                        n.group = 1,
                        n.repl  = 1)
> str(w.index)
```

```
List of 3
 $ w       : int [1:731] 1 2 3 4 5 6 7 8 9 10 ...
 $ w.group: int [1:731] 1 1 1 1 1 1 1 1 1 1 ...
 $ w.repl : int [1:731] 1 1 1 1 1 1 1 1 1 1 ...
```

The object `w` inside this list contains the numbers 1 to 731 (the number of vertices in the mesh). These are the w_ks from Equation (12.3). The `w.group` and `w.repl` will be discussed later in this book when we analyse spatial-temporal data.

12.10 Define the stack

In Section 12.6 we informed R-INLA at which vertices of the mesh we have sampling locations, and this gave us the projector matrix `A5`. In Section 12.7 we defined the technical details of the SPDE. We need to inform R-INLA at which sampling locations we have data for the response variable and where we have the covariate data. This step is necessary as covariates may be available at different locations than the response variable. In R we use the function `inla.stack` to inform R-INLA about the setup. It is perhaps the most confusing part of the R coding process. To understand `inla.stack`, we rewrite the model using some matrix notation.

$$\mu_i = \alpha + \sum_{j=1}^{7} \beta_j \times X_{ij} + u_i \tag{12.4}$$

X_{ij} represents a main or interaction term. For example, X_{i1} contains the SDI_i value, and X_{i7} is the three-way interaction term $SDI_i \times LogAltitude_i \times Forested_i$. As explained earlier in this section R-INLA estimates the posterior mean value for each w_k, and using the weights in the matrix `A5` we can calculate u_i via Equation (12.3). We might as well substitute Equation (12.3) into (12.4), which results in the following expression.

$$\mu_i = \alpha + \sum_{j=1}^{7} \beta_j \times X_{ij} + \sum_{k=1}^{731} a_{ik} \times w_k \tag{12.5}$$

The model in Equation (12.5) can also be written in matrix notation as

$$\begin{aligned} \boldsymbol{\mu} &= \mathbf{1} \times \alpha + \boldsymbol{X} \times \boldsymbol{\beta} + \boldsymbol{A} \times \boldsymbol{w} \\ &= \text{Intercept + Covariates + Spatial random effect} \end{aligned} \tag{12.6}$$

There are three components on the right-hand side in Equation (12.6). The **1** is a vector with ones and is used for the intercept. The component with the covariates is written as $\mathbf{X} \times \boldsymbol{\beta}$. The seven regression parameters are in $\boldsymbol{\beta}$. The spatial correlated random effect is written as $\mathbf{A} \times \mathbf{w}$, where the 731 w_ks are in $\mathbf{w}$. The matrix $\mathbf{A}$ contains the known weighting factors as discussed in Section 12.6. We called it `A5` in Section 12.5.

The hyperparameters consist of σ from the normal distribution for pH, and the two Matérn correlation parameters κ and σ_u that are obtained via the finite element and SPDE approaches applied on the $\mathbf{w}$.

To run the model in Equation (12.6) in R-INLA we need to provide the three components to R-INLA. We can do this with the `inla.stack` function.

Let us start with the second component, the covariates. Using the `model.matrix` function we create a matrix $\mathbf{X}$. Categorical covariates and interactions may cause some confusion. The interaction between two covariates is simply a new covariate that is equal to the product of the two covariates. For example, the interaction between altitude and SDI is implemented in any statistical model as a new covariate $Altitude_i \times SDI_i$. The `model.matrix` function calculates this product for us and stores it in the fifth column.

```
> Xm <- model.matrix(~ -1 + LogAlt * SDI *
                     fForested, data = iph)
> colnames(Xm) #Results not shown here
> X <- data.frame(Alt         = Xm[,1],
                  SDI         = Xm[,2],
                  fFor        = Xm[,3],
                  Alt.SDI     = Xm[,4],
                  Alt.fFor    = Xm[,5],
```

```
                    SDI.fFor     = Xm[,6],
                    Alt.SDI.fFor = Xm[,7])
```

Now we have the **X** in Equation (12.6). We have all the ingredients to combine the data via the `inla.stack` function.

```
> StackFit <- inla.stack(
                tag = "Fit",
                data = list(y = iph$pH),
                A = list(1, 1, A5),
                effects = list(
                    Intercept = rep(1, N),
                    X = X,
                    w = w.index))
```

The resulting object is called a 'stack'. The `data` argument in the `inla.stack` function contains the response variable. The `tag` is just a string that allows us to give the stack a name. This name can be used at a later stage of the analysis if we want to combine stacks (e.g. for fitting the model and doing predictions at the same time), or if we want to extract information.

The `A` and the `effects` arguments are the important parts. The confusing part is perhaps that there is an 'A5' in the `list(1, 1, A5)` and an 'A' as an argument for the `inla.stack` function. The `A5` is the projector matrix for the spatial random field that we defined in Section 12.6 with the `inla.spde.make.A` function. It has the dimensions 210 by 731. The `1` in the list is short-hand notation for an identity matrix and each argument in the `list(1, 1, A5)` corresponds to an argument in the `effects` list, and the order matches! Hence, the first `1` in the `list(1, 1, A5)` corresponds to the `Intercept` term, the second `1` to the `X` term, and the `A5` in the list belongs to the `w`; see also Figure 12.7.

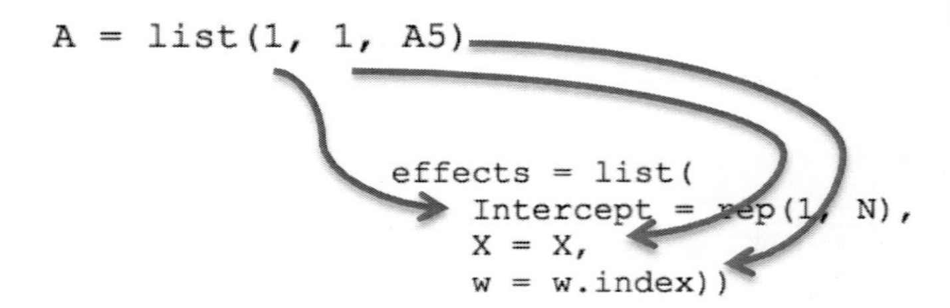

Figure 12.7. Elements of the projector matrix A are linked to the elements of the effects argument.

If more terms are added to the `effects` component (e.g. a smoother or another random effect) then the list for `A` has to be expanded. The purpose of the `A = list(1, 1, A5)` and the `effects` terms is to link the covariates and the spatial random intercept to the observed data. And the three components in the list correspond to the three components in Equation (12.6). If more covariates are added to the `X` matrix, then no changes to the `effects` and `A` matrix are required.

12.11 Define the formula for the spatial model

We will fit two models: a model without spatial correlation and a model with spatial correlation. Because we have a relatively large number of covariates, we use the formula. The model with all main terms, two-way interactions, three way-interaction, and without spatial dependency has the following model formula:

```
> f2a <- y ~ -1 + Intercept + Alt  + SDI + fFor +
             Alt.SDI + Alt.fFor + SDI.fFor  +
             Alt.SDI.fFor
```

The '–1' means that the default intercept is being dropped from the model, and instead we use our own `Intercept` component. This approach is optional here, but becomes a requirement when multiple response variables are fitted in the same model. It also allows us to add the intercept to the spatial random field in case there are no covariates in the model.

The model with the spatial dependency contains an extra term:

```
> f2b <- y ~ -1 + Intercept + Alt  + SDI + fFor +
             Alt.SDI + Alt.fFor + SDI.fFor  +
             Alt.SDI.fFor + f(w, model = spde)
```

The `spde` was made in Section 12.7. Inside the `spde` object is variable `w` with values 1 to 731. We are now ready to run the two models.

12.12 Execute the spatial model in R

To fit a model without spatial dependency we can either use the code from Section 12.3, or we can use the `StackFit` data with the formula that does not contain the spatial random effect.

```
> IM2a <- inla(f2a,
               family = "gaussian",
               data = inla.stack.data(StackFit),
               control.compute = list(
                                  dic = TRUE,
                                  waic = TRUE),
               control.predictor = list(
                     A = inla.stack.A(StackFit)))
```

The model with the spatial dependency model requires similar code, except that the formula `f2b` now contains the spatial random effect.

```
> IM2b <- inla(f2b,
             family = "gaussian",
             data = inla.stack.data(StackFit),
             control.compute = list(dic = TRUE.
                                    waic = TRUE),
             control.predictor = list(
                    A = inla.stack.A(StackFit)))
```

We compare the two models via the DIC and WAIC. The code below extracts the DIC and WAIC from both models and presents the values in a table.

```
> dic    <- c(IM2a$dic$dic, IM2b$dic$dic)
> waic   <- c(IM2a$waic$waic, IM2b$waic$waic)
> Z.out <- cbind(dic, waic)
> rownames(Z.out) <- c("Gaussian lm",
                       "Gaussian lm + SPDE")
> Z.out

                        dic     waic
Gaussian lm        194.5433 196.9315
Gaussian lm + SPDE 124.1180 127.7217
```

Both the DIC and WAIC values indicate that the model with the spatial correlated random effects is better. To investigate whether the size of the mesh has any effects on the results, we also ran the model with mesh 2 (see Figure 12.4). Computing time was a few minutes longer but the DIC and WAIC values were nearly the same as those obtained with mesh 5. Hence, we might as well keep on using mesh 5.

12.13 Results

The posterior mean values and the 95% credible intervals of the fixed parameters can be obtained for both models with the `$summary.fixed`.

```
> IM2a$summary.fixed[, c("mean", "0.025quant",
                         "0.975quant")]
> IM2b$summary.fixed[, c("mean", "0.025quant",
                         "0.975quant")]
```

We can either compare the two sets of numbers or plot the results side by side; see Figure 12.8.

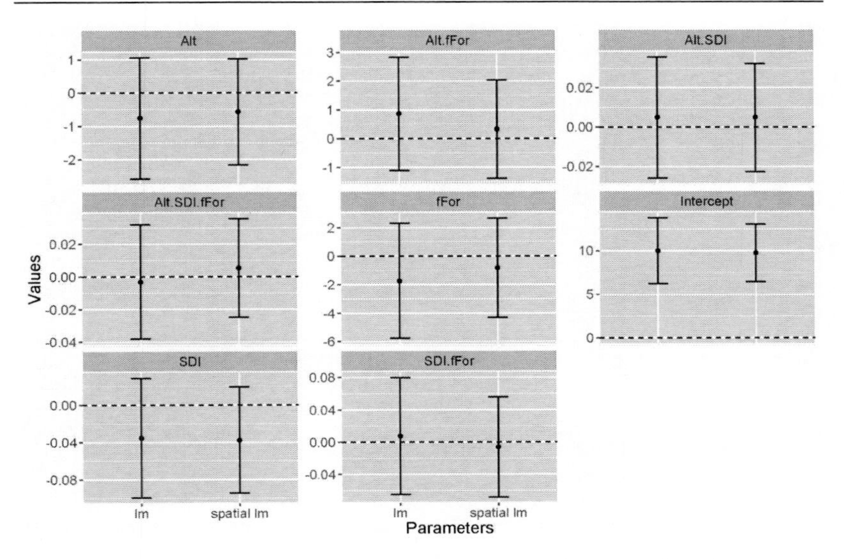

Figure 12.8. Results of the model without ('lm') and with the spatial correlated random effect ('spatial lm'). The R code to create this graph is on the website for this book.

There are only minimal differences between the fixed parts of the two models. This is perhaps an indication that the spatial random effect is not that important. It may well be that for a different data set the results for a model without and with a spatial random effect are much larger. We also noted that for the model without the three-way interaction differences between the two models are larger.

Instead of the posterior mean values and 95% credible intervals it may also be interesting to draw the posterior distributions and compare these between the models.

Next we focus on the hyperparameters. Here, we just show how to access them with the function `inla.spde2.result`.

```
> SpFi.w <- inla.spde2.result(inla = IM2b,
                              name = "w",
                              spde = spde,
                              do.transfer = TRUE)
> Kappa <- inla.emarginal(function(x) x,
                     SpFi.w$marginals.kappa[[1]])
> sigmau <- inla.emarginal(function(x) sqrt(x),
             SpFi.w$marginals.variance.nominal[[1]])
> r <- inla.emarginal(function(x) x,
                SpFi.w$marginals.range.nominal[[1]])
```

The posterior mean of κ (`Kappa` in the code above) is 0.0314 and the posterior mean of σ_u (`sigmau` in the code) is 0.282. We can use the `Kappa` value to calculate the Matérn correlation values; see Figure 12.9.

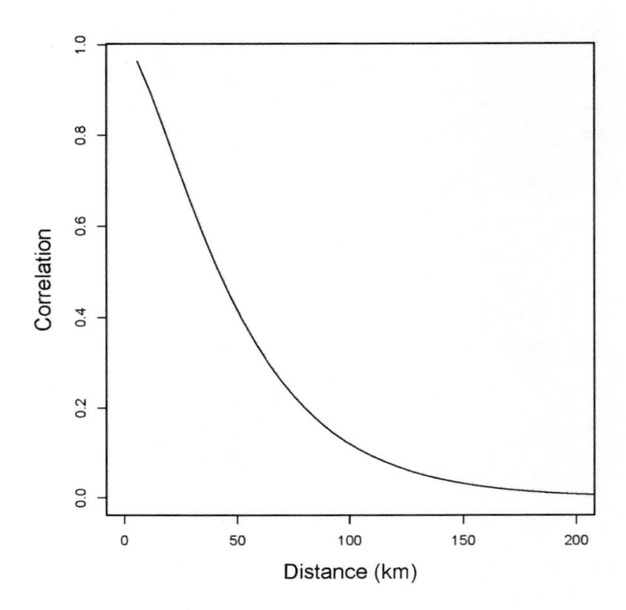

Figure 12.9. Matérn correlation function. The online R code shows how to add a 95% credible interval around the line.

We have strong spatial correlation up to about 40 km. In Chapter 11 we mentioned the phrase 'diminishing correlation' and quantified it as a correlation of 0.1. We also mentioned that the range is the distance at which this happens. Eyeballing the value of the range from the figure gives a value of the range of about 100 km.

In Chapter 11 we also showed an equation that linked the Matérn correlation parameters to the range. R-INLA used this equation when we calculated `r`. This value is the range! Its value is 106.66 km. This means that spatial dependency diminishes for distances larger than 106.66 km!

The R code to create Figure 12.9 is as follows. First we calculate the distances between each sampling location with the `dist` function and determine the maximum distance between any two sampling locations. We then create a vector with 100 distance values between 0 and the maximum distance. Using the posterior mean value of `Kappa` and the Matérn correlation function (see Chapter 11), we calculate the correlation for each distance. The rest is elementary plotting tools.

```
> D     <- as.matrix(dist(mesh5$loc[,1:2]))
> d.vec <- seq(0, max(D), length = 100)
> Cor.M <- (Kappa*d.vec)*besselK(Kappa * d.vec,1)
> Cor.M[1] <- 1
> plot(x = d.vec / 1000,
       y = Cor.M,
       type = "l",
```

```
       xlab = "Distance (km)",
       ylab = "Correlation",
       xlim = c(0, 200))
```

Instead of looking at posterior mean values of the hyperparameters it is also useful to look at the posterior distribution of the hyperparameters, as quite often these are not symmetric.

We finally present the spatial component, the w_ks. Their posterior mean values can be obtained via

```
> w.pm <- IM2b$summary.random$w$mean
```

This is a vector of length 731 by 1. Each value in `w.pm` belongs to a specific vertex on mesh 5. We can either obtain the coordinates of the 731 vertices, match these with `w.pm` (in the correct order), and use existing graphical functions to plot the spatial random field, or we can use R-INLA functions to do this for us. We will opt for the second approach. There are various ways to plot the spatial field. One option is to use the function `inla.mesh.projector`; it creates a lattice using the mesh and specified ranges.

```
> w.proj <- inla.mesh.projector(mesh5,
                                xlim = range(Loc[,1]),
                                ylim = range(Loc[,2]))
```

The function `inla.mesh.project` can then be used to project the 731 posterior mean values on this grid. By default a lattice of 100 by 100 is used. The boundaries of this lattice are given by the extremes of the observed coordinates. It is also possible to omit the `xlim` and `ylim` arguments and use:

```
> w.proj <- inla.mesh.projector(mesh5)
```

In this case the bounding box for the projections is defined by the mesh.

```
> w.pm100_100 <- inla.mesh.project(w.proj, w.pm)
```

This `w.pm100_100` has the dimensions 100 by 100 and is a projection (interpolation and extrapolation) of the random field **w**. We can use the `levelplot` function from the `lattice` package to plot `w.pm100_100`; see Figure 12.10. The code to create this graph uses elementary `lattice` code (see below).

```
> grid <- expand.grid(x = w.proj$x / 1000,
                      y = w.proj$y / 1000)
> grid$z <- as.vector(w.pm100_100)
> levelplot(z ~ x * y,
          data = grid,
          scales = list(draw = TRUE),
          xlab = list("Easting (km)", cex = 1.5),
          ylab = list("Northing (km)", cex = 1.5),
          main = list("Posterior mean spatial
```

```
                    random field", cex = 1.5),
   panel=function(...){
    panel.levelplot(...)
    grid.points(x = iph$Easting / 1000,
                y = iph$Northing / 1000,
                pch = 1,
                size = unit(0.5, "char"))})
```

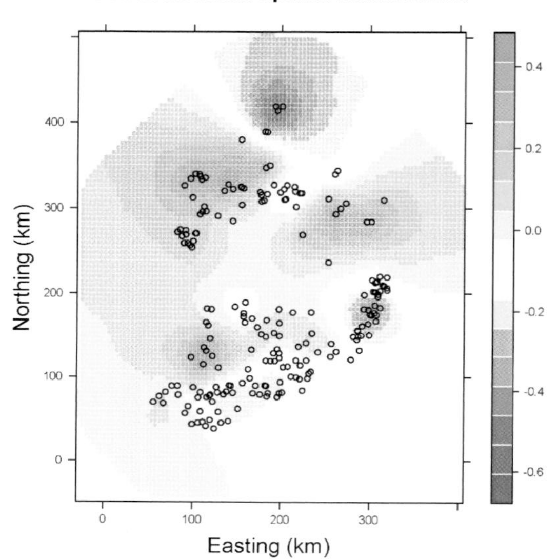

Figure 12.10. Posterior mean values of the spatial field. The dots represent the sampling locations.

The R code on the website for this book also contains `ggplot2` code to create a similar graph; see Figure 12.11 for the final graph. It is also possible to superimpose the spatial random field in Figure 12.10 on the Irish map.

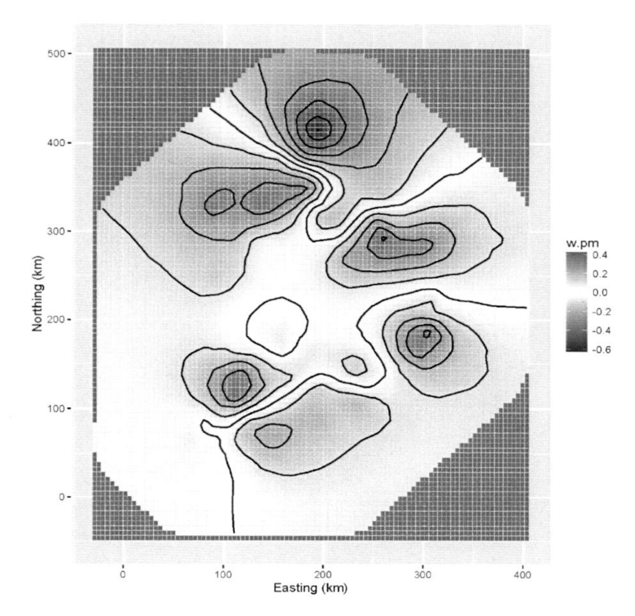

Figure 12.11. Posterior mean values of the spatial field. The graph is made with ggplot2.

12.14 Model selection

We dropped the three-way interaction and reran the model. The DIC and WAIC indicated that the model without the three-way interaction is better. We then dropped each of the two-way interactions and compared the DICs and WAICs. This process was repeated until no further model improvement was found. This model selection process is essentially the same as implemented in the `step` function, except that we have to do it manually and use the DIC and/or WAIC.

The optimal model contains all three main terms and the interaction between forested and (log transformed) altitude. The R code to fit the optimal model is as follows.

```
> f6 <- y ~ -1 + Intercept + Alt  + SDI + fFor +
           Alt.fFor + f(w, model = spde)
> X6 <- data.frame(Alt      = Xm[,2],
                   SDI      = Xm[,3],
                   fFor     = Xm[,4],
                   Alt.fFor = Xm[,6])
> StackFit <- inla.stack(
               tag = "Fit",
               data = list(y = iph$pH),
               A = list(1, 1, A5),
               effects = list(
                    Intercept = rep(1, N),
```

```
                          X = X6,
                          w = w.index))
> IM6 <- inla(f6,
             family = "gaussian",
             data = inla.stack.data(StackFit),
             control.compute = list(dic = TRUE),
             control.predictor = list(A =
                         inla.stack.A(StackFit)))
```

We suggest that for a paper you present the DIC and WAIC values of all the models fitted during the model selection.

12.15 Model validation

We need to get the residuals for model validation. In order to get the residuals we need the (posterior mean) fitted values. And we can get these as follows.

```
> Fit6.a <- IM6$summary.fitted.values[1:210,
                                      "mean"]
```

The first 210 rows in `summary.fitted.values` are the fitted values. But for another data set with different numbers of observations you need to change the 210. It is also possible to obtain the information which rows in the object `summary.fitted.values` correspond to the fitted values via

```
> FitIndex <- inla.stack.index(StackFit,
                               tag = "Fit")$data
> Fit6.a <- IM6$summary.fitted.values[FitIndex,
                                      "mean"]
```

For this data set the `FitIndex` variable contains the values 1 to 210 and we use it to extract the fitted values.

We will discuss in a moment what the remaining 420 rows in `IM6$summary.fitted.values` are.

The fitted values obtained in this way include the spatial random effects. We can now calculate the residuals via

```
> E6 <- iph$pH - Fit6.a
```

These residuals can be plotted versus fitted values, versus each covariate in the model and each covariate not in the model, and they should be assessed for any remaining spatial correlation. We leave this as an exercise for the reader.

12.16 Model interpretation

In this section we focus on the interpretation of model `IM6`. We first present the output for the covariates.

```
> Out6 <- IM6$summary.fixed
```

```
> print(Out6, digits = 2)

           mean   sd 0.025q  0.5q 0.975q  mode  kld
Intercept  9.14 0.39   8.36  9.14   9.91  9.14 0.00
Alt       -0.35 0.18  -0.71 -0.35  -0.00 -0.35 0.00
SDI       -0.02 0.00  -0.02 -0.02  -0.02 -0.02 0.00
fFor      -1.03 0.38  -1.79 -1.03  -0.26 -1.03 0.00
Alt.fFor   0.53 0.18   0.17  0.53   0.89  0.53 0.00
```

This output shows that there is a negative effect of altitude in non-forested sites, there is a negative effect of SDI, sites that are forested have a lower pH, and the altitude effect is positive for forested sites. The numerical output can be translated into

$$pH_i \sim N(\mu_i, 0.29^2)$$
$$E(pH_i) = \mu_i \quad \text{and} \quad \text{var}(pH_i) = 0.29^2$$

$$\mu_i = \begin{cases} 9.14 - 0.02 \times SDI_i - 0.35 \times LogAltitude_i + u_i & \text{If non-forested} \\ 8.11 - 0.02 \times SDI_i + 0.18 \times LogAltitude_i + u_i & \text{If forested} \end{cases}$$

The variance in the normal distribution was obtained as follows.

```
> tau <- IM6$marginals.hyperpar$`Precision for the
                   Gaussian observations`
> sigma <- inla.emarginal(function(x) (1/sqrt(x)),
                                  tau)
```

For a non-technical audience it may be wise to provide a visual interpretation of this model. We have two equations for the mean (see above) and each equation contains two continuous covariates. If we plot this in a three-dimensional space then we have two planes; see Figure 12.12.

It is also interesting to visualise the spatial correlated random effects. This can be done as in Figure 12.10, or we can superimpose the spatial field on a map of Ireland. A third option is presented in Figure 12.13. The points reflect the sampling locations and the size of a point is proportional to the value of the spatial random intercept u_i. We used different symbols (and colours) for the positive and the negative values. Note that all positive values are on the southeastern part of Ireland.

Figure 12.14 shows the same information but in this case we also superimposed the w_ks.

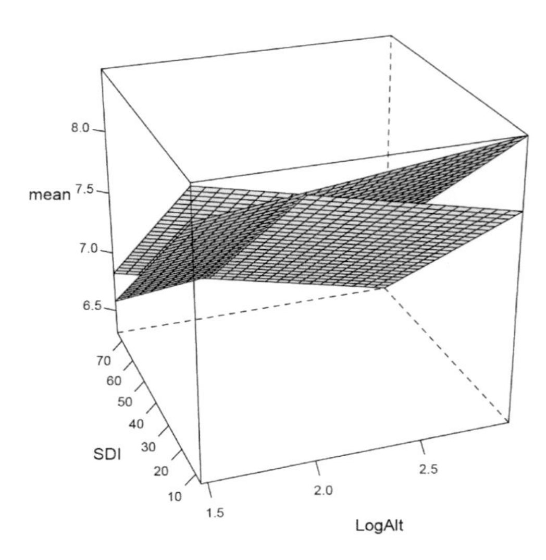

Figure 12.12. Fit of the optimal model. We have two planes, one for the forested data and one for the non-forested data. The *z*-axis shows the posterior mean values. It is also possible to add 95% credible intervals and the observed data.

Figure 12.13. Map of Ireland. Positive values are represented by circles and negative values by triangles. The R code to create this graph is on the website for this book.

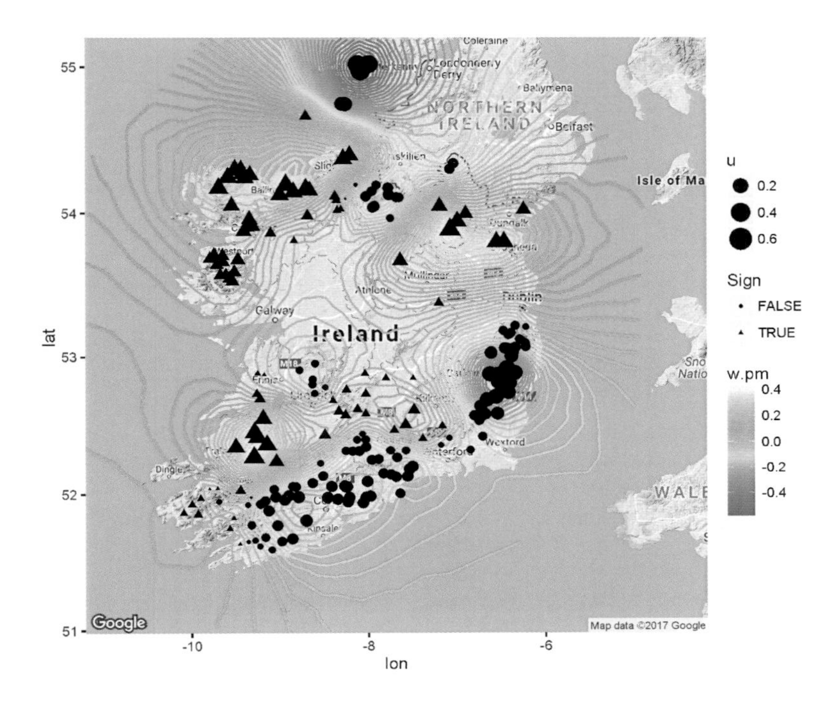

Figure 12.14. As Figure 12.13 but we also added the w_k information as coloured contour lines. The R code to create this graph is on the website for this book.

It is not easy to create Figure 12.12. The challenge is not the actual coding of the three-dimensional graph, but more so to get the data for this graph. For this we create a grid of covariates and predict the pH values for these specific covariate values. The model has two continuous covariates and one categorical covariate. We create 25 values for log altitude and also for SDI, and combine these with the two levels of forested. The `expand.grid` function below creates 1250 covariate values for which we want to obtain predicted pH values (posterior mean values and 95% credible intervals).

```
> MyData <- expand.grid(
         LogAlt     = seq(1.47, 2.94, length = 25),
         SDI        = seq(6.3, 74.25, length = 25),
         fForested = levels(iph$fForested))
> dim(MyData)

1250    3
```

```
> head(MyData)

   LogAlt SDI fForested
1 1.47000 6.3       Yes
2 1.53125 6.3       Yes
3 1.59250 6.3       Yes
4 1.65375 6.3       Yes
5 1.71500 6.3       Yes
6 1.77625 6.3       Yes
```

In Chapters 7 and 8 we discussed two options to get posterior mean values and 95% credible intervals for such a grid of covariate values. The first option was to add the 1250 covariate values to the existing set of covariates and add 1250 NAs for the response variable, and let R-INLA estimate the posterior mean values of pH wherever there is an NA. The second option used `lincomb = inla.make.lincombs`.

Because of the presence of the mesh and the spatial random effect, things work slightly differently here. We will apply the equivalent of the first approach, adding the grid of covariates and NAs. But instead of adding this data to the `iph` object, we will use different stacks. This approach is worked out in the next section.

12.17 Detailed information about the stack*

Upon first reading of this book, this section may be skipped (hence the * in the title). But if you want to do predictions or sketch the fitted values plus credible intervals (which is also prediction) for your own model and data, then this section is relevant.

In the previous section we typed `head(MyData)`, which showed the first six lines of 1250 covariate values. The only aim of this section is to explain how to add three columns to `MyData`, namely the posterior mean and the two values making up the 95% credible interval for each predicted pH value. Once we have this information, creating Figure 12.12 requires simple `wireframe` coding.

12.17.1 Stack for the fitted model again

We already created the stack for the model in Section 12.9. We reproduce it here so that you don't have to hunt back through the book to find the details.

```
> StackFit <- inla.stack(
                 tag = "Fit",
                 data = list(y = iph$pH),
                 A = list(1, 1, A5),
                 effects = list(
                       Intercept = rep(1, N),
                       X = X,
                       w = w.index))
```

The A5 is the projector matrix that links the **w**s at the 731 vertices to the spatial random intercept **u** at the 210 sampling locations; see also Equation (12.3). The dimensions of A5 are 210 by 731.

```
> dim(A5)
```

```
210 731
```

As explained in Section 12.6, each row i in A5 contains weighting factors a_{ik}. The sum of these weighting factors equals 1. Just type rowSums(A5) and you will see 210 times a 1. Let us now focus on the columns of A5. If all sampling locations are on vertices, then the a_{ik}s are either 1 or 0. If a sampling location is inside a triangle, then per row there are three non-zero a_{ik}s. If you type

```
> colSums(A5)
```

then you will see a lot of zeros and also some non-zero numbers. The zeros correspond to vertices with no sampling locations around. If all sampling locations are on vertices then all the non-zero numbers in the colSums(A5) output will be 1. This is not the case if some vertices are inside triangles. For mesh 5 we have 160 columns where the column sum equals 1.

```
> table(colSums(A5) == 1)
```

```
FALSE  TRUE
  571   160
```

And there are 212 columns in A5 where there is somewhere a non-zero weighting factor a_{ik}, as can be seen from the following table command.

```
> table(colSums(A5)>0)
```

```
FALSE  TRUE
  519   212
```

Out of the 713 columns in A5, there are 519 columns with only zeros. The model that we are using is of the form

$$\boldsymbol{\mu} = \text{Intercept} + \text{Covariates} + \text{Spatial random effect} \tag{12.7}$$

We can also write this (in quasi-mathematical notation) as

$$\boldsymbol{\mu} = \mathbf{A}_1 \times \text{Intercept} + \mathbf{A}_2 \times \text{Covariates} + \boldsymbol{A}_3 \times \text{Spatial field} \tag{12.8}$$

We can also write this as

$$\boldsymbol{\mu} = \begin{pmatrix} \mathbf{A}_1 & \mathbf{A}_2 & \mathbf{A}_3 \end{pmatrix} \times \begin{pmatrix} \text{Intercept} \\ \text{Covariates} \\ \text{Spatial field} \end{pmatrix} = \mathbf{A} \times \mathbf{Z} \tag{12.9}$$

The $\mathbf{A}_1$, $\mathbf{A}_2$, and $\mathbf{A}_3$ are the three arguments in the list for `A=list(1,1,A5)` in the `inla.stack` function. So, in this example $\mathbf{A}_1$ is just a matrix with 210 rows and one column, $\mathbf{A}_2$ is an identity matrix with the dimensions 210 by 210, and $\mathbf{A}_3$ is equal to `A5` (which had 210 rows and 713 columns). The $\mathbf{A}$ is a concatenation of $\mathbf{A}_1$, $\mathbf{A}_2$, and $\mathbf{A}_3$, and the whole purpose of the `inla.stack` function is to set up this construction. Let us have a look at the dimensions of the stack.

```
> dim(inla.stack.A(StackFit))
```

```
210 423
```

As expected there are 210 rows. It is confusing that the number of columns of the stack is not 1 + 210 + 713 = 924; instead it is 423. The reason that it is 423 and not 924 is that when R-INLA calculates the fitted values, the `inla.stack` function removes any columns in $\mathbf{A}$ that do not contain weighting factors. Hence, it has 1 + 210 + 212 = 423 columns.

Note that the entire mesh is needed when calculating the **w**s and the spatial random intercepts **u**. But once we have these and want to calculate the fitted values at the sampling locations, we don't need the entire mesh anymore. By dropping the columns with only zeros from the stack the calculations to obtain the fitted values get shorter.

12.17.2 Stack for the new covariate values

In Section 12.15 we defined a grid of covariate values for which we want to predict the pH. We need to convert these data into an $\mathbf{X}$ matrix with the help of the `model.matrix` function.

```
> Xmm <- model.matrix(~ LogAlt + SDI + fForested +
                        LogAlt : fForested,
                        data = iph)
```

But because R-INLA requires some extra work if there are categorical covariates or interactions we prepare a data frame in which each covariate has its own name (including the interaction, which is called `Alt.fFor`).

```
> Xp <- data.frame(Alt         = Xmm[,2],
                   SDI         = Xmm[,3],
                   fFor        = Xmm[,4],
                   Alt.fFor    = Xmm[,5])
```

We create a stack for this specific data set. In the `effects` part we specify the intercept and `Xp`, but no spatial term because we only want to visualise the covariate effect. There are two terms in `effects`, hence the `A` has also two arguments.

```
> StackPred <- inla.stack(
                  tag = "Predict",
                  data = list(y = NA),
                   A = list(1, 1),
                   effects = list(
                      Intercept = rep(1, nrow(Xp)),
                      Xp = Xp))
```

The crucial part is that we specify NA for the response variable. As a result the data in this stack will not affect the estimated parameters or spatial field.

The fitted values linked to this stack represent the fitted values due to the intercept and the covariates that are specified in `Xp`. Note that the `StackCov` is the second stack that we have created, as we already have the stack defined in Section 12.9. It is perhaps wise to use superscripts 1 and 2 to differentiate between the two stacks. In Equation (12.9) we defined the stack for the observed data, intercept, and spatial random field. We will use a superscript 1 for this stack.

$$\begin{aligned} \boldsymbol{\mu}^1 &= \begin{pmatrix} \mathbf{A}_1^1 & \mathbf{A}_2^1 & \boldsymbol{A}_3^1 \end{pmatrix} \times \begin{pmatrix} \text{Intercept} \\ \text{Covariates} \\ \text{Spatial field} \end{pmatrix} \\ &= \mathbf{A}^1 \times \mathbf{Z}^1 \end{aligned} \quad (12.10)$$

Then there is the new stack for the 1250 covariates values that we just generated. This one can be written as

$$\begin{aligned} \boldsymbol{\mu}^2 &= \begin{pmatrix} \mathbf{A}_1^2 & \mathbf{A}_2^2 \end{pmatrix} \times \begin{pmatrix} \text{Intercept} \\ \text{Covariates} \end{pmatrix} \\ &= \mathbf{A}^2 \times \mathbf{Z}^2 \end{aligned} \quad (12.11)$$

The dimensions of this stack are as follows.

```
> dim(inla.stack.A(StackPred))
1250 1251
```

We have 1250 covariates combinations, hence there are 1250 rows. $\mathbf{A}^2{}_1$ is a matrix with 1250 rows and one column and is used for the intercept. $\mathbf{A}^2{}_2$ is an identity matrix with 1250 rows and 1250 columns and is used for the 1250 covariate values. The diagonal values in $\mathbf{A}^2{}_2$ are all 1; there are no columns in $\mathbf{A}^2$ that have a column sum of 0. As a result $\mathbf{A}^2$ has the dimensions 1250 by 1251.

12.17.3 Combine the two stacks

We now have two stacks and these can be combined into a new stack.

```
> All.stacks <- inla.stack(StackFit, StackPred)
```

We are combining Equations (12.10) and (12.11), resulting in the following expression (using quasi-mathematical notation).

$$\begin{pmatrix} \boldsymbol{\mu}^1 \\ \boldsymbol{\mu}^2 \end{pmatrix} = \begin{pmatrix} \mathbf{A}^1 & 0 \\ 0 & \mathbf{A}^2 \end{pmatrix} \times \begin{pmatrix} \mathbf{Z}^1 \\ \mathbf{Z}^2 \end{pmatrix} = \mathbf{A} \times \mathbf{Z} \tag{12.12}$$

The response variable is now equal to $\mathbf{y} = (\text{pH}_1, \ldots, \text{pH}_{210}, \text{NA}, \ldots, \text{NA})$. There are 1250 NAs. The dimensions of the combined stack are as follows.

```
> dim(inla.stack.A(All.stacks))

1460 1673
```

The number of rows in the combined stack is the number of rows of $\mathbf{A}^1$ and $\mathbf{A}^2$ combined, which is 210 + 1250 = 1460. The number of columns in the combined stack is 1673, which is one less than the sum of 423 + 1251 = 1674.

12.17.4 Run the model

We use the same set of covariates as in `IM6`, but for convenience we call the formula `f7` so that it matches the model object name.

```
> f7 <- y ~ -1 + Intercept + Alt  + SDI + fFor +
             Alt.fFor + f(w, model = spde)
```

The model is executed with the same R-INLA code as `IM6`, except that we now use `All.stacks` instead of `StackFit`.

```
> IM7 <- inla(f7,
              family = "gaussian",
              data=inla.stack.data(All.stacks),
              control.compute = list(dic = TRUE),
              control.predictor = list(
                   A = inla.stack.A(All.stacks)))
```

Let's inspect the dimension of the fitted values.

```
> dim(IM7$summary.fitted.values)

3133    6
```

The question is why we have 3133 rows in the fitted values. The answer is that the `summary.fitted.values` object contains the fitted values for each component of the stack. To make life easier, R-INLA has the inla.stack.index function that gives the index showing what is what in the summary.fitted.values object. The crucial point is to use the name of the individual stack when using inla.stack.index.

```
> Index.Fit <- inla.stack.index(All.stacks,
                                 tag = "Fit")$data
```

This is a vector with the numbers $1, 2, 3, \ldots, 210$, which means that the first 210 rows in summary.fitted.values are the fitted values of the model. If we type

```
> F7a <- IM7$summary.fitted.values[Index.Fit,
                                    c(1,3,5)]
```

then we have the posterior mean values and 95% credible intervals for the fitted values. This is the part we already had, and it is not linked to the predicted values for the 1250 new rows.

Then we have the index for the extra data. These are taken from the 'Covariates' stack.

```
> Index.Predict <- inla.stack.index(All.stacks,
                              tag = "Predict")$data
```

The values in this index are 211, 212, ..., 1460. Hence rows 211 to 1460 give us the predicted values for the 1250 covariate values that we created.

```
> F7b <- IM7$summary.fitted.values[Index.Predict,
                                    c(1,3,5)]
```

We now have the posterior mean and 95% credible intervals for the covariates specified in the MyData object in Section 12.15. We can combine the MyData and F7b.

```
> MyData2 <- cbind(MyData, F7b)
```

The wireframe code to create Figure 12.12 is as follows.

```
> wireframe(mean ~ LogAlt + SDI,
            data = MyData2,
            group = fForested,
            shade = FALSE,
            scales = list(arrows = FALSE),
            drape = FALSE, colorkey = FALSE,
            screen = list(z = 20, x = -60 - 20/5))
```

If you wonder what is in the remaining 1673 rows in summary.fitted.values then please read on, though it is not relevant for the discussion on how to create Figure 12.12. We can get the stack matrix **A** of the two combined stacks via

```
> Astk <- as.matrix(inla.stack.A(All.stacks))
```

Using this matrix and the expression in Equation (12.12) we can calculate the fitted values and predicted values again.

```
> F7c <- Astk %*%
              IM7$summary.fitted.values[1461:3132,
                                              "mean"]
```

The fitted values are given by `F7c[index.Fit]` and the predicted value by `F7c[index.Cov]`.

13 Spatial Poisson models applied to plant diversity

In all previous chapters we introduced theory in an applied context. In this chapter there is no new theory. Instead we present a detailed example of the analysis of a plant data set. Techniques employed are Poisson generalised linear models (GLM) without and with spatial dependency. We will include most of the R code and also discuss what to present in a paper.

Prerequisite for this chapter: You need to be familiar with all the material discussed in Chapter 12.

13.1 Introduction

The data used in this chapter were taken from Irl et al. (2015), who studied the effect of climatic and topographic variables on plant species richness and endemic richness on the oceanic island La Palma, Canary Islands, Spain. Endemic (i.e. unique to a defined geographic location) species and especially single-island endemic species are a result of isolated biological populations evolving to become distinct species. La Palma is relatively far away from the mainland and therefore species diversity is low, but it has high endemism.

Irl et al. (2015) explained that the spatial distribution of endemism within an oceanic island gives valuable information about the origin and drivers of biodiversity patterns on the landscape scale, and this has important implications for conservation. The question that we will focus on in this chapter is whether climatic and topographic variables drive single-island endemic richness.

At 890 plots on the island the absence and presence of perennial (a plant living for several years) vascular species (land plants that have lignified tissues for conducting water and minerals) was sampled. Total richness, endemic richness, single-island richness, and archipelago richness per plot were determined.

13.2 Data exploration

13.2.1 Sampling locations

We first import the data (available from Dryad; Irl et al. 2015) using the `read.table` function.

```
> LP <- read.table(file = "LaPalma.txt",
                   header = TRUE,
                   dec = ".")
```

The sampling locations in the text file are denoted by latitude and longitude, but these are in fact UTM coordinates. We can easily convert these into real latitude and longitude coordinates with some `sp` (Pebesma and Bivand 2005) magic.

```
> library(sp)
> utmcoor <- SpatialPoints(
                    coords = cbind(LP$Longitude,
                                   LP$Latitude),
                    proj4string =
                          CRS("+proj=utm +zone=28N"))
> longlat <- spTransform(utmcoor,
                           CRS("+proj=longlat"))
> LP$Longitude2 <- coordinates(longlat)[,1]
> LP$Latitude2  <- coordinates(longlat)[,2]
```

We now have the real latitude and longitude coordinates of each plot, and we can use these to visualise where the plots are on La Palma; see Figure 13.1. We have a large number of plots close to one another and spatial dependency is likely to be present. We suggest that you also make a histogram of the distances between sites (see Chapter 12 for the R code and an example, and also Figure 13.6).

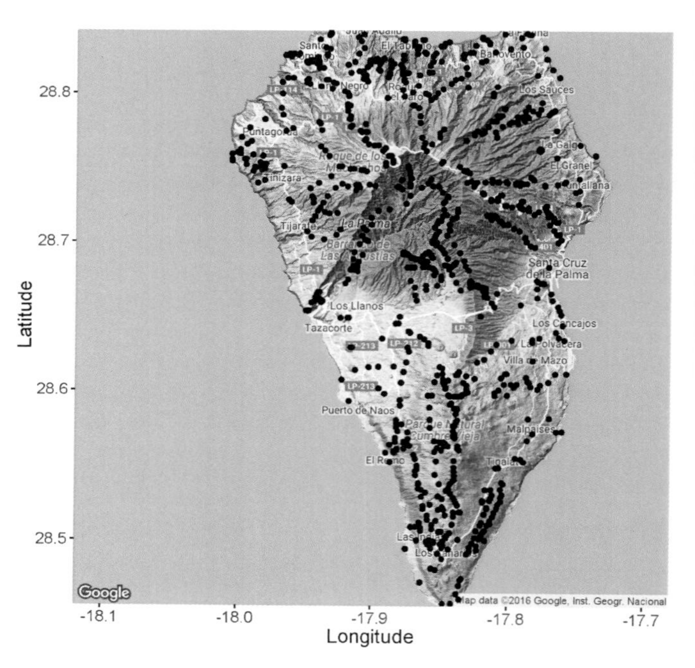

Figure 13.1. Map of La Palma (west of Morocco) and the sampling locations (dots).

Figure 13.1 was created with the following R code. First we load the ggmap package (Kahle and Wickham 2013) and use the get_map function to obtain the map of La Palma. Finding the right coordinates and zooming factor so that the island fits on the graph requires some fiddling with the location values. The rest of the code is a matter of adding the plots and some cosmetics.

```
> library(ggmap)
> glgmap <- get_map(location = c(-18.1, 28.4,
                                 -17.7,28.9),
                    maptype = "terrain",
                    zoom = 7)
> p <- ggmap(glgmap)
> p <- p + geom_point(aes(Longitude2, Latitude2),
                     data = LP)
> p <- p + theme(text = element_text(size = 15))
> p <- p + xlab("Longitude") + ylab("Latitude")
> p
```

13.2.2 Outliers

We continue the data exploration with Cleveland dotplots to investigate the presence of potential outliers; see Figure 13.2. None of the variables has observations that are relative small or large, as compared to the majority of the observations. Hence, there are no obvious outliers.

CR_CAN and CR_LP are climate rarity indices for the Canary Islands and La Palma respectively. INTRA_VAR and INTER_VAR are intra- and inter-annual precipitation variability respectively. MAT and MAP are mean annual temperature and mean annual precipitation. RSI stands for rainfall seasonality index. Macro is the macro-aspect. TCI is a topographic index. All these variables are covariates and their coding and interpretation are explained and justified in Irl et al. (2015). The response variable is nSIE, which is the number of single-island endemic species per plot.

The R code to create Figure 13.2 uses our support file, which can be downloaded from the website for this book.

```
> library(lattice)
> source("HighstatLibV10.R")
> MyVar <- c("CR_CAN", "CR_LP", "Elevation",
             "INTRA_VAR", "INTER_VAR", "MAT",
             "MAP", "RSI", "ASR", "Easterness",
             "Age", "Macro", "Northerness",
             "Slope", "TCI", "nSIE")
> Mydotplot(LP[, MyVar])
```

Note that the scales of the covariates differ considerably, and to avoid numerical estimation problems later in this chapter it is advisable to standardise the covariates.

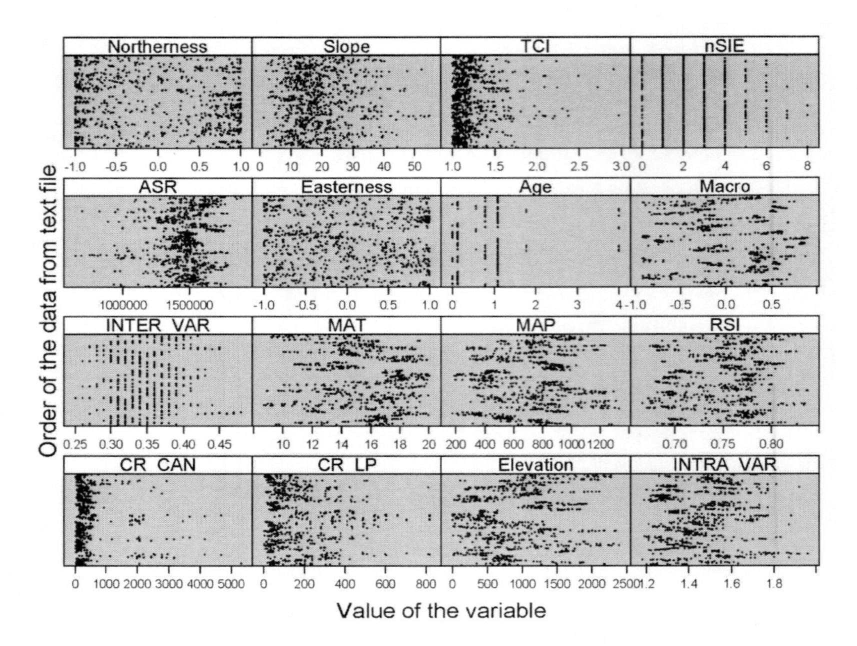

Figure 13.2. Cleveland dotplots for each variable.

13.2.3 Collinearity

Based on the nature of the covariates we expect collinearity to be present (e.g. elevation and temperature tend to be correlated in most ecosystems). We use variance inflation factors (VIF) to investigate this. The function `corvif` is taken from our support file.

```
> MyVar <- c("CR_CAN", "CR_LP", "Elevation",
             "INTRA_VAR", "INTER_VAR", "MAT",
             "MAP", "RSI", "ASR", "Easterness",
             "Age", "Macro", "Northerness",
             "Slope", "TCI")
> corvif(LP[,MyVar])

                   VIF
CR_CAN        2.757690
CR_LP         3.064728
Elevation    63.931930
INTRA_VAR     5.456039
INTER_VAR     3.065116
MAT          63.569856
MAP          11.092176
RSI          14.413193
ASR           7.384237
Easterness    4.111226
Age           1.445933
```

```
Macro          9.243082
Northerness    3.169867
Slope          3.042687
TCI            1.915045
```

There is indeed collinearity. We dropped elevation as it has the highest VIF value, recalculated the VIFs, dropped the covariate with the highest VIF, and continued this process until all VIFs were smaller than 3. This process resulted in dropping elevation, MAT, Macro, ASR, RSI, and INTRA_VAR. The selected covariates for the analysis are CR_CAN, CR_LP, INTER_VAR, MAP, easterness, Age, northerness, slope, and TCI. A multi-panel scatterplot of the selected covariates still shows some odd patterns, and further selection of covariates is recommended. We will continue with the current set of covariates and hope that we do not encounter any numerical estimation problems. A more sensible selection of covariates can be made using expertise knowledge of these covariates instead of VIFs (or using both VIFs and biological knowledge).

13.2.4 Relationships

As part of data exploration we also present scatterplots of each covariate versus the response variable nSIE; see Figure 13.3. There are no strong patterns present. A few covariates show a weak non-linear effect and we need to pay attention to this once we reach the model validation stage.

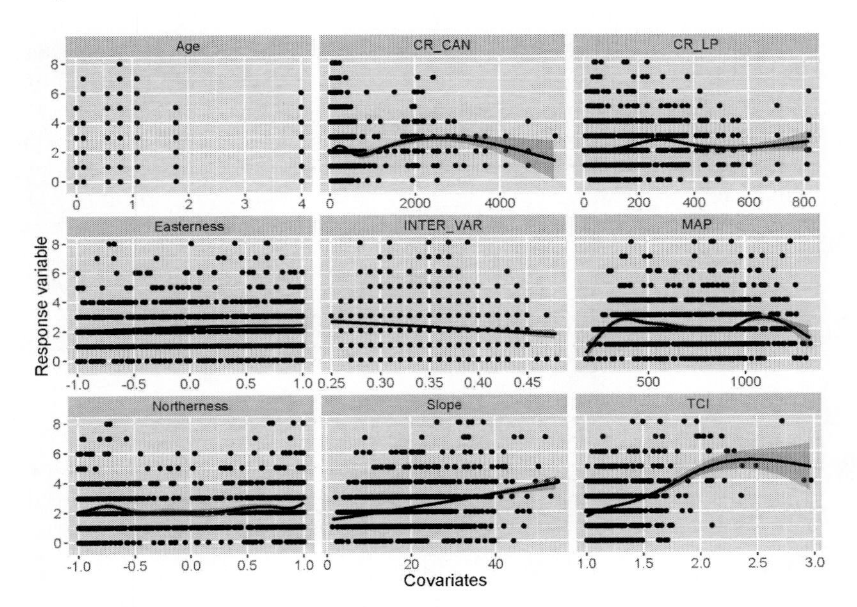

Figure 13.3. Scatterplot of selected covariate versus nSIE. A smoother is added to each panel to aid visual interpretation.

The following code was used to produce Figure 13.3. The function MyMultipanel.ggp2 is in our support file.

```
> MyVar <- c("CR_CAN", "CR_LP", "INTER_VAR",
             "MAP", "Easterness", "Age",
             "Northerness", "Slope", "TCI")
> MyMultipanel.ggp2(Z = LP,
                    varx = MyVar,
                    vary = "nSIE",
                    ylab = "Response variable",
                    addSmoother = TRUE,
                    addRegressionLine = FALSE,
                    addHorizontalLine = FALSE)
```

13.2.5 Numbers of zeros

Quite often biological count data contain a lot of zeros, but this data set seems to be an example without such problems. Only 85 (9.5%) of the 890 plots had no single-island endemic species present. Hence, we do not expect any major problems with zero inflation. See Zuur et al. (2016a) for a detailed volume on zero-inflated models.

13.2.6 Conclusions data exploration

Data exploration indicated that there is collinearity and we took a sub-selection of the covariates based on VIF values. A more sensible approach for selecting covariates is possible. There are no obvious outliers in the covariates or in the response variable. Relationships between covariates and the response variable are weak. We do not expect problems with the number of zeros in the response variable. Spatial dependency in the response variable is likely to be present. Due to the large differences in the scales of the covariates we decide to standardise the covariates, which we will do now using the function MyStd, which is in our support file.

```
> LP$CRCAN.std    <- MyStd(LP$CR_CAN)
> LP$CRLP.std     <- MyStd(LP$CR_LP)
> LP$INTERVAR.std <- MyStd(LP$INTER_VAR)
> LP$MAP.std      <- MyStd(LP$MAP)
> LP$Age.std      <- MyStd(LP$Age)
> LP$Slope.std    <- MyStd(LP$Slope)
> LP$TCI.std      <- MyStd(LP$TCI)
```

We decided not to standardise easterness and northerness as these two variables try to capture the spatial position. In fact, we are not sure whether these two variables should be included in the model at all.

13.3 Model formulation

Single-island species richness is a count. The question is whether we should start with an ordinary Poisson GLM or whether we should add a spatial correlation term to the model to avoid pseudoreplication. Some of the covariates are spatially correlated and we also use easterness and northerness, which may capture some (or all) of the spatial correlation (if present). We decided to start simple and ignore the spatial dependency for the moment. The reason for this is that it is easier to justify the application of complicated statistical methods (e.g. a Poisson model with spatial correlation) if standard models (e.g. a Poisson GLM) fail. Too often we have encountered the following question during a refereeing process: "Why are you applying these complicated methods?" Hence, we will start with a simple Poisson GLM; see Equation (13.1).

$$\begin{aligned} & nSIE_i \sim P(\mu_i) \\ & E(nSIE_i) = \text{var}(nSIE_i) = \mu_i \\ & \log(\mu_i) = CRCAN + CRLP + INTERVAR + MAP + Easterness\ + \\ & \qquad\qquad Age + Northerness + Slope + TCI \end{aligned} \tag{13.1}$$

Note that most covariates are standardised.

13.4 GLM results

To execute the model in Equation (13.1) in R-INLA we use the following code.

```
> I1 <- inla(nSIE ~ CRCAN.std + CRLP.std +
                   INTERVAR.std + MAP.std +
                   Easterness + Age.std +
                   Northerness + Slope.std +
                   TCI.std,
            family = "poisson",
            control.predictor = list(
                              compute = TRUE),
            data = LP)
```

The `summary(I1)` output is not presented here but it shows that plenty of the covariants are important, but before drawing any conclusions we first need to assess the dispersion statistic and apply model validation. The dispersion statistic indicates no major over- or underdispersion problems.

```
> ExpY <- I1$summary.fitted.values[,"mean"]
> E1   <- (LP$nSIE - ExpY) / sqrt(ExpY)
> N    <- nrow(LP)
> p    <- length(I1$names.fixed)
> Dispersion <- sum(E1^2) / (N - p)
> Dispersion

0.8751238
```

The value of 0.87 is slightly smaller than 1 but we are sure that if we do a simulation study (see Chapter 11) then 0.87 is something that one can expect based on the model and the size of this data set.

We also plotted the Pearson residuals versus each covariate in the model, and each covariate not in the model; see Figure 13.4. There are one or two covariates that are rather skewed and these would benefit from a transformation (e.g. CR_CAN and TCI). We should have picked this up when investigating the Cleveland dotplots. There are no clear non-linear patterns, though the shape of the smoother in the CR_CAN and TCI panels caused us to raise our eyebrows a bit.

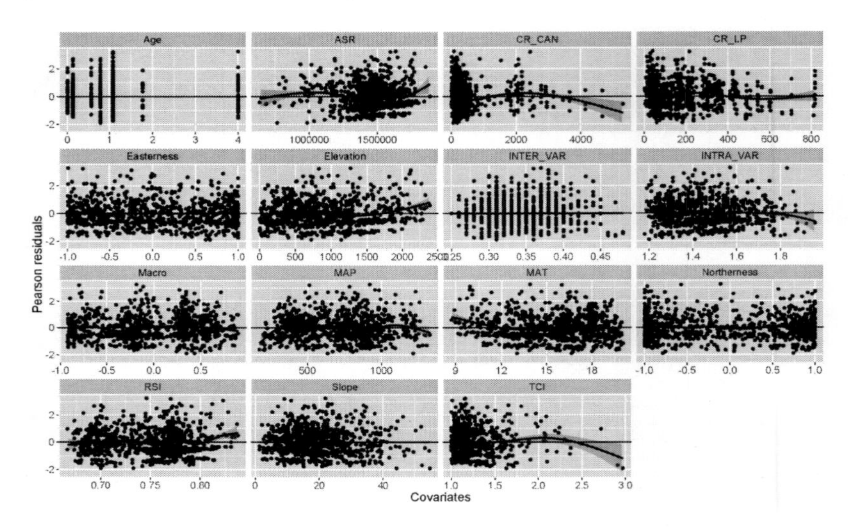

Figure 13.4. Pearson residuals plotted versus each covariate in the model and each covariate not in the model.

The following R code was used to create Figure 13.4. It is similar to the code we used for Figure 13.3.

```
> LP$E1 <- E1
> MyVar <- c("CR_CAN", "CR_LP",
             "Elevation", "INTRA_VAR","INTER_VAR",
             "MAT", "MAP", "RSI", "ASR",
             "Easterness", "Age", "Macro",
             "Northerness", "Slope", "TCI")
> MyMultipanel.ggp2(Z = LP,
                    varx = MyVar,
                    vary = "E1",
                    ylab = "Pearson residuals",
                    addSmoother = TRUE,
                    addRegressionLine = FALSE,
                    addHorizontalLine = TRUE)
```

Finally we investigate spatial dependency in the Pearson residuals. One option is to plot the residuals versus their spatial positions and use different point sizes depending on the value of the residuals. Such a plot should not show any patterns. Alternatively, we can make a sample variogram; see Figure 13.5. The sample variogram shows clear spatial correlation up to around 2½ km. This means that we cannot proceed with the Poisson GLM in Equation (13.1).

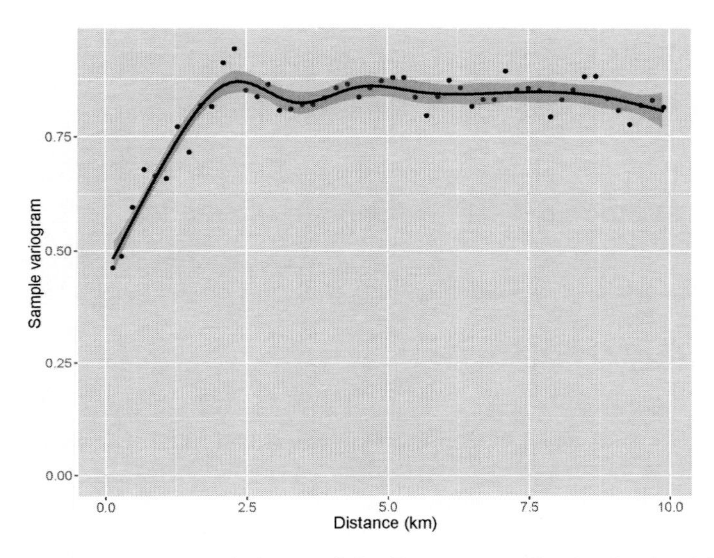

Figure 13.5. Sample variogram of the Pearson residuals obtained from the Poisson GLM in Equation (13.1).

The R code to create the sample variogram in Figure 13.5 is as follows. First we make a data frame `MyData` that contains the Pearson residuals and the UTM coordinates. Then we use the function `variogram` from the `gstat` package (Pebesma 2004) to calculate the sample variogram. Because we have a relative large sample size we decreased the default value for `width` (it controls the width of distances intervals for the variogram calculations) and as a result we get more points in the graph.

```
> library(sp)
> library(gstat)
> LP$Xkm <- LP$Longitude / 1000
> LP$Ykm <- LP$Latitude  / 1000
> mydata <- data.frame(E1, LP$Xkm, LP$Ykm)
> coordinates(mydata) <- c("LP.Xkm", "LP.Ykm")
> Vario <- variogram(object = E1 ~ 1,
                     data = mydata,
                     cressie = TRUE,
                     cutoff = 10, width = 0.2)
```

The rest is a matter of some trivial ggplot2 code.

```
> p <- ggplot(data = Vario,
              aes(x = dist, y = gamma))
> p <- p + geom_point()
> p <- p + geom_smooth(method = "gam",
                       formula = y ~ s(x,
                                       bs = "cs"),
                       colour = "black")
> p <- p + ylim(0,1)
> p <- p + theme(text = element_text(size = 15))
> p <- p + xlab("Distance (km)")
> p <- p + ylab("Sample variogram")
> p
```

13.5 Adding spatial correlation to the model

In this section we will use R-INLA to execute a model that contains a spatial correlated random effect. We will follow the outline presented in Figure 13.2.

13.5.1 Model formulation

The model in Equation (13.2) is nearly identical to the one in Equation (13.1), except that it contains an extra term u_i at the end. This is the spatial correlated random term. The SPDE approach is used by R-INLA to estimate it.

$$\begin{aligned}
& nSIE_i \sim P(\mu_i) \\
& E(nSIE_i) = \text{var}(nSIE_i) = \mu_i \\
& \log(\mu_i) = CRCAN + CRLP + INTERVAR + MAP + Easterness\ + \\
& \qquad Age + Northerness + Slope + TCI + u_i
\end{aligned} \tag{13.2}$$

We use the same set of covariates (which are standardised).

13.5.2 Mesh

As explained in Chapter 12 we need to define a mesh for the spatial correlated random effects. To get an idea what values of `max.edge`, `cutoff`, and `offset` to choose in the `inla.mesh.2d` function, we make a histogram of the distances between the 890 plots and we also inspect the cumulative distribution of these distances; see Figure 13.6. The majority of the sites are separated by less than 20 km. Based on these two graphs and some trial and error we initially choose the mesh in Figure 13.7. We did not use an outer area in the mesh because these plants do not grow in the sea.

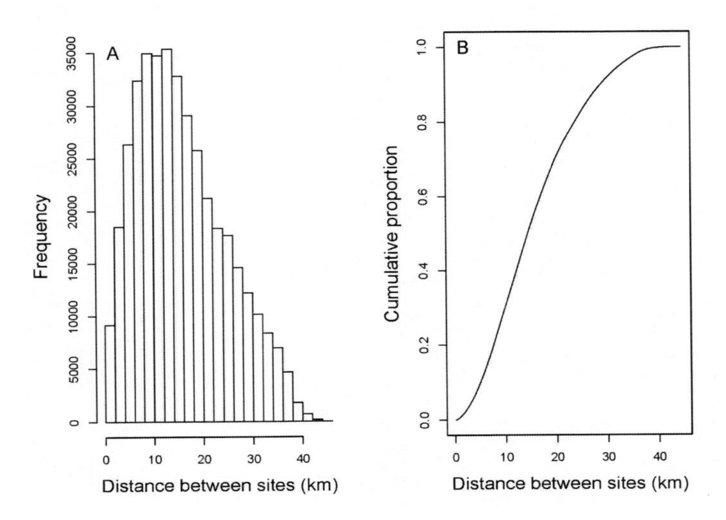

Figure 13.6. A: Histogram of distances between the 890 sites on La Palma. B: Cumulative proportion of distances versus distance.

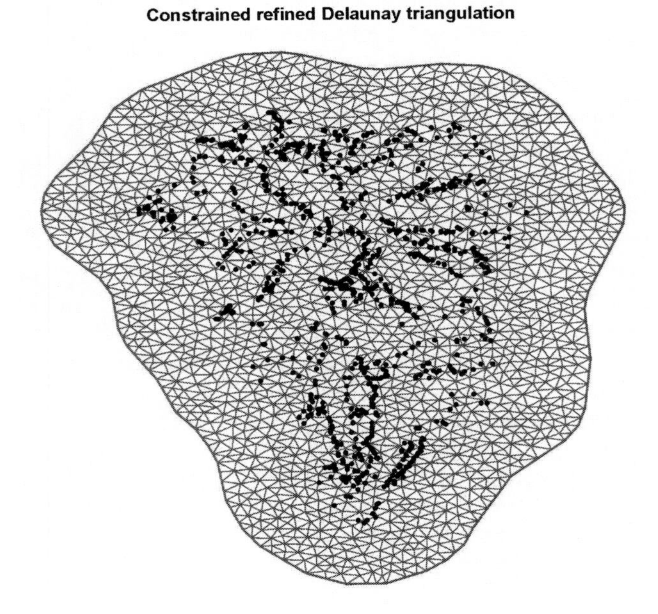

Figure 13.7. Mesh selected initially.

The R code to produce Figure 13.6 uses the `dist` function to calculate distances between the sites and elementary plotting tools.

```
> Loc <- cbind(LP$Xkm, LP$Ykm)
```

```
> D <- dist(Loc)
> hist(D,
       freq = TRUE,
       main = "",
       xlab = "Distance between sites (km)",
       ylab = "Frequency")
> plot(x = sort(D),
       y = (1:length(D))/length(D),
       type = "l",
       xlab = "Distance between sites (km)",
       ylab = "Cumulative proportion")
```

For the mesh in Figure 13.7 we first define a non-convex hull. We only specify one value for `max.edge` so that no outer area is used.

```
> ConvHull <- inla.nonconvex.hull(Loc)
> mesh1    <- inla.mesh.2d(boundary = ConvHull,
                           max.edge = c(1.5))
> plot(mesh1)
> points(Loc, col = 1, pch = 16, cex = 1)
```

This mesh has 1,829 vertices, as can be seen from `mesh1$n`. We initially ran all the models in this chapter with this mesh. The end result was that the spatial field and the fitted values extend beyond the coastline and cover the sea. This does not make sense as we are not sampling sites in the sea. Obviously, we can crop the picture containing the fitted values with the contour lines of La Palma (and that is what we did). But it is also possible to define a boundary for a mesh, and therefore constrain the spatial field to be on land. Figure 13.8 shows the mesh obtained by using the coastline of La Palma as boundary. As can be seen from the figure, this mesh does not extend into the sea. It has 1,192 vertices.

To make this graph you need to have a shapefile. Shapefiles can be downloaded from various sources on the internet (e.g. http://www.arcgis.com/features/index.html) or you can create them using specialized software like GIS. We created the shapefile `lapalma.shp` by downloading a general shapefile for Spain, and subsetting it by using the spatial coordinates of the sampling locations on La Palma.

The code below imports the shapefile in R using the rgdal package (Bivand et al. 2016).

```
> library(rgdal)
> DSN <- "lapalma.shp"
> ShapeF.utm <- readOGR(dsn = DSN,
                        layer = "lapalma")
```

If you type `plot(ShapeF.utm)` in R then you get a picture with the contour lines of the island. If you type `ShapeF.utm` in R then the output will tell you that this is a spatial polygon data frame.

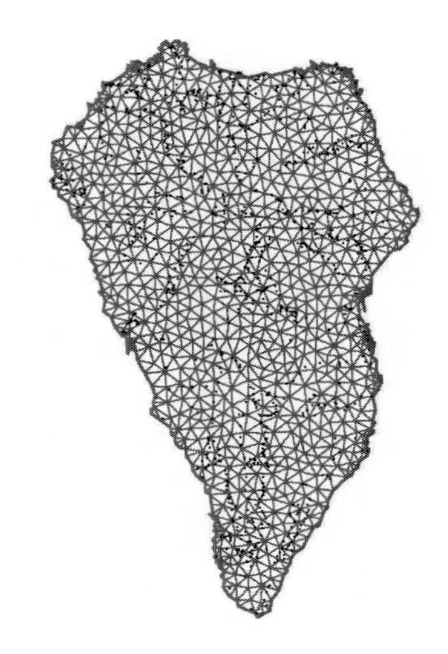

Figure 13.8. Mesh obtained by using a shapefile as boundary.

The `fortify` function from the `ggplot2` package can be used to convert the data in `ShapeF.utm` into an ordinary data frame, which is easier to manipulate.

```
> LaPalma_df <- fortify(ShapeF.utm)
> head(LaPalma_df, 4)
      long     lat order  hole piece id group
1 233949.4 3182045     1 FALSE     1  0   0.1
2 233661.6 3181397     2 FALSE     1  0   0.1
3 233217.2 3181025     3 FALSE     1  0   0.1
4 233254.1 3180755     4 FALSE     1  0   0.1
```

If we plot the `long` and `lat` columns versus each other then we get a picture of the contours of La Palma (Figure 13.9). It is crucial to realise that the rows in the `LaPalma_df` data frame are sorted clockwise. Counter-clockwise values are interpreted as a boundary by R-INLA whereas clockwise orientation is seen as a hole.

The coordinates in the shapefile are also in UTM and we convert them to kilometres.

```
> LaPalma_df$Xkm <- LaPalma_df$long / 1000
> LaPalma_df$Ykm <- LaPalma_df$lat / 1000
```

We extract the relevant columns and store them in an object with the name CoastLine.

```
> CoastLine <- LaPalma_df[,c("Xkm", "Ykm")]
```

Simple plotting tools can be used to make a scatterplot of Xkm versus Ykm; see Figure 13.9. We used the xyplot function from the lattice package to create this graph. Changing type = "p" to type = "l" in this function gives the coastline as a line instead of points.

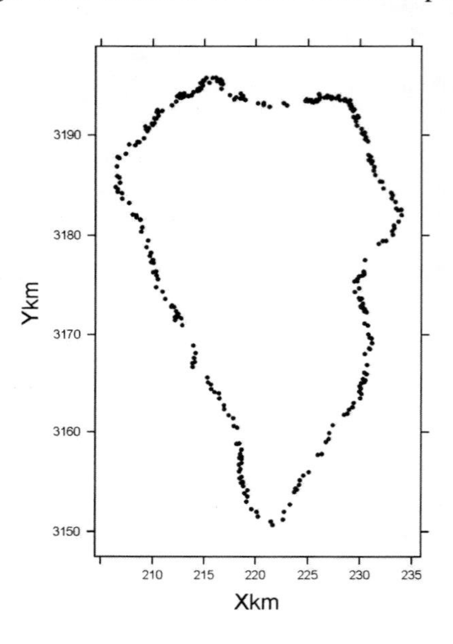

Figure 13.9. Simple scatterplot of the Xkm and Ykm columns in the CoastLine object.

Because the coordinates should form a boundary in R-INLA we reverse the order of the rows of the CoastLine object.

```
> N <- nrow(CoastLine)
> Coast.Rev <- CoastLine[N:1, c("Xkm", "Ykm")]
```

We are now ready to make the mesh in Figure 13.8.

```
> mesh2 <- inla.mesh.2d(loc.domain = CoastLine,
                          max.edge = 1.5,
                          boundary =
                        inla.mesh.segment(Coast.Rev))
> plot(mesh2, asp = 1)
> points(x = LP$Xkm, y = LP$Ykm,
         col = 1, pch = 16, cex = 0.5)
```

We also show what happens if the orientation of the boundary points is clockwise; see Figure 13.10. The island is now interpreted as a hole in the mesh. This may be useful for dealing with fjords in case fisheries data are analysed, though alternative options are available; see Bakka et al. (2016) who discuss how to deal with physical barriers when doing analysis with R-INLA. The mesh in Figure 13.10 was created with the following code.

```
> mesh3 <- inla.mesh.2d(loc.domain = CoastLine,
                        max.edge = 1.5,
                        boundary =
                      inla.mesh.segment(CoastLine))
```

We will use mesh2 in the remainder of this chapter. However, when we compared the results obtained by models using mesh1 and mesh2, we were surprised to see minimal differences between them.

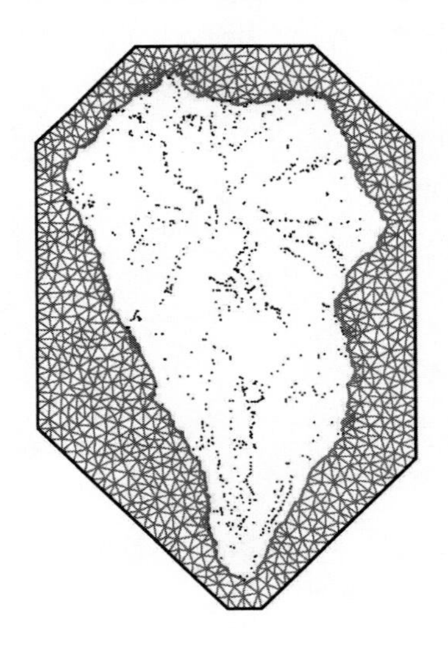

Figure 13.10. Mesh obtained by using clockwise orientation of the boundary points.

13.5.3 Projector matrix

In the next step we need to define the weighting factors a_{ik}. Recall from Chapter 12 that these are used to link the spatial random field w_k to the random effects u_i via the following expression.

$$u_i = \sum_{k=1}^{1829} a_{ik} \times w_k \tag{13.3}$$

The matrix A2 below has the dimensions 890 by 1192 and contains the weighting factors a_{ik}.

```
> mesh <- mesh2  #Use mesh2
> A2   <- inla.spde.make.A(mesh2, loc = Loc)
> dim(A2)
```

```
890 1192
```

Each row in A2 contains either one or three non-negative values. If a sampling location i is on vertex k then only one value in row i in A2 is non-negative (and a_{ik} is equal to 1). If location i is not on a vertex then there are three non-negative values (and their sum is 1); see also Figure 12.6.

13.5.4 SPDE

Next we define the SPDE model. This means that we specify some of the parameters for the Matérn correlation function.

```
> spde   <- inla.spde2.matern(mesh, alpha = 2)
```

13.5.5 Spatial field

We define the spatial field w_k.

```
> w.index <- inla.spde.make.index(
                  name     = 'w',
                  n.spde   = spde$n.spde,
                  n.group  = 1,
                  n.repl   = 1)
```

The `n.group` and `n.repl` options are explained in Chapter 15.

13.5.6 Stack

Next we define the stack. We first define a data frame that contains the intercept and all the covariates.

```
> N <- nrow(LP)
> X <- data.frame(Intercept     = rep(1, N),
                  CRCAN.std     = LP$CRCAN.std,
                  CRLP.std      = LP$CRLP.std,
                  INTERVAR.std  = LP$INTERVAR.std,
                  MAP.std       = LP$MAP.std,
                  Age.std       = LP$Age.std,
                  Slope.std     = LP$Slope.std,
                  TCI.std       = LP$TCI.std)
> X <- as.data.frame(X) #Avoids potential trouble
```

Now we reach the challenging point, the definition of the stack. We call it 'Fit', so that in the next section we can add a 'simulation' part. In the effects part we have two components, the spatial random field w_k and the covariates. The first one needs to match to the locations of the observed data on the mesh via the projector matrix A2. The covariates are measured at the sampling locations, hence the '1' in the list for A.

```
> stk2 <- inla.stack(
              tag  = "Fit",
              data = list(y = LP$nSIE),
              A = list(A2, 1),
              effects = list(w = w.index,
                             X = X))
```

If you swap the order of the w and the X in the effects list then you also need to swap the order of the A2 and the 1 in the list for A.

13.5.7 Formula

We define two models in R, one without a spatial component (f1) and one with the spatial random effect (f2). We already applied the first model earlier in this chapter, but we did not execute it with R-INLA.

```
> f2 <- y ~ -1 + Intercept + CRCAN.std +
          CRLP.std + INTERVAR.std +
          MAP.std + Age.std + Slope.std +
          TCI.std
> f3 <- y ~ -1 + Intercept + CRCAN.std +
          CRLP.std + INTERVAR.std + MAP.std +
          Age.std + Slope.std + TCI.std +
          f(w, model = spde)
```

13.5.8 Run R-INLA

We are now ready to execute both models in R-INLA and we also compare the DICs and WAICs. Except for the formula, the code for both models is identical.

```
> I2 <- inla(f2,
             family = "poisson",
             data = inla.stack.data(stk2),
             control.compute = list(dic = TRUE,
                                    waic = TRUE),
             control.predictor = list(
                         A = inla.stack.A(stk2)))
> I3 <- inla(f3,
             family = "poisson",
             data = inla.stack.data(stk2),
             control.compute = list(dic = TRUE,
                                    waic = TRUE),
             control.predictor = list(
                         A = inla.stack.A(stk2)))
```

Here are the DICs and WAICs. Both indicate that the model with spatial correlation is better.

```
> dic  <- c(I2$dic$dic, I3$dic$dic)
> waic <- c(I2$waic$waic, I3$waic$waic)
> Z    <- cbind(dic, waic)
> rownames(Z) <- c("Poisson GLM",
                   "Poisson GLM + SPDE")
> Z

                        dic     waic
Poisson GLM        3043.055 3042.724
Poisson GLM + SPDE 2990.545 2972.809
```

13.5.9 Inspect results

There are a series of things we need to do: inspect posterior mean values and 95% credible intervals of the fixed parameters and the hyperparameters. Model validation: Do we still have spatial correlation? Model selection. Sketching of results. Present the spatial field. We will start with the interesting part and compare the results of the two models; see Figure 13.11. Note that the 95% credible intervals obtained by the model with the spatial correlated random effect are all larger than their ordinary Poisson GLM counterparts. This is to be expected if spatial dependency is present in the data; if spatial dependency is wrongly ignored then you are likely to state that things are important whereas in reality they are not. In this example, if we ignore spatial dependency then we conclude that slope is important. But after correcting for spatial dependency it is not important anymore. The same holds for the climate rarity index for the Canary Islands (CRCAN.std).

We also need to check whether there is still spatial dependency in the residuals. We can calculate the fitted values and Pearson residuals of both models via

```
> Index <- inla.stack.index(stk2,tag = "Fit")$data
> mu2 <- I2$summary.fitted.values[Index, "mean"]
> mu3 <- I3$summary.fitted.values[Index, "mean"]
> E2 <- (LP$nSIE - mu2) / sqrt(mu2)
> E3 <- (LP$nSIE - mu3) / sqrt(mu3)
```

We can use the Pearson residuals of the Poisson GLM (`E2`) and of the model with spatial correlation (`E3`) to make variograms again; see Figure 13.12. The graph shows two curves. The left curve is the sample-variogram for the Poisson GLM and is the same sample-variogram as in Figure 13.5. The curve in the right panel is for the model with spatial correlation. Although the spatial correlation is less strong, there is still a small amount of spatial correlation present! We will return to this in the last section of this chapter.

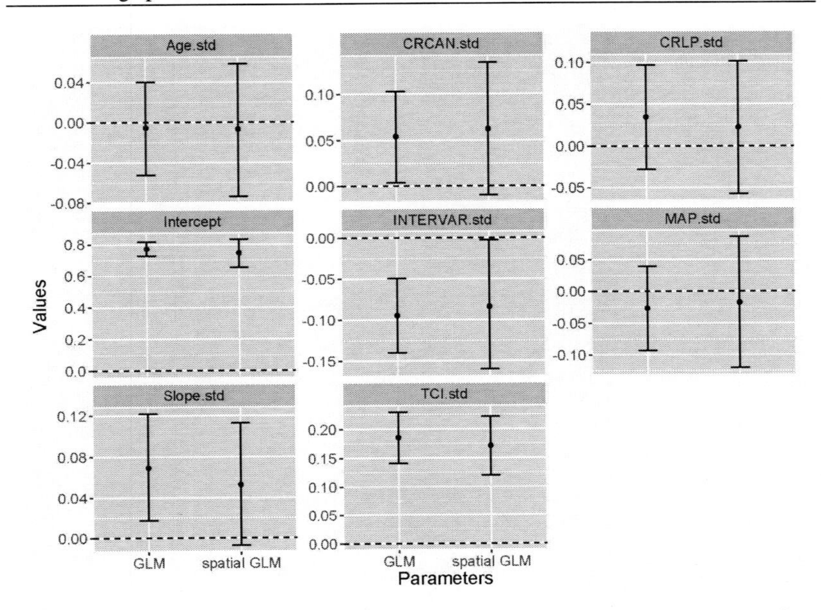

Figure 13.11. Results of the Poisson GLM without spatial correlation and the model with spatial correlation.

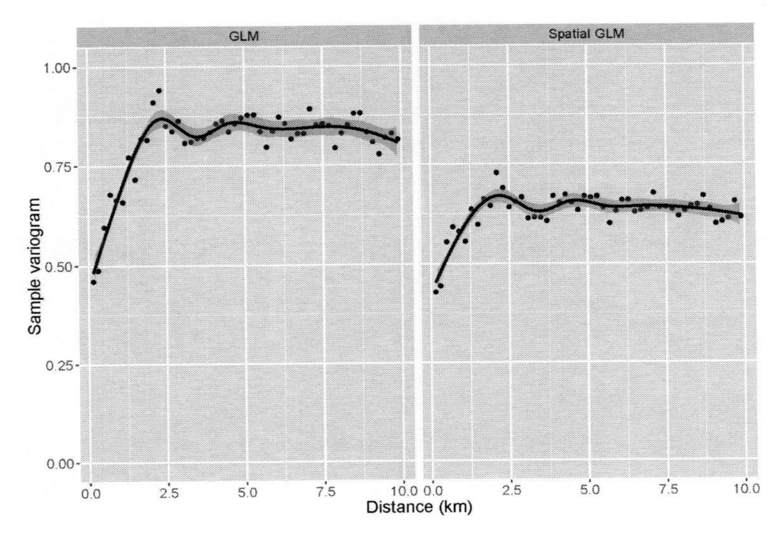

Figure 13.12. Sample-variograms of the Pearson residuals for the Poisson GLM (left panel) and the model with spatial correlation (right panel).

It is also interesting to present the spatial component information. There are multiple ways of doing this. One option is to interpolate the w_ks and superimpose these on a map of La Palma (Figure 13.13). An alternative option is to calculate the posterior mean values of the u_is and superimpose these on the map of La Palma (Figure 13.14). We will work out both options. The color versions of these graphs are much nicer than the grey-scale version in the printed book. You can reproduce the graphs with help of the R code.

Plotting interpolated w_ks

The posterior mean values of the 1192 w_ks are obtained via

```
> w.pm <- I3$summary.random$w$mean
```

These are defined on the 1192 vertices of the mesh. We can also calculate the matching standard deviations to get some idea about the uncertainty around the posterior mean.

```
> w.sd <- I3$summary.random$w$sd
```

To make a picture of the w_ks we interpolate them on a grid, say a grid consisting of 100 by 100 cells. R-INLA has functions to do this.

```
> wproj        <- inla.mesh.projector(mesh)
> w.pm100_100 <- inla.mesh.project(wproj, w.pm)
> w.sd100_100 <- inla.mesh.project(wproj, w.sd)
```

The `w.pm100_100` object has the dimensions 100 by 100. The help file of `inla.mesh.projector` shows how to change various settings. To visualise the *interpolated* values in `w.pm100_100` we can use standard graphical tools from the `lattice` (Sarkar 2008) or `ggplot2` (Wickham 2009) packages, or any package that allows for plotting of three-dimensional data. Some functions are able to plot `w.pm100_100` directly, whereas other functions require the data in a long vector. We present an example of the latter. First we combine the UTM coordinates and interpolated w_k values. The `as.vector` function converts the 100 by 100 matrix in a long vector with 100,000 values.

```
> Grid <- expand.grid(Xkm = wproj$x,
                      Ykm = wproj$y)
> Grid$w.pm <- as.vector(w.pm100_100)
> Grid$w.sd <- as.vector(w.sd100_100)
```

The object `Grid` has 100,000 rows and the four columns contain the spatial coordinates (UTM), the interpolated w_k values, and the interpolated standard deviation values. Some simple `levelplot` code visualised the interpolated spatial random field; see Figure 13.13. It is easy to change the colours, for example using

```
col.r  <- colorRampPalette(c('red', 'green'))(50)
```

instead of the rainbow colors gives much darker colours between red (low values) and green (high values).

```
> col.r <- rev(rainbow(30, alpha = 0.35))
```

```
> levelplot(w.pm ~ Xkm * Ykm,
            data = Grid,
            aspect = "iso",
            col.regions = col.r,
            scales = list(draw = TRUE),
            xlab = "X-coordinates (km)",
            ylab = "Y-coordinates (km)",
            main = "Posterior mean spatial random
                    field")
```

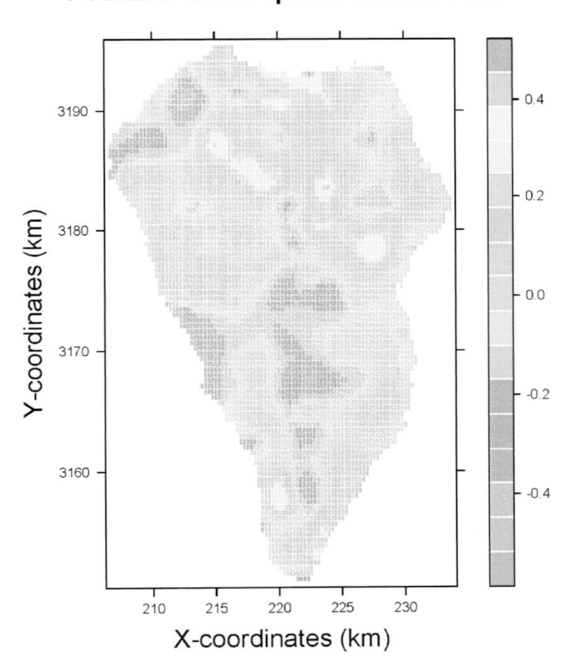

Figure 13.13. Interpolated spatial random field.

The same graph can be made for the interpolated posterior standard deviations. The graph is not shown here, but most values are around 0.2.

Instead of showing the spatial random field w_k, it is interesting to show its exponential. The motivation for doing this is that the expected values of the expected endemic richness values are given by μ_i = exp(Covariates + u_i) = exp(Covariates) × exp(u_i). Hence, the exponential of the spatial random field is the multiplication factor of the fitted values of the covariate effects. If the spatial random field is 0, then the spatial term has no effect as exp(0) = 1. Results in Figure 13.13 indicate that in some regions the spatial random effect causes an increase in expected endemic richness values of exp(0.4) = 50%, and in other areas there is a reduction up to around 30%.

Plotting the random intercepts u_i

Besides the spatial random field w_k we can also visualise the spatial random effects u_i. There are different ways to obtain the u_is. One option is to let R-INLA do the hard work and use the `inla.mesh.projector` function. Internally it uses Equation (13.3).

```
> u.proj <- inla.mesh.projector(mesh, loc = Loc )
> u.mean <- inla.mesh.project(u.proj,
                          I2$summary.random$w$mean)
```

We now have the 890 u_is. We can plot them using standard graphical tools; see Figure 13.14. We use filled circles for negative residuals and open circles for positive u_i values. The negative random effects u_i are clustered in the northwestern and central southern areas, whereas the positive random effects are in the northern and southern areas. It is now up to you as a biologist to interpret and explain this.

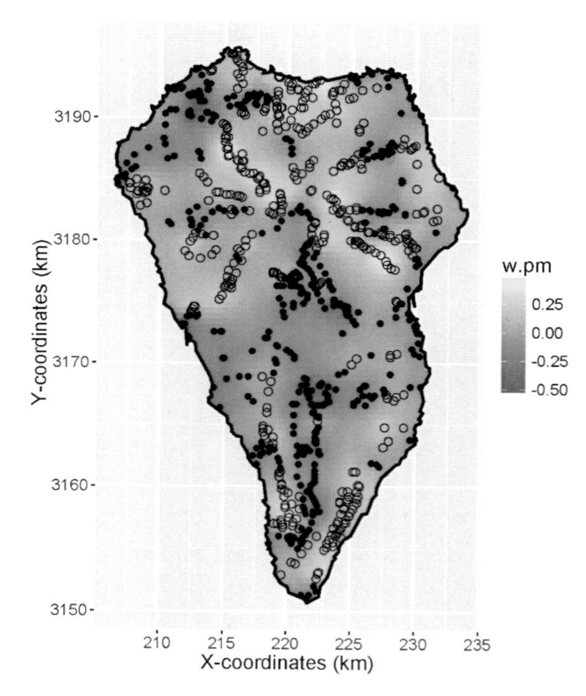

Figure 13.14. Map of La Palma with posterior mean values of the spatial random effects u_i, and the spatial random field w_k. The open circles represent positive u_i values and the filled circles are negative u_is. We used `ggplot2` code to create this figure; it is available on the website for this book.

Another interesting graph is the imposed correlation via the Matérn correlation function; see Figure 13.15. It shows that strong correlation is up to about 1 km, and the range is about 2.5 km. The code below calculates the Matérn correlation parameters and it also calculates the range (i.e. the distance at which the correlation is lower than 0.1-ish).

```
> SpFi.w <- inla.spde2.result(inla = I3,
                               name = "w",
                               spde = spde,
                               do.transfer = TRUE)
> Kappa <- inla.emarginal(function(x) { x },
                  SpFi.w$marginals.kappa[[1]])
> sigmau <- inla.emarginal(function(x) {sqrt(x)},
            SpFi.w$marginals.variance.nominal[[1]])
> r <- inla.emarginal(function(x) { x },
               SpFi.w$marginals.range.nominal[[1]])
> c(Kappa, sigmau, r)

1.3465744 0.2835614 2.3557519
```

There are many more things we can do, for example dropping covariates that are not important (model selection) and applying a detailed model validation. We have provided the code to get Pearson residuals, fitted values, and DICs, and we leave these steps as an exercise for the reader.

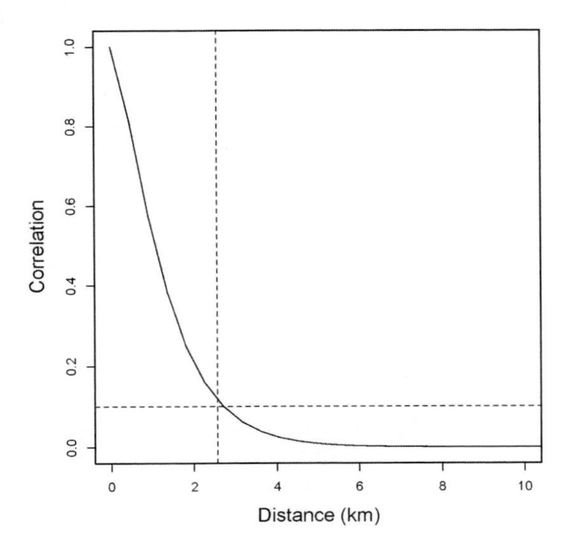

Figure 13.15. Matérn correlation function. The R code to create this graph is presented in Chapter 12.

13.6 Simulating from the model

We have applied a model validation and noticed that there are no major problems. But how do we know that the model is good? In Subsection 2.2.2 we presented a 10-step protocol and in the last step we proposed doing a simulation study. The idea behind this step is as follows. When we fit a model on a data set then we would hope that when we simulate data from the model that the simulated data are comparable to the observed data. If the simulated data and the observed data are rather different then we have a major problem. There are many possible definitions of 'comparable'. Before delving more deeply into one of them we first show how to simulate data from our model. We first rerun the model with the `control.compute = list(config =TRUE)` option added.

```
> I4 <- inla(f3,
             family = "poisson",
             data = inla.stack.data(stk2),
             control.compute = list(config =TRUE),
             control.predictor = list(
                         A = inla.stack.A(stk2)))
```

We can now use the `inla.posterior.sample` function to simulate a value for each regression parameter and each w_k of the spatial random field.

```
> set.seed(12345)
> Sim <- inla.posterior.sample(n = 1, result = I4,
                               seed = 12345)
```

We have set the random seed so that your results are identical to ours. The information that we seek is hidden in `Sim[[1]]$latent`. If you type

```
> Sim[[1]]$latent
```

then you will see a lot of numbers floating over the screen. The relevant part is given below.

```
...
w:1191           0.4016490349
Intercept        0.8153216028
CRCAN.std        0.0350298041
CRLP.std        -0.0096557649
INTERVAR.std    -0.1579886787
MAP.std         -0.0710062611
Age.std          0.0543022426
Slope.std        0.0967356809
TCI.std          0.1834819518
```

If you run the `inla.posterior.sample` function again then you get slightly different values. Each time you run this function, it samples from the posterior distributions. This is a look at bootstrap analysis in frequentist analysis.

Let us extract the numbers. First we need to determine the row numbers for the eight betas. We can either figure out that these are in rows 3634 to 3641, or do it semi-automatically using the `grep` function.

```
> MyParams <- c("Intercept", "CRCAN.std",
                "CRLP.std",  "INTERVAR.std",
                "MAP.std",   "Age.std",
                "Slope.std", "TCI.std")
> rownum <- lapply(MyParams,
              function(x)
                grep(x,
                     rownames(Sim[[1]]$latent),
                     fixed = TRUE))
> rownum <- as.numeric(rownum)
> rownum
```

```
3634 3635 3636 3637 3638 3639 3640 3641
```

All this `lappy` and `grep` stuff is used to determine that the eight simulated values for the regression parameters are on rows 3634 to 3641. To access the simulated values of the eight regression parameters from `Sim[[1]]$latent` we use

```
> Betas <- SimData[[1]]$latent[as.numeric(rownum)]
```

Just type `Betas` in R to convince yourself that we indeed have the correct eight values. We also need the simulated w_k values. These are in rows 2442 to 3633 (these numbers change for a different mesh or a different data set).

```
> wk <- SimData[[i]]$latent[2442:3633]
```

We still have the matrix with covariates and the projector matrices. To ensure that we can use them for matrix multiplications we convert them into matrices using `as.matrix`.

```
> Xm <- as.matrix(X)
> Am <- as.matrix(A2)
```

We can now calculate the fitted values based on the simulated parameter values.

```
> FixedPart   <- Xm %*% Betas
> SpatialPart <- Am %*% wk
> mu          <- exp(FixedPart + SpatialPart)
```

We now have fitted values that correspond to the simulated betas and w_ks; these are stored in `mu`. Using these fitted values we can simulate 890 endemic diversity values using the `rpois` function.

```
> Ysim <- rpois(n = nrow(LP), lambda = mu)
```

We now have 890 simulated endemic richness values and Figure 13.16 shows a visualisation of the frequency table. The graph was obtained with

```
> plot(table(Ysim),
       xlab = "Simulated endemic richness values",
       ylab = "Frequencies")
```

The figure shows how many of the 890 simulated endemic richness values are equal to 0, 1, 2, etc. Rather than do this process once we can do it a thousand times with a loop (see the online R code), resulting in 1,000 frequency tables. Instead of plotting 1,000 frequency tables we calculate an average frequency table; see Figure 13.17. We also included the frequency table of the observed endemic richness. It seems that our model is generating too many zeros and too few ones.

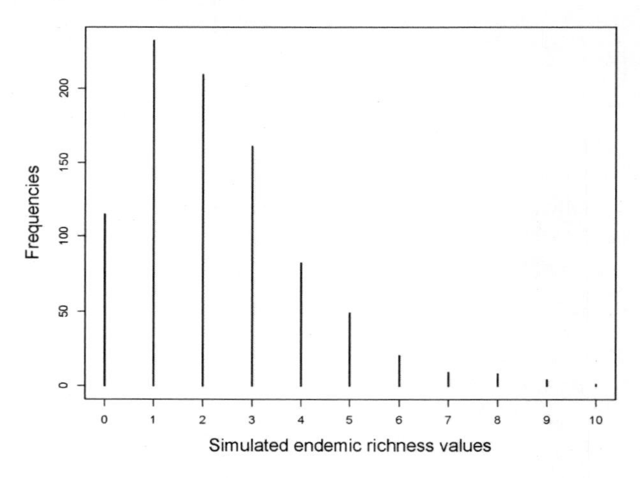

Figure 13.16. Plot of frequency table of simulated endemic richness values.

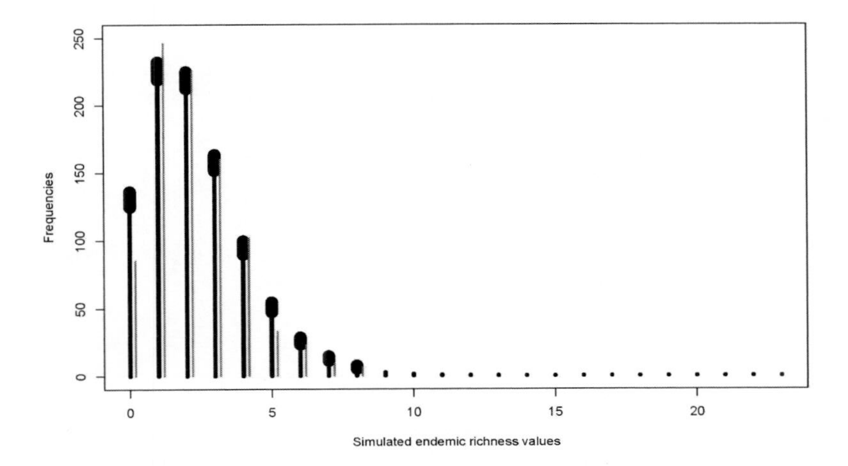

Figure 13.17. Average frequency table and frequency table of the observed endemic richness values. The thick lines on the left sides are for the simulated data and the thinner lines on the right sides are for the observed endemic richness values. The thicker line on top of the lines are standard deviations.

13.7 What to write in a paper

When writing this chapter we tried to stick close to the protocol specified in Chapter 2. For a publication we suggest following the same steps. Specify the underlying questions in the introduction, show a map of La Palma with the sampling locations, summarise the data exploration with 2–3 lines of text, list the models, and explain why R-INLA is used. The methods in R-INLA are relatively new in the biological literature and we recommend explaining some of the key steps such as making a mesh, explaining its role, and providing a two- to three-line summary of what the method does. Figure 13.11 is a nice graph for a paper, perhaps combined with a table showing the results of model `I2`. And certainly you need to include a graph like Figure 13.13 or Figure 13.14 (or an improved version of it). In the Discussion section you first need to discuss the fixed parameters (and state which covariates are important) and then hypothesise what the spatial random field may represent.

We have one remaining point to discuss. The spatial correlated random effect clearly improves the model, but there is still a small amount of spatial correlation present in the Pearson residuals of the model in Equation (13.2). The spatial correlation imposed by the Matérn correlation function is assumed to be isotropic; correlation in all spatial directions is assumed to be identical. Figure 13.18 shows sample variograms split up in four directions. Isotropy means that all four sample variograms are identical, but the results in Figure 13.18 seem to suggest that residual patterns in the northern and northeastern directions differ from the eastern and southeastern directions, which does not come as a surprise for an island with two big volcanoes.

It is possible in R-INLA to allow for anisotropic correlation by allowing the parameters of the Matérn correlation to vary spatially (Lindgren and Rue 2015).

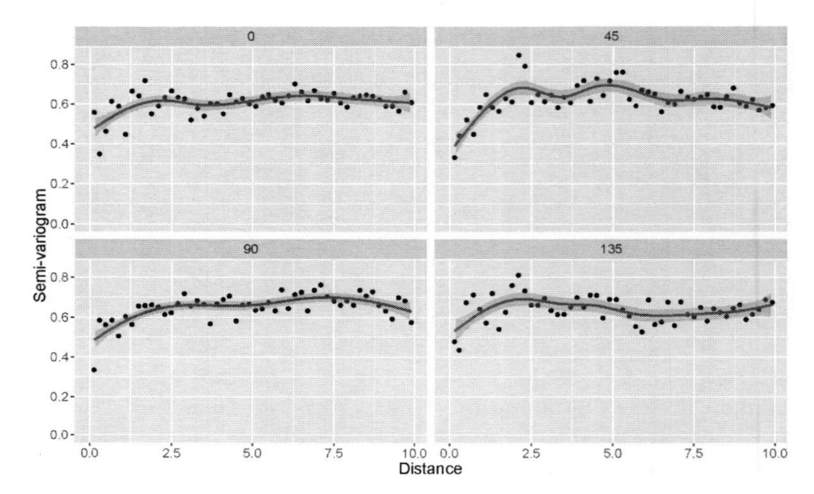

Figure 13.18. Sample variogram of the Pearson residuals obtained by the model in Equation (13.2). The panels with the labels '0', '45', '90', and '135' represent the sample variograms in northern, northeastern, eastern, and southeastern directions respectively. This is the same as 180, 225, 270, and 315 degrees respectively.

14 Time-series analysis in R-INLA

The scientific field of time-series analysis consists of such a wide variety of techniques that we could easily fill an entire book about this topic; see for example Harvey (1989), Chatfield (2003), Shumway and Stoffer (2017), and Durbin and Koopman (2012), among many others.

In this chapter we do not want to delve too deeply into time-series analysis techniques. Instead, we will touch upon methods available in R-INLA that can be used to identify trends. Some of these techniques will also be used in Chapters 15 and 16 which deal with spatial-temporal models.

Prerequisite for this chapter: You need to be familiar with the time-series analysis topics that were discussed in Chapter 3 and with linear regression and R-INLA as discussed in Chapter 8.

14.1 Simulation study

Let Y_t be a response variable that has been measured repeatedly over time, e.g. the biomass of a fish at time t or the number of birds in year t. We will assume that Y_t follows a certain distribution, given its mean value. For example we can use a normal distribution for the biomass of fish or a Poisson distribution for the number of birds. Actually, depending on the nature of the response variable we can use any distribution. It is the mean value of the distribution that is relevant for this chapter. We will model it as a function of a trend over time, and possibly also covariate effects. In its general form, using a normal distribution, the model is given by

$$\begin{aligned} Y_t &= \text{Intercept} + \text{Covariates}_i + \text{Trend}_t + \varepsilon_t \\ \varepsilon_t &\sim N\left(0, \sigma_\varepsilon^2\right) \end{aligned} \tag{14.1}$$

This looks very much like an ordinary linear regression model. But it depends on what we are going to do with the Trend_t component, whether this is indeed a linear regression model or whether it is something else.

Using a model in which the trend component is modelled as $Time \times \beta$ means that we indeed have an ordinary regression model, and this ultimately results in pseudoreplication when we apply the model to time-series data. To avoid this we will use a so-called random walk to model the trend. The simplest random walk model is given in Equation (14.2); we dropped the Covariates_i term.

$$\begin{aligned} &Y_t = \text{Intercept} + \mu_t + \varepsilon_t \\ &\mu_t = \mu_{t-1} + v_t \\ &\varepsilon_t \sim N\left(0, \sigma_\varepsilon^2\right) \quad \text{and} \quad v_t \sim N\left(0, \sigma_v^2\right) \end{aligned} \tag{14.2}$$

Let us for the sake of simplicity assume that the time index t refers to years. According to the model the response variable Y_t in year t is equal to a trend μ_t plus independent, identical, and normal distributed noise ε_t. The random walk trend itself is modelled as the trend from last year plus some independent, identical, and normal distributed noise v_t. The latter term has mean 0 and variance σ_v^2. In words, using the birds as an example, we have

Birds in year t = Trend in year t + pure noise ε_t
Trend in year t = Trend from last year + pure noise v_t

The model has two residual terms (the 'pure noise' components), namely the ε_t and the v_t, and each term has its own variance; these are σ_ε and σ_v respectively. To understand the roles of σ_ε and σ_v in the model we simulated 15 data sets; see Figure 14.1. These simulated data sets use different values for σ_ε and σ_v.

Based on the shapes of the curves we can conclude that a small value for σ_v implies that the v_t values tend to be small and consequently μ_t is close to μ_{t-1}. The resulting random walk trend μ_t is a smooth curve; see for example the simulated data sets 4–6. On the other hand, a large σ_v means that the v_ts can be large and μ_t can be rather different from μ_{t-1}. As a result we obtain trends that can change quite a lot (see simulated data sets 10–15).

The random walk trend is given by $\mu_t = \mu_{t-1} + v_t$, where $v_t \sim N(0, \sigma_v^2)$. The smaller the σ_v, the smoother the trend.

A small value of σ_ε means that the ε_t values are likely to be small, and as a result the variation of the response variable Y_t around the trend is small. Formulated differently, we have a good fit (see simulations 1–3 and 10–12). The opposite also holds; a large value for σ_ε means larger ε_ts, and therefore a large scatter around the trend (simulations 4–6 and 13–15).

It is relatively easy to add covariates to the model. These are added on the first line of the model. It is also possible to add daily, quarterly, seasonal, or cyclic components to the model.

Models with random walk trends are described in detail in Harvey (1989) and Durbin and Koopman (2012). More advanced applications can be found in Zuur et al. (2003a; 2003b; 2004) who applied a multivariate extension and dynamic factor analysis.

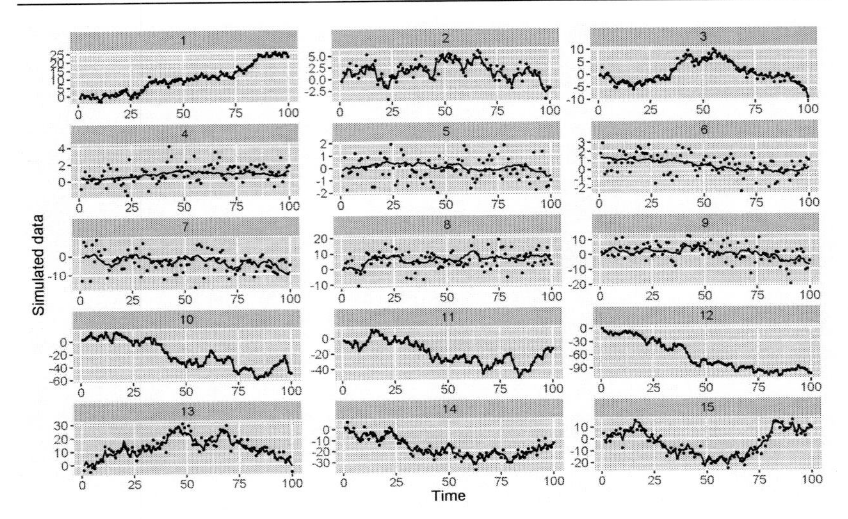

Figure 14.1. Examples of simulated data sets with different random walk trends. Each of the three panels on the same row represents a different simulation from the same model. The number above a panel is the simulation number. In simulations 1–3 (top row) we used $\sigma_\varepsilon = 1$ and $\sigma_v = 1$; in simulations 4–6 (second row) we used $\sigma_\varepsilon = 1$ and $\sigma_v = 0.1$; in simulations 7–9 we used $\sigma_\varepsilon = 5$ and $\sigma_v = 1$; in simulations 10–12 we used $\sigma_\varepsilon = 1$ and $\sigma_v = 5$; and in simulations 13–15 we used $\sigma_\varepsilon = 3$ and $\sigma_v = 3$. The R code for these simulations is available on the website for this book.

14.2 Trends in migration dates of sockeye salmon

14.2.1 Applying a random walk trend model

The data used in this section were taken from Crozier et al. (2011), who studied the migration date in sockeye salmon (*Oncorhynchus nerka*) in the Columbia River, Canada. Figure 14.2 shows a time-series plot of the (median) arrival day versus year. The first question is whether there is a trend in these data. We will apply the following model.

$$\begin{aligned} MigDay_t &= \text{Intercept} + \mu_t + \varepsilon_t \\ \mu_t &= \mu_{t-1} + v_t \\ \varepsilon_t &\sim N\left(0, \sigma_\varepsilon^2\right) \quad \text{and} \quad v_t \sim N\left(0, \sigma_v^2\right) \end{aligned} \tag{14.3}$$

where $MigDay_t$ is the (median) migration day in year t. Of prime interest is the trend μ_t; does it show a pattern that diverges from a horizontal line (which would mean an increase or decrease in the trend)? Figure 14.3A shows the estimated random walk trend and 95% credible intervals, and panel B shows the fitted values of the model (= intercept + trend). There seems to be a decrease from 1970 onwards. The trend is clearly non-linear.

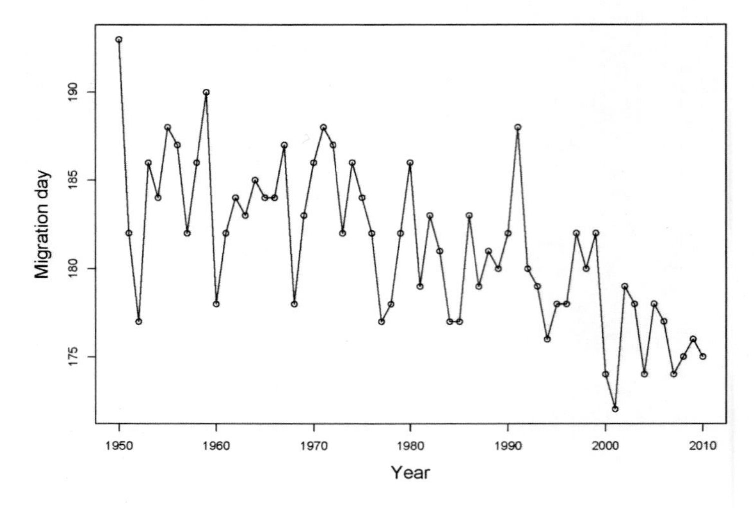

Figure 14.2. Time-series plot of (median) arrival day versus year.

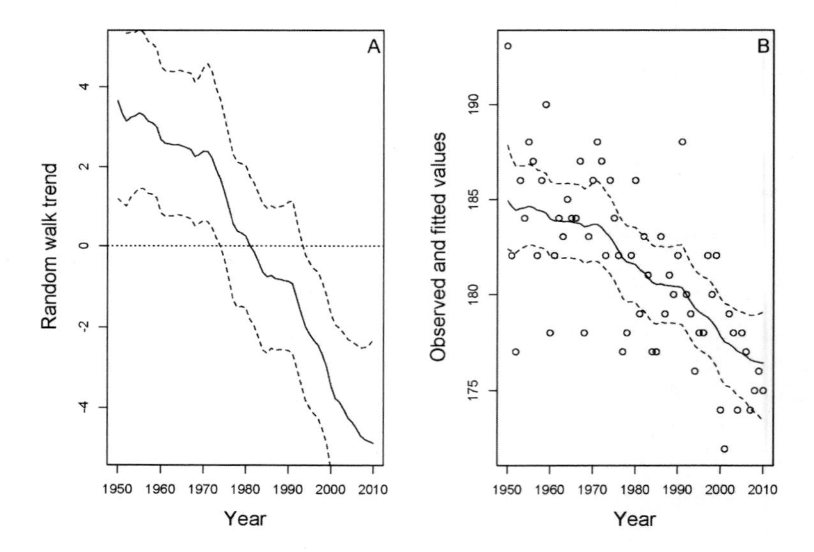

Figure 14.3. A: Random walk trend and 95% credible interval. B: Observed data with fitted values (plus 95% credible intervals).

The following R code was used to fit the model in R-INLA. We first import the data using the `read.csv` function.

```
> Salmon <- read.csv(file = "sockeye.csv",
                     header = TRUE)
```

Migration date is modelled as a function of year, and the rw1 model imposes the random walk trend.

```
> library(INLA)
> I1 <- inla(MigDay ~ f(Year, model = "rw1"),
             control.compute = list(dic = TRUE),
             family = "gaussian",
             data = Salmon)
```

To get the numerical information on the intercept and the hyperparameters we can use the summary function.

```
> summary(I1) #Results not shown here
```

The trend μ_t is a random effect that changes over time and the posterior information is stored in the summary.random part of the output.

```
> Yearsm <- I1$summary.random$Year
```

The second column in Yearsm contains the posterior mean values, and the fourth and sixth columns contain the values for the 95% credible intervals. The first column contains the years. It is then a matter of using simple plotting tools to create Figure 14.3A.

```
> plot(Yearsm[,1:2], type='l',
       xlab = 'Year',
       ylab = 'Smoother',
       ylim = c(-5, 5) )
> abline(h = 0, lty = 3)
> lines(Yearsm[, c(1, 4)], lty = 2)
> lines(Yearsm[, c(1, 6)], lty = 2)
```

The fitted values are accessed via

```
> Fit1 <- I1$summary.fitted.values
```

These can be used to create Figure 14.3B.

```
> plot(x = Salmon$Year,
       y = Salmon$MigDay,
       xlab = 'Year',
       ylab = 'Observed and fitted values')
> lines(Salmon$Year, Fit1$mean)
> lines(Salmon$Year, Fit1$"0.025quant", lty = 2)
> lines(Salmon$Year, Fit1$"0.975quant", lty = 2)
```

Once we have the fitted values we can calculate the residuals ε_t via

```
> E1 <- Salmon$MigDay - Fit1$mean
```

An auto-correlation function of E1 does not indicate any major dependency problems, so in principle we have a working model.

14.2.2 Posterior distribution of the sigmas

We have two variances in the model, but R-INLA uses precision parameters, $\tau_\varepsilon = 1 / \sigma_\varepsilon^2$ and $\tau_v = 1 / \sigma_v^2$. All the numerical output is for these precision parameters, but we can easily convert it back to posterior information on σ_ε and σ_v using existing R-INLA functions.

The marginal distributions of the two precision parameters are obtained via the `marginals.hyperpar` part in the output.

```
> hp <- I1$marginals.hyperpar
> tau.eps <- hp$`Precision for the Gaussian
                 observations`
> tau.v   <- hp$`Precision for Year`
```

We are not interested in the distributions of the precision parameters. Instead we want the two standard deviation parameters, σ_ε and σ_v. First we calculate the posterior mean values of σ_ε and σ_v.

```
> Tau2Sigma <- function(x) { sqrt(1/x) }
> sigma.eps <- inla.emarginal(fun = Tau2Sigma,
                              marg = tau.eps)
> sigma.v   <- inla.emarginal(fun = Tau2Sigma,
                              marg = tau.v)
> c(sigma.eps, sigma.v)

3.2769911 0.5933391
```

The posterior mean values of σ_ε and σ_v are 3.27 and 0.59 respectively. The marginal posterior distributions for σ_ε and σ_v are given in Figure 14.4A and Figure 14.4B. These were obtained with the following code.

```
> md.eps <- inla.tmarginal(fun = Tau2Sigma,
                           marg = tau.eps)
> md.v <- inla.tmarginal(fun = Tau2Sigma,
                         marg = tau.v)
```

To get the figures we need to plot the information in `md.eps` and `md.v`; see Section 8.3.3.

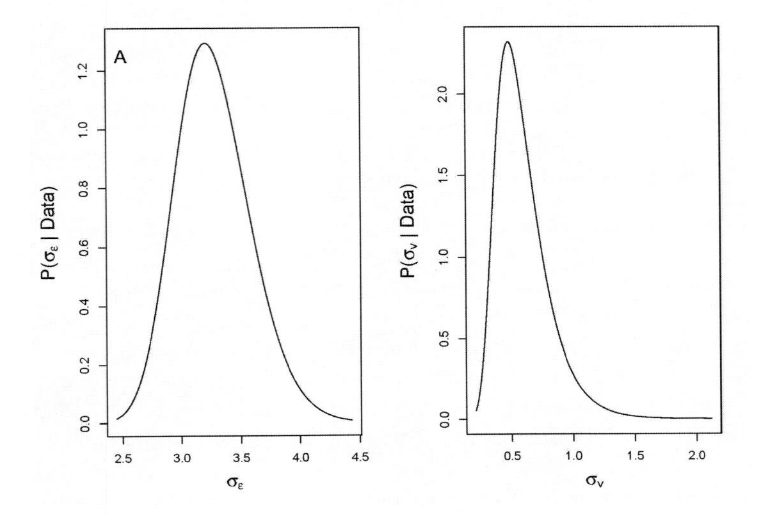

Figure 14.4. A: Marginal posterior distribution of σ_ε. B: Marginal posterior distribution for σ_v. The R code to create this figure is available on the website for this book.

14.2.3 Covariates and trends

The model in Equation (14.3) can be extended by adding a covariate, for example the mean June temperature measured at a nearby water dam.

$$\begin{aligned} &MigDay_t = \text{Intercept} + Temp_t \times \beta + \mu_t + \varepsilon_t \\ &\mu_t = \mu_{t-1} + v_t \\ &\varepsilon_t \sim N(0, \sigma_\varepsilon^2) \quad \text{and} \quad v_t \sim N(0, \sigma_v^2) \end{aligned} \tag{14.4}$$

Note that the trend in Equation (14.3) represents the overall temporal pattern in the migration dates, which may reflect a covariate. The trend in Equation (14.4) represents the trend in the migration dates while taking into account the temperature effect. That is a different interpretation! Also note that similar long-term trends in temperature and the component μ_t may cause collinearity problems.

The model in Equation (14.4) can be fitted in R-INLA by adding Temp to the formula.

```
> f2 <- MigDay ~ Temp + f(Year, model = "rw1")
> I2 <- inla(f2,
             control.compute = list(dic = TRUE),
             family = "gaussian",
             data = Salmon)
> c(I1$dic$dic, I2$dic$dic)

326.3199 321.4010
```

Adding temperature to the model gives a reasonable improvement in the DIC. The trend of the new model still shows the same non-linear pattern, and the 95% credible interval for the regression parameter of temperature does not contain 0, indicating that temperature is an important covariate for migration dates.

14.2.4 Making the trend smoother

It took about 10 data sets before we managed to find the salmon example that showed a trend over time that was easy on the eyes. Ending the chapter at this point would have given the wrong impression that random walk models are easy-to-work-with models in R-INLA.

The reality is that random walk trends in R-INLA are extremely sensitive to the specification of the priors of the hyperparameters. Let us formulate this differently: The shape of the trend is very much determined by the prior distribution for σ_v. In many of the examples that we looked at when writing this chapter we ended up with trends that either had $\sigma_v \approx \infty$ (the trend goes from point to point) or $\sigma_v \approx 0$ (the trend is a straight line). Readers familiar with generalised additive models (GAM) will recognise this problem as 'how to choose the degrees of freedom of a smoother'. In fact, the `rw1` model is one of the two tools for implementing a smoother in R-INLA (the other one is `rw2`, which will be discussed in a moment). Luckily, better tools for GAMs in R-INLA are available, as we will see in Volume II of this book.

We will discuss three approaches to control the amount of wiggliness of the trend, namely changing the values of the gamma prior for σ_v, replacing the gamma distribution by another distribution, and adopting a different random walk model. Each of these approaches is discussed next.

Changing the values of the gamma prior

The random walk trend can also be written as

$$\begin{aligned} \mu_t - \mu_{t-1} &= v_t \\ v_t &\sim N\left(0, \sigma_v^2\right) \end{aligned} \tag{14.5}$$

R-INLA considers σ_v as a hyperparameter and the $\mu_1,\ldots,\ \mu_T$ variables are part of the latent field (together with the betas). A normal distribution is used for the differences $\mu_t - \mu_{t-1}$, and a gamma prior with parameters 1 and 0.00001 is used for the precision parameter τ_v, which is equal to 1 / σ_v^2. See Chapter 9, Section 9.11 for a detailed discussion on gamma priors. In that section we also argued that the 1 and 0.0001 values are not always the best options for a prior. We mentioned a paper by Carroll et al. (2015) who recommended using a gamma(1, 0.5) prior. To change the gamma prior from the default gamma(1, 0.00001) to gamma(1, 0.5), we can use the following code.

```
> f3 <- MigDay ~ f(Year,
                   model = "rw1",
                   hyper = list(
                         theta = list(
                           prior = "loggamma",
                           param = c(1, 0.5))))
> I3 <- inla(f3,
             control.compute = list(dic = TRUE),
             family = "gaussian",
             data = Salmon)
```

Note the extra `hyper` argument in the model specification for Year. This is the prior recommended by Carroll et al. (2015) for mixed-effects models, but without conducting a similar simulation study as carried out in their paper we have no idea whether the gamma(1, 0.5) prior is also appropriate for random walk models.

In Chapter 9, Section 9.11 we showed that the gamma(1, 0.5) for a precision parameter τ_v corresponds to a σ_v of around 0–5. That is a very large range for the standard deviation for a random walk trend. Suppose that we have an idea about the possible range of values for σ_v. How does this information translate into the parameters of the gamma distribution? The owl below summarises the problem that we now face.

Suppose that we have prior knowledge and we know the range of likely values of σ_v. R-INLA uses $\tau_v = 1 / \sigma_v^2$ and assumes that τ_v is gamma distributed. But the gamma distribution requires the specification of two parameters; the shape a_1 and the scale a_2. What values for a_1 and a_2 should we use?

Blangiardo and Cameletti (2015) show how information on σ_v can be translated into the two gamma parameters (also called the shape and scale). This can be done via a small simulation. Suppose that from a similar research study we know that σ_v is somewhere between 0.3 and 0.7. Let's simulate 10,000 values for such a σ_v.

```
> set.seed(12345)
> sigma.v <- runif(10000, min = 0.3, max = 0.7)
```

The precision τ_v is defined as 1 divided by σ_v^2.

```
> tau.v <- 1 /sigma.v^2
```

So, we have converted the 10,000 simulated σ_v values into 10,000 τ_v values. We have plotted the density curve for these 10,000 τ_v values in Figure 14.5. Next we do something that requires a fair amount of statistical knowledge of a gamma distribution. This distribution is specified with the shape (a_1) and scale (a_2) parameters. These are in fact the 1 and 0.00001 parameters when specifying the default gamma prior. Here comes the difficult part. Using standard rules for a variable X that follows a gamma distribution with mean $E(X)$ and variance var(X) the following expressions hold for the shape a_1 and scale a_2 parameters of the gamma distribution: a_1 = E(X)^2 / var(X) and a_2 = a_1 / E(X). Take these expressions as givens if you are not familiar with the gamma distribution.

Following these standard statistical expressions for the shape and scale parameters, we can calculate them using the simulated τ_v values.

```
> a1 <- mean(tau.v)^2 / var(tau.v)
> a2 <- a1 / mean(tau.v)
```

We now have the shape (a1) and scale (a2) parameters for specifying the gamma prior for τ_v, if we know that σ_v is somewhere between 0.3 and 0.7. So, instead of `param = c(1, 0.5)` we can use `param = c(a1, a2)` for the prior of τ_v. The dotted line in Figure 14.5 represents a gamma distribution with the given values of a1 and a2. The gamma density curve and the density curve of the simulated τ_v values are vaguely similar.

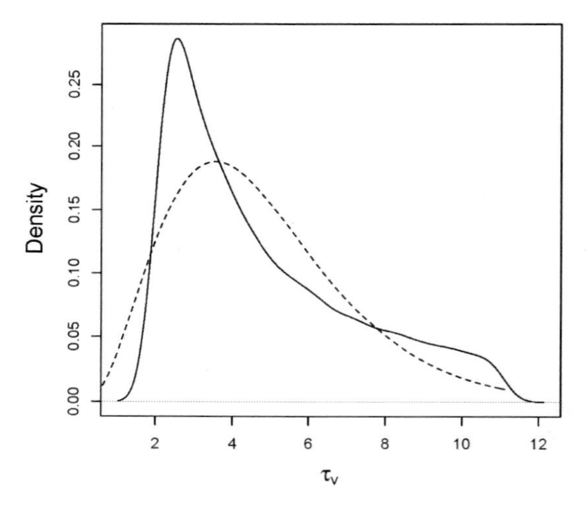

Figure 14.5. Density curve (solid line) of 10,000 τ_v values and a gamma distribution with parameters a1 and a2. The R code to create this figure is available on the website for this book.

Changing the gamma prior to the PC prior

So far we have used default prior distributions for the hyperparameters; these were mainly gamma distributions. Such gamma distributions require the specification of two parameters (the shape and scale) and R-INLA uses default values for them, namely 1 and 0.00001. In Chapter 8 we discussed the Carroll et al. (2015) paper which advises us to use a gamma(1, 0.5) distribution for hyperparameters instead of gamma(1, 0.00001). This may have worked for the simulated data in Carroll et al. (2015), but for your own data and model this is still a 'let's hope it also works for me' approach, where 'it works' is defined in terms of not crashing the software or not overfitting the data. And if the gamma(1, 0.5) or the gamma(1, 0.00001) priors don't work, then finding better values for the shape and scale parameters becomes a 'how does the wind blow' exercise.

Simpson et al. (2014) published an interesting paper in which they classified priors into objective priors (no prior information is used), weakly informative priors, and computationally convenient priors, among various other types. They also introduced a new type of prior distribution, namely the Penalised Complexity (PC) prior. The idea is that the user specifies the prior information from strong to weak with a single parameter. The mathematical expression for the density function of the PC prior for a precision parameter is not presented here, but it can be found in the Simpson et al. (2014) paper.

We will use the PC prior for the precision parameter in the random walk trend $\tau_v = 1 / \sigma_v^2$. The PC prior density function is for the precision parameter τ_v, but luckily we only need to provide information for the σ_v, which is slightly easier as it represent the change in the trend. A small value for σ_v means a trend that is nearly a straight line, whereas a large value for σ_v allows for a 'wigglier' curve (more change in the trend). When using the PC prior we need to specify a value for U in the following expression.

$$\text{Prob}\left(\sigma_v > U\right) = \alpha$$

For the value of α we can use 0.05. The U and α are used as known parameters in the density function of τ_v. If we choose a large value for U, say 100, then we are stating that the probability that σ_v is larger than 100 is very small. This is like stating that the majority of the people do not get older than 100 years. That is not very informative; we all know this. On the other hand, suppose that we use a small value for U, say 0.1. We are now stating that the probability that σ_v is larger than 0.1 is 0.05, so most likely σ_v is somewhere between 0 and 0.1. That is an informative prior!

Figure 14.6 shows a visualisation of the underlying idea of the PC prior. We have chosen five different values for U, from small to large. If $U = 5$ then we are stating that the probability that σ_v is larger than 5 is 0.05. The corresponding density function for τ_v is mainly concentrated at small values, and this means that σ_v can be anywhere (a small value of τ_v implies a large value of σ_v). At the other extreme, if $U = 0.1$, then the distribution covers large values for τ_v, which means that σ_v is rather small.

When using the PC prior, you need to specify the U value; the smaller the value you choose for U, the smaller the value for σ_v.

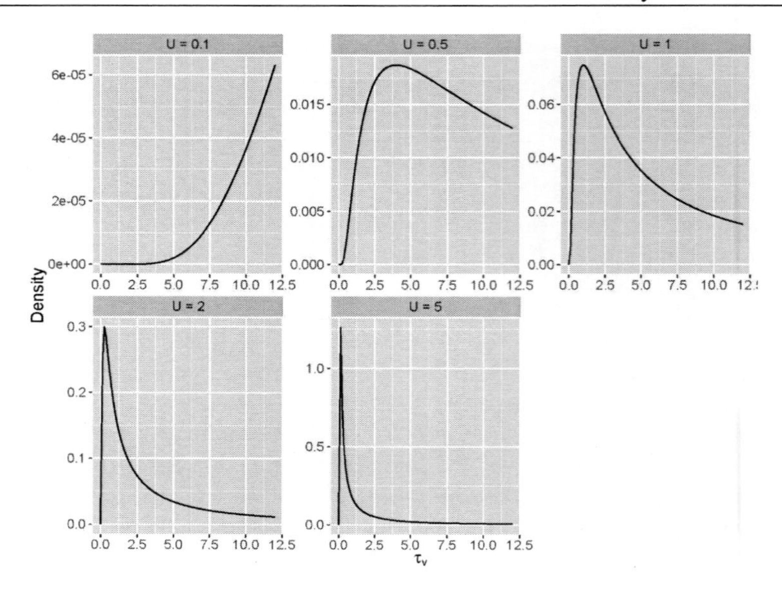

Figure 14.6. PC density function for τ_v obtained with different values of U and α = 0.05. The R code to create this figure is available on the website for this book.

Running a model with PC priors in R-INLA is rather simple, as can be seen from the code below.

```
> U <- 0.1
> hyper.prec <- list(theta = list(
                        prior = "pc.prec",
                        param = c(U, 0.05)))
> f4 <- MigDay ~ f(Year,
                   model = "rw1",
                   hyper = hyper.prec)
> I4 <- inla(f4,
             control.compute = list(dic = TRUE),
             family = "gaussian",
             data = Salmon)
```

To show the effect of the U value on the random walk trend, we ran the code above for six different U values. The fitted values are shown in Figure 14.7. Small values of U result in nearly straight lines of the rw1 trend (these have small σ_v values).

It is now your task to pick the trend that you like! Inspecting residual patterns is one tool that may help you.

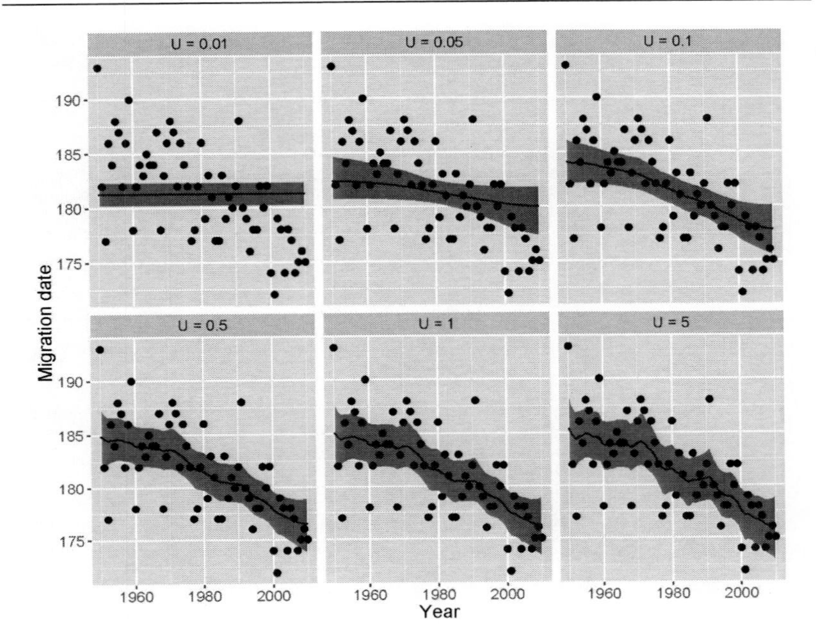

Figure 14.7. Fitted values for the rw1 model applied on the salmon arrival dates data. A PC prior was used for the precision parameter τ_v of the variance of the random walk. Each panel corresponds to a different value of U. The R code to create this figure is available on the website for this book.

Changing the random walk model: rw2

Besides the rw1 model that we have used so far, there is also the rw2 model in R-INLA. It is also a random walk trend, but it tends to produce smoother curves. This is achieved by using the following model.

$$\begin{aligned} &\mu_t - (2 \times \mu_{t-1} - \mu_{t-2}) = v_t \\ &v_t \sim N(0, \sigma_v^2) \end{aligned} \tag{14.5}$$

The justification for this expression goes along the lines that the rw1 model can be written as $\Delta\mu_t = \mu_t - \mu_{t-1}$, and the rw2 model is then defined by $\Delta^2\mu_t$, where $\Delta^2\mu_t = \Delta\Delta\mu_t$; see Bolin (2015) for a technical explanation. There is actually no urgent need to memorise or understand the expression in Equation (14.5). All you need to remember is that an rw2 model results in a smoother trend.

Implementation of this model is just a matter of changing `rw1` by `rw2`; the rest of the code stays the same. Results are not shown here, but the resulting trend is indeed smoother.

It should also be noted that both the rw1 and rw2 models are able to cope with irregularly spaced data; see Bolin (2015).

An rw1 model is a smooth curve if differences between sequential trend values are small. In an rw2 trend we work with the differences of the differences.

14.3 Trends in polar bear movements

In this section we use data originally published in Mauritzen et al. (2003), who studied the passive and active movement of polar bears (*Ursus maritimus*). Between 1988 and 1999, 86 adult female polar bears were tagged with a satellite telemetry transmitter. The transmitters gave information on position every sixth day for up to 3 years. Sampling took place in two areas, the Svalbard archipelago and the Barents Sea. In this section we only use the Barents Sea data. The data were also analysed in Zuur et al. (2014).

We have multiple observations from the same polar bear and therefore we will use a random intercept 'polar bear'. We would like to know whether there is a long-term trend in the movement data, but we also have to take into account the seasonal patterns. Only females were tagged and the reproductive state of a polar bear is also likely to affect movement.

Movement is strictly positive (only dead polar bears don't move), hence a gamma distribution with a log link seems to be a sensible starting point. This leads to the following model.

$$
\begin{aligned}
&Movement_{ij} \sim Gamma(\mu_{ij}, r) \\
&E(Movement_{ij}) = \mu_{ij} \\
&\log(\mu_{ij}) = Intercept + f_1(Year_{ij}) + f_2(DayInYear_{ij}) + Repro_{ij} + Bear_i \\
&Bear_i \sim N(0, \sigma^2_{Bear})
\end{aligned}
$$

$Movement_{ij}$ is the *j*th movement observation on bear *i*. We use a long-term trend $f_1(Year_{ij})$ and a seasonal component $f_2(DayInYear_{ij})$. We will use random walk models for both components. $Repro_{ij}$ is the reproductive state of a polar bear (a categorical variables with three levels: 'With cubs of the year', 'With one-year old cubs', 'With 2-year old cubs/weaning and reproducing'), and $Bear_i$ is the random effect for bears.

We execute the gamma generalised linear mixed-effects model (GLMM) with the R code below. The first two commands import the data and define the categorical covariate. The random effect is defined as a variable with the values 1 1 1 … 2 2 2 …3 3 3 …, etc. We now have three components in the formula; two times an rw1 random walk trend, and one random effect.

The model has four hyperparameters; the parameter *r* from the gamma distribution used for the movement data, a σ_{Year} for the first rw1 trend, a $\sigma_{DayInYear}$ for the second rw1 trend, and a σ_{Bear} for the random effect. In the first instance we will use the default gamma(1, 0.00005) prior for each hyperparameter.

```
> PB <- read.table(file = "PolarBearsV2.txt",
                   header = TRUE,
                   dec = ".")
> PB$fRepro <- factor(PB$Repro,
               levels = c("0", "1", "2"),
               labels = c("With cubs of the year",
                          "With one-year old cubs",
                          "With 2-year old cubs /
                           weaning & reproducing"))
> PB$BearN <- as.integer(as.factor(PB$BearID))
> f1 <- Movement ~ fRepro +
                   f(Year, model = "rw1") +
                   f(DayInYear, model = "rw1") +
                   f(BearN, model = "iid")
> I1 <- inla(f1,
        control.predictor = list(compute = TRUE),
        family = "gamma",
        data = PB)
```

The two random walk trends are presented in Figure 14.8. We are not too impressed with the wiggly behavior of the day in the season trend. We would have preferred it to be slightly smoother. And the random walk trend for year is rather too smooth.

The first thing we do to improve the patterns of the trends is to replace the rw1 trend for day in the season with an rw2 trend. The resulting curve was still wiggly. We could have tried to modify the shape and scale values of the gamma priors, but we did not do this. The reason for this is that we would have to do this for multiple hyperparameters, each with its own shape and scale values. An additional issue is that the hyperparameters are on different scales, which makes it even more difficult to find appropriate shape and scale values for each hyperparameter. Sørbye (2013) recommends applying a scaling so that priors for the precision parameters for certain models (e.g. rw1, rw2, besag, bym) become comparable. This is achieved by either adding `scale.model = TRUE` to the rw1 and rw2 models, or by typing

```
> inla.setOption(scale.model.default = TRUE)
```

We can now specify one set of scale and shape values (instead of different ones for the rw1 and rw2 models), and that should in principle work better.

In the R code below we define the formula and run the model again. The resulting two random walk trends are presented in Figure 14.9 and are easier on the eye. Note that we use the PC prior. The value of U was obtained with trial and error; if your choice is too large the day in the year trend is too wiggly. If your choice is too small it becomes a straight line.

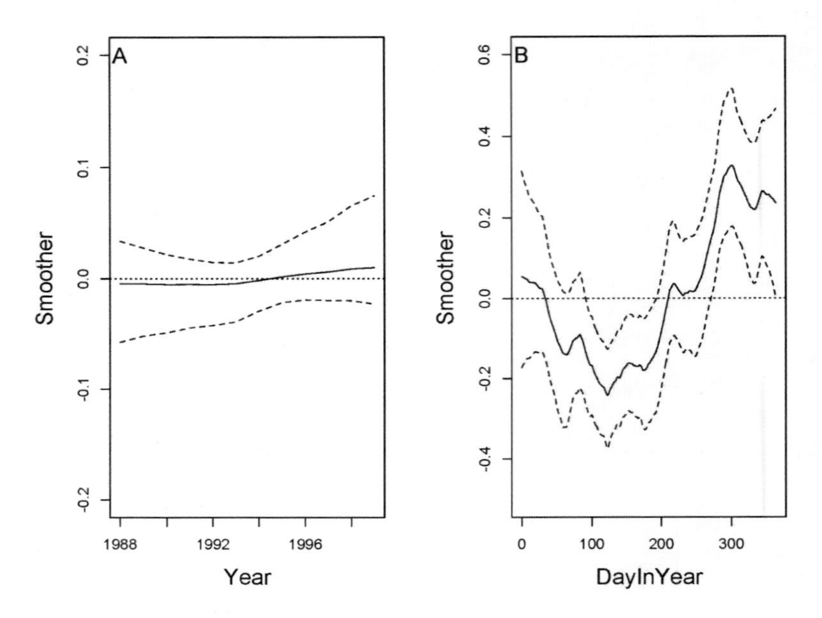

Figure 14.8. A: Random walk trends for year. B: Day in the season (right) with 95% credible intervals. The R code to create this figure is available on the website for this book.

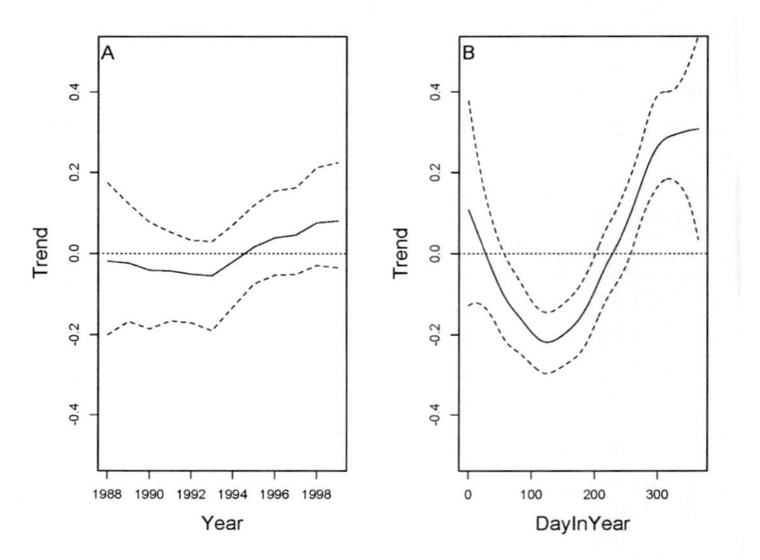

Figure 14.9. Random walk trends for year (left) and day in the year (right). The R code to create this figure is available on the website for this book.

```
> U <- 1.5
> hyper.prec = list(theta =
                 list(prior = "pc.prec",
                      param = c(U, 0.01)))
> f1 <- Movement ~ fRepro +
                   f(Year,
                     model = "rw1",
                     hyper = hyper.prec) +
                   f(DayInYear,
                     model = "rw2",
                     hyper = hyper.prec) +
                   f(BearN, model = "iid")
```

14.4 Trends in whale strandings

In this section we use data from Smeenk (1997; 1999) and Pierce et al. (2007) who analysed trends in sperm whale (*Physeter macrocephalus*) strandings on North Sea coasts. Information on strandings of sperm whales in the North Atlantic has been preserved in historical records since the 16th century. Whale strandings have been linked to climate change, sonar activity, food availability, migration patterns, sunspots, and various other factors. Obviously, one can (and should) question the quality and reliability of the data. Nowadays, whale strandings attract lots of onlookers, selfie-taking tourists, and news coverage, but that is no different from hundreds of years ago (when a stranded whale meant free food and bones).

Whatever your opinion about the quality of the data, it provides an interesting data set to compare Poisson and negative binomial models. The data were also used in Zuur et al. (2012), who employed zero-inflated generalised additive models with a CAR (temporal) correlation. The main question we focus on in this section is whether there are temporal patterns in the number of annual strandings.

We have annual number of strandings per year from 1563 until 2001, which means that we have 439 observations. In the first instance we applied a Poisson model with a random walk trend; see Equation (14.6).

$$\begin{aligned}
&Whales_t \sim Poisson(\mu_t) \\
&E(Whales_t) = \mu_t \\
&\log(\mu_t) = Intercept + \mu_t \\
&\mu_t = \mu_{t-1} + v_t \\
&v_t \sim N(0, \sigma_v^2)
\end{aligned} \qquad (14.6)$$

where $Whale_t$ is the number of strandings in year t. The random walk trend is presented in Figure 14.10A and it is not a useful curve at all. We refitted the model with a negative binomial GLMM and the resulting random walk trend is presented in Figure 14.10B; it is far more pleasing to the eye. Note that to get the fitted values we have to add the intercept and take the exponential function.

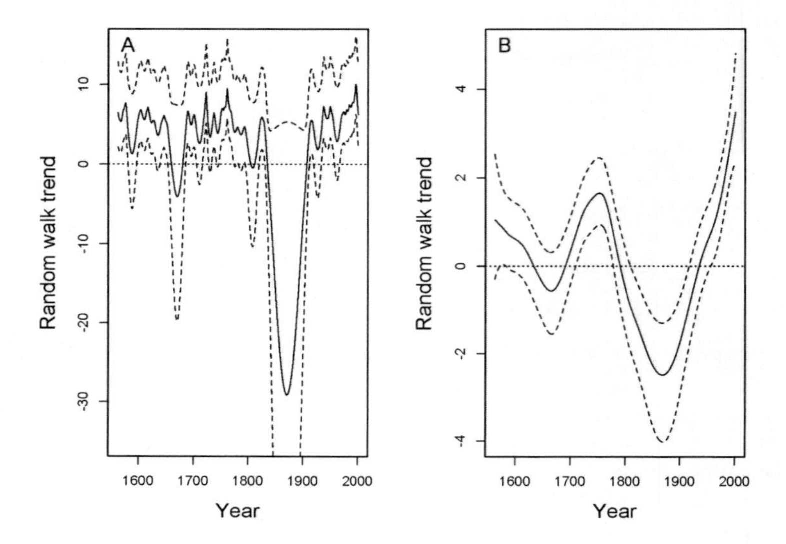

Figure 14.10. A: random walk trend of the Poisson GLMM. B: Random walk trend of a negative binomial GLMM. The R code to create this figure is available on the website for this book.

The reason for the odd behavior of the random walk trend in the Poisson GLMM is that 82% of the years have 0 recorded strandings, which means that the data is zero inflated. A large number of these zeros were recoded between 1820 and 1920. The Poisson GLMM models these zeros with a large negative trend and due to the log link these large negative trend values result in fitted values close to 0. The negative binomial distribution is better capable to deal with the zeros and as a result its random walk trend is smoother.

What do we learn from this? That choosing the wrong distribution may result in odd behaviour of trends.

The Poisson GLMM and NB GLMM were fitted with the following R code.

```
> SW <- read.table("Spermwhales.txt",
                   header = TRUE)
> SW1 <- subset(SW, Year >= 1563 & Year <= 2001)
> f1 <- Whales ~ f(Year, model = "rwl")
> I1 <- inla(f1,
             control.compute = list(dic = TRUE),
             family = "poisson",
             data = SW1)
> f2 <- Whales ~ f(Year, model = "rwl")
```

```
> I2 <- inla(f2,
             control.compute = list(dic = TRUE),
             family = "nbinomial",
             data = SW1)
```

Extracting the random walk trends requires R code identical to that presented in Section 14.2 and is not repeated here. A detailed model validation is required for the NB GLMM.

14.5 Multivariate time series for Hawaiian birds

In this section we use a data set that we analysed in Zuur et al. (2007) and in Reed et al. (2011). It was also used in Ieno and Zuur (2015) for explaining `ggplot2` code. The aim of the study is to analyse long-term survey data for three endangered waterbirds that are found only in the Hawaiian Islands: Hawaiian stilt (*Himantopus mexicanus knudeseni*), Hawaiian coot (*Fulica alai*), and Hawaiian moorhen (*Gallinula chloropus sandvicensis*). The survey data come from biannual waterbird counts that are conducted in both winter and late summer. In this chapter we will use the winter survey data as it reduces the contribution of recently hatched birds to survey numbers, which can increase count variability.

14.5.1 Importing and preparing the data

We import the data using the `read.table` function.

```
> HB <- read.table(file = "HawaiiBirdsV2.txt",
                   header = TRUE,
                   dec = ".")
```

The `names` function shows that we have six variables.

```
> names(HB)

"Year"      "Birds"      "ID"      "Rainfall"
"Species"   "Island"
```

The variable `Birds` contains the observed number of birds and `Year` contains the year of sampling. The data are from three different bird species and from three different islands in Hawaii. The `Species` variable contains the values 1, 2, and 3, referring to the bird species stilts, coots, and moorhens respectively. The variable `Island` has the values 1, 2, and 3, referring to the islands Oahu, Maui, and Kauai and Niihau respectively.

Instead of the values 1, 2, and 3 for the variables `Species` and `Island` we would like to use the bird species names and the names of the islands. To do this we define two variables that contain the names for which we would like to see graphs.

```
> SpeciesNames <- c("Stilts", "Coots", "Moorhens")
> IslandsNames <- c("Oahu", "Maui",
                    "Kauai and Niihau")
```

Next we use some indexing in R. For every 1 in `HB$Species` we put `Stilts`; for every 2 we put `Coots`; and for every 3 we put `Moorhens`. For every 1 in `HB$Island` we put `Oahu`; for every 2 we put `Maui`; and for every 3 we put `Kauai and Niihau`.

```
> HB$Species2 <- SpeciesNames[HB$Species]
> HB$Island2  <- IslandsNames[HB$Island]
> HB$SpeciesIsland <- paste(HB$Species2,
                            HB$Island2, sep = ".")
```

The `paste` function joins together the species and island names and inserts a dot between the two names. We define `SpeciesIsland` as a factor:

```
> HB$fSpeciesIsland <- factor(HB$SpeciesIsland)
```

The categorical variable `fSpeciesIsland` has eight levels, not nine, because one bird species (moorhen) was not observed on a specific island (Kauai and Niihau).

```
> levels(HB$fSpeciesIsland)
```

```
"Coots.Kauai and Niihau"
"Coots.Maui"
"Coots.Oahu"
"Moorhens.Maui"
"Moorhens.Oahu"
"Stilts.Kauai and Niihau"
"Stilts.Maui"
"Stilts.Oahu"
```

14.5.2 Data exploration

We will keep the data exploration short. Figure 14.11 shows a time-series plot for each species–island combination. Some time series have missing values and the abundances are rather different per species. Coots and moorhen had high abundances at about the same time on Oahu.

Figure 14.11 was created using the following `ggplot2` code.

```
> p <- ggplot(data = HB, aes(y = Birds, x = Year))
> p <- p + xlab("Year") + ylab("Bird abundance")
> p <- p + geom_point(shape = 16, size = 1.5)
> p <- p + geom_line()
> p <- p + facet_grid(Species3 ~ Island3,
                      scales = "free")
> p
```

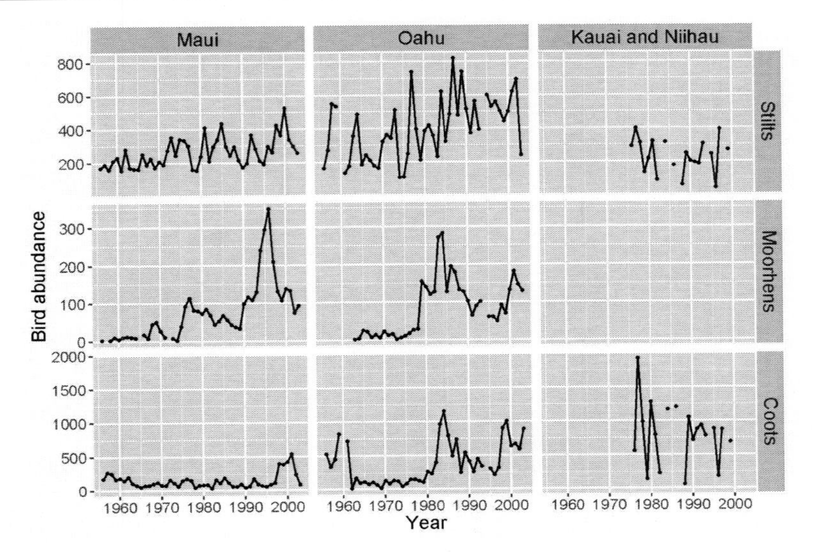

Figure 14.11. Time-series plot of the bird abundance at three islands and three species.

14.5.3 Model formulation

We initially used models with a Poisson distribution, but for some of the models this resulted in a trend that gave a perfect fit (a line that goes from point to point). Using the negative binomial distribution solved this issue.

The underlying question is whether there is a trend over time, but if we think a little bit more deeply about this question, we can also ask whether all eight time series have one trend, or whether the trend differs by island, by species, or by island–species combination. To answer these questions means that we can formulate, execute, and compare the following four models. For all models we use the same distributional assumption, which is given below.

$$
\begin{aligned}
&Birds_{tij} \sim NB\left(\mu_{tij}, \theta\right) \\
&E\left(Birds_{tij}\right) = \mu_{tij} \\
&\text{var}\left(Birds_{tij}\right) = \mu_{tij} + \mu_{tij}^2 / \theta
\end{aligned}
$$

$Birds_{tij}$ is the number of birds in year t for species i on island j, and we assume that it is negative binomial distributed. The four models that we will apply differ in how we model the mean value μ_{tij}. In the first model that we consider we assume that all species at all islands have the same trend.

$$
\log\left(\mu_{tij}\right) = a_{tij} + Trend_t \qquad (14.7)
$$

The a_{tij} is the categorical variable `fSpeciesIsland` and has eight levels. It allows for a different mean abundance per time series. The trend is modeled as an rw1 random walk trend. Note that it has only a subscript t for time.

To allow for three trends, one per island, we add a subscript j to the trend component.

$$\log\left(\mu_{tij}\right) = a_{tij} + Trend_{tj} \tag{14.8}$$

This model assumes that all species on an island follow the same temporal trend.

To investigate whether the trends differ per species (but are the same for all the islands) we use a trend for each species. We now use the subscripts t and i in the random walk trend.

$$\log\left(\mu_{tij}\right) = a_{tij} + Trend_{ti} \tag{14.9}$$

Finally, we specify a model that has a different random walk trend for each species and each island.

$$\log\left(\mu_{tij}\right) = a_{tij} + Trend_{tij} \tag{14.10}$$

Now the random walk trend has three subscripts: time, species, and island. We will compare the models with the DIC.

14.5.4 Executing the models

One trend for all time series

We remove all the rows with missing values (which are present in the response variable) from the data set.

```
> HB2 <- na.exclude(HB)
```

We first focus on the model in Equation (14.7). The R code to execute this model in R-INLA is as follows. We use a PC prior with $U = 1$, though we did not really experiment with different values of U. The selected value of 1 seems to work for this example.

```
> U <- 1
> hyper.prec = list(theta = list(
                                    prior = "pc.prec",
                                    param = c(U, 0.01)))
> f1 <- Birds ~ fSpeciesIsland  +
                f(Year,
                  model = "rw1",
                  scale.model = TRUE,
                  hyper = hyper.prec)
> I1 <- inla(f1,
             control.compute = list(dic = TRUE),
```

```
                  family = "nbinomial",
                  data = HB2)
```

We obtain the fitted values, posterior mean of the θ (this is the parameter in the variance of the negative binomial distribution), and the Pearson residuals via

```
> mu    <- I1$summary.fitted.values[,"mean"]
> theta <- I1$summary.hyperpar[1, "mean"]
> E1    <- (HB2$Birds - mu)/sqrt(mu + mu^2 / theta)
```

and these can be used for model validation purposes. We extract the fitted values and the boundaries for the 95% credible intervals, and these are all added to the data object so that we can plot the fitted values in a moment using ggplot2.

```
> Fit1           <- I1$summary.fitted.values
> HB2$Fitted1    <- Fit1$"mean"
> HB2$Fit1.025   <- Fit1$"0.025quant"
> HB2$Fit1.975   <- Fit1$"0.975quant"
```

Once we have the fitted values and 95% credible intervals in the same object as the original data, we use ggplot2 code to produce Figure 14.12. All trends have the same pattern, because that is how we specified it. The only difference between the fitted values in eight panels is the contribution from the categorical covariate. That is a vertical shift of the lines.

```
> p <- ggplot(data = HB2,
              aes(y = Birds, x = Year))
> p <- p + xlab("Year") + ylab("Bird abundance")
> p <- p + geom_line(aes(x = Year, y = Fitted1))
> p <- p + geom_ribbon(aes(x = Year,
                           ymax = Fit1.975,
                           ymin = Fit1.025),
                       fill = grey(0.5),
                       alpha = 0.4)
> p <- p + facet_grid(Species2 ~ Island2 ,
                      scales = "free")
> p
```

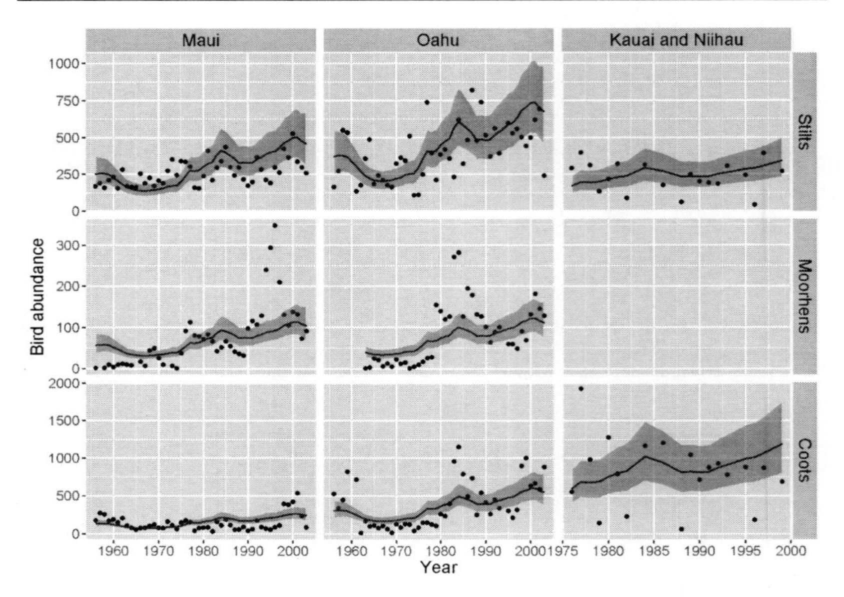

Figure 14.12. Fitted values of the model in Equation (14.7). The model has one trend component.

One trend for each island

Next we apply the model in Equation (14.8). This model has three random walk trends, one per island. This means that we assume that on each island all the species behave similarly over time. To execute such a model in R-INLA we first apply the internal standardisation for the priors because we will be using multiple rw1 models.

```
> inla.setOption(scale.model.default = TRUE)
```

The trick to fit this model in R-INLA is to create three new dummy variables with the names Oahu, Maui, and KN. The variable Oahu has the value 1 if an observation is from Oahu, and otherwise it is 0. We do the same with the other two variables.

```
> HB2$Oahu <- as.numeric(HB2$Island2 == "Oahu")
> HB2$Maui <- as.numeric(HB2$Island2 == "Maui")
> HB2$KN   <- as.numeric(HB2$Island2 == "Kauai and
                                         Niihau")
```

Because we will be using the rw1 models of year we need to duplicate the year variable for each model. This is just an R-INLA syntax issue.

```
> HB2$Year2 <- HB2$Year
> HB2$Year3 <- HB2$Year
```

We are now ready to define the model. As before we use the categorical covariate `fSpeciesIsland` that allows for a different mean value per time series. There are three rw1 models in the formula `f2`. The only difference between these three models is the variable immediately after `Year`. R-INLA multiplies `Year` with the Oahu dummy variable, and as a result we end up with a trend that is purely for the Oahu data. We do exactly the same for the other two rw1 components. For readers who are familiar with the `gam` function from the `mgcv` package (Wood 2006), this is equivalent to the command `gam(Birds ~ s(Year, by = Island2, data = HB2)`.

```
> f2 <- Birds ~ fSpeciesIsland  +
                f(Year, Oahu,
                  model = "rw1",
                  hyper = hyper.prec) +
                f(Year2, Maui,
                  model = "rw1",
                  hyper = hyper.prec) +
                f(Year3, KN,
                  model = "rw1",
                  hyper = hyper.prec)
```

We are now ready to execute the model in R-INLA and compare the DICs of the two models.

```
> I2 <- inla(f2,
             control.compute = list(dic = TRUE),
             family = "nbinomial",
             data = HB2)
> c(I1$dic$dic, I2$dic$dic)

3823.289 3783.985
```

Allowing for a different temporal pattern per island has improved the model as compared to using one overall trend.

We can use nearly the identical R code (just change `I1` by `I2`) that was used for Figure 14.12 to plot the fitted values; see Figure 14.13. The shapes of the trends indicate that there are clear differences between the islands.

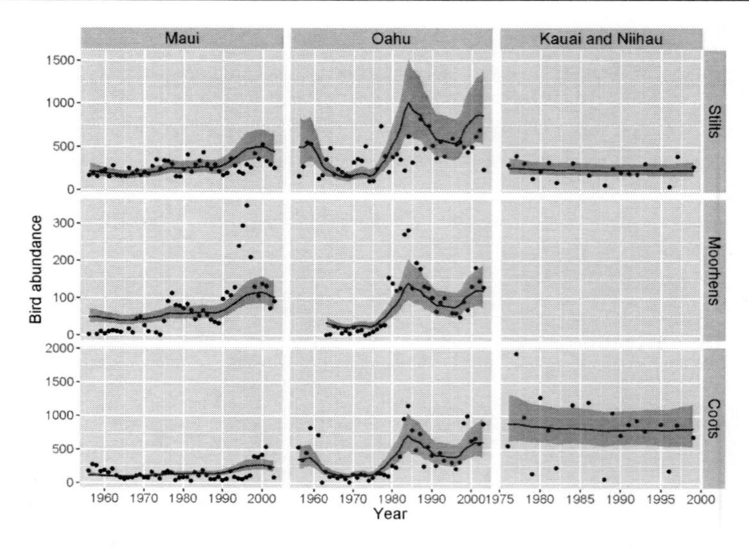

Figure 14.13. Fitted values of the model in Equation (14.8). The model has three random walk trends, one per island. The R code to create this figure is available on the website for this book.

One trend for each species

The same process can be repeated to fit the model in Equation (14.9). We again make three dummy variables.

```
> HB2$Stilts <- as.numeric(HB2$Species2=="Stilts")
> HB2$Coots   <- as.numeric(HB2$Species2=="Coots")
> HB2$Moorhens <- as.numeric(HB2$Species2 ==
                                      "Moorhens")
```

And we make the formula again with three rw1 models, one per species in this case.

```
> f3 <- Birds ~ fSpeciesIsland  +
                f(Year, Stilts,
                  model = "rw1",
                  hyper = hyper.prec) +
                f(Year2, Coots,
                  model = "rw1",
                  hyper = hyper.prec) +
                f(Year3, Moorhens,
                  model = "rw1",
                  hyper = hyper.prec)
```

Executing the model and comparing the DICs shows that the model with three species-specific trends is better than the model with three island-specific trends. The graph of the three trends is not presented here but the online R code can be used to plot them.

```
> I3 <- inla(f3,
             control.compute = list(dic = TRUE),
             family = "nbinomial",
             data = HB2)
> c(I1$dic$dic, I2$dic$dic, I3$dic$dic)

3823.289 3783.985 3699.825
```

One trend for each species and island combination

We finally apply the model in Equation (14.10); it has eight rw1 trends. To run this model in R-INLA we need to create eight dummy variables.

```
> HB2$Coots_KN       <- as.numeric(HB2$SpeciesIsland == "Coots.Kauai and
                                                         Niihau")
> HB2$Coots_Maui     <- as.numeric(HB2$SpeciesIsland == "Coots.Maui")
> HB2$Coots_Oahu     <- as.numeric(HB2$SpeciesIsland == "Coots.Oahu")
> HB2$Moorhens_Maui  <- as.numeric(HB2$SpeciesIsland == "Moorhens.Maui")
> HB2$Moorhens_Oahu  <- as.numeric(HB2$SpeciesIsland == "Moorhens.Oahu")
> HB2$Stilts_KN      <- as.numeric(HB2$SpeciesIsland == "Stilts.Kauai
                                                         and Niihau")
> HB2$Stilts_Maui    <- as.numeric(HB2$SpeciesIsland == "Stilts.Maui")
> HB2$Stilts_Oahu    <- as.numeric(HB2$SpeciesIsland == "Stilts.Oahu")
```

We also need eight copies of the year variable. We already have two copies so we only have to create six extra ones.

```
> HB2$Year4 <- HB2$Year
> HB2$Year5 <- HB2$Year
> HB2$Year6 <- HB2$Year
> HB2$Year7 <- HB2$Year
> HB2$Year8 <- HB2$Year
```

We define the model with eight trends and execute the model in R-INLA.

```
> f4 <- Birds ~ fSpeciesIsland  +
                f(Year, Coots_KN,
                  model = "rw1",
                  hyper = hyper.prec) +
                f(Year2, Coots_Maui,
                  model = "rw1",
                  hyper = hyper.prec) +
                f(Year3, Coots_Oahu,
                  model = "rw1",
                  hyper = hyper.prec) +
                f(Year4, Moorhens_Maui,
                  model = "rw1",
                  hyper = hyper.prec) +
                f(Year5, Moorhens_Oahu,
                  model = "rw1",
                  hyper = hyper.prec) +
                f(Year6, Stilts_KN,
                  model = "rw1",
```

```
                    hyper = hyper.prec) +
                  f(Year7, Stilts_Maui,
                    model = "rw1",
                    hyper = hyper.prec) +
                  f(Year8, Stilts_Oahu,
                    model = "rw1",
                    hyper = hyper.prec)
> I4 <- inla(f4,
             control.compute = list(dic = TRUE),
             family = "nbinomial",
             data = HB2)
> c(I1$dic$dic, I2$dic$dic, I3$dic$dic,I4$dic$dic)

3823.289 3783.985 3699.825 3613.540
```

The model with eight trends is the best model among the four models that we have applied. Using nearly identical code to what we used for Figure 14.12 we plot the fitted values in Figure 14.14.

Pearson residuals and fitted values of the model in Equation (14.9) can easily be obtained and used for model validation. We also suggest applying a simulation study to investigate whether the negative binomial distribution is appropriate; see Chapter 10.

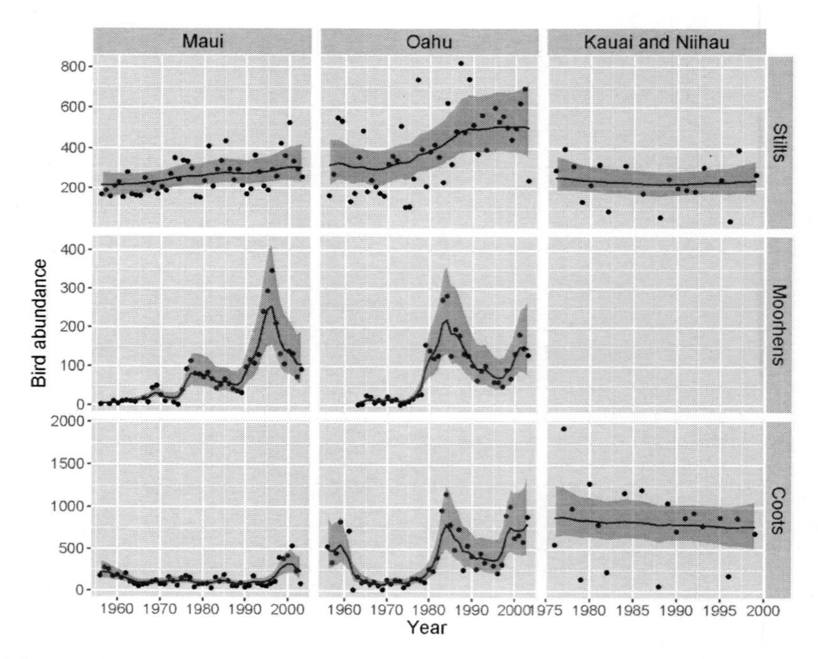

Figure 14.14. Fitted values of the model in Equation (14.10). The model has eight random walk trends, one per island and species combination. The R code to create this figure is available on the website for this book.

14.5.5 Mixing Poisson and negative binomial distributions

In the previous subsections we used the negative binomial distribution. The motivation for this was that the model with the Poisson distribution resulted in fitted values that were identical to the observed values when using the model in Equation (14.10). We could have played around with the U value of the PC prior, but when we saw the time-series plots in Figure 14.11 for the first time we already had the impression that for some time series the negative binomial distribution may be needed due to the large abundance values.

We deliberately used the word 'some' in the previous sentence. Not all time series have large values. A question that we hear often during statistical courses is: 'Why don't you analyse the time series separately?' The answer is that we want to know whether the time series are different from one another, not so much in absolute values but more whether the trends over time differ. And in order to investigate this we need to fit models on all the data, and how different models (e.g. with and without common trends) perform. So splitting up the data is not an option. But if one species follows a Poisson distribution and another species is negative binomial distributed, then the models that we have fitted so far on the Hawaiian data set are faulty. Turns out that R-INLA can apply models with different distributions on the same data! To keep the code compact we only use two time series from the Hawaiian data set, but you can easily do it for more time series and more distributions.

First we take a subset of only two time series.

```
> HB3 <- subset(HB2,
           fSpeciesIsland == "Moorhens.Oahu" |
           fSpeciesIsland == "Stilts.Oahu")
> HB3$fSpeciesIsland <-
                 droplevels(HB3$fSpeciesIsland)
```

Suppose that we want to apply a model in which the Moorhens data follow a Poisson distribution and the stilts data have a negative binomial distribution. To run such a model in R-INLA we need to know a few things about the sample sized.

```
> dim(HB3)

 86 50
```

We have 86 rows in total, the Moorhens time series contains 40 observations, and the stilts time series is 46 years.

```
> table(HB3$fSpeciesIsland)

Moorhens.Oahu   Stilts.Oahu
           40            46
```

We need to create a new object with the bird abundances that has the following format.

$$\begin{pmatrix} Moorhens_1 & NA \\ Moorhens_2 & NA \\ \vdots & \vdots \\ Moorhens_{40} & NA \\ NA & Stilts_1 \\ NA & Stilts2 \\ \vdots & \vdots \\ NA & Stilts_{46} \end{pmatrix}$$

The first column contains the 40 Moorhens abundance values and 46 NAs. The second column contains 40 NAs and the 46 stilts abundance values. Perhaps now you understand why we only picked two species to explain this technique. If you have eight time series then there are eight columns like this.

The matrix above is going to be the response variable. We also need to ensure that we have matching covariates and a column identifying which rows belong to the same group. This is the response variable.

```
> Y <- matrix(NA, nrow = 86, 2)
> Y[1:40,1]  <- HB3$Birds[HB3$fSpeciesIsland ==
                          "Moorhens.Oahu"]
> Y[41:86,2] <- HB3$Birds[HB3$fSpeciesIsland ==
                          "Stilts.Oahu"]
```

If you inspect the values in `Y` then you will see that they matched the matrix that we sketched above. We are now ready to combine all the data. This has to be in a list; otherwise we have trouble with the two column names in `Y`.

```
> MyData <- list(
      Birds = Y,
      Year  = HB3$Year,
      Year2 = HB3$Year,
      Moorhens = as.numeric(HB3$fSpeciesIsland ==
                           "Moorhens.Oahu"),
      Stilts   = as.numeric(HB3$fSpeciesIsland ==
                           "Stilts.Oahu"),
      fSpeciesIsland = HB3$fSpeciesIsland)
```

The Birds contains the matrix with the abundances and the NAs. Year and Year2 are two copies required for the two rw1 trends. ID identifies which rows in Y belong to the same time series and can be used by `ggplot2` for plotting the results. Moorhens is a dummy variable with values 1 and 0, and is needed for the first rw1 trend. Stilts is also a dummy variable and identifies which rows belong to the Stilts time series. It is needed for the second rw1 trend.

We define the model with two rw1 trends and execute the model in R-INLA. The crucial point is the `family` argument. It has 'poisson' and 'nbinomial' as arguments. That means that a Poisson distribution is used for the first column in Birds, and a negative binomial distribution is used for the second column.

```
> f5 <- Birds ~ fSpeciesIsland  +
                f(Year, Moorhens.Oahu,
                  model = "rw1",
                  hyper = hyper.prec) +
                f(Year2, Stilts.Oahu,
                  model = "rw1",
                  hyper = hyper.prec)
> I5 <- inla(f5,
             control.compute = list(dic = TRUE),
             family = c("poisson", "nbinomial"),
             data = MyData)
```

How do you know whether these distributions are correct? We suggest applying a simulation as discussed in Chapter 10. If the simulated and observed data match, then the selected distributions are appropriate. Using DIC may be a hazardous exercise if there is also a random walk trend in the model that potentially produces a perfect fit for one of the columns in case the wrong distribution is selected.

The facility to fit models with different distributions also enables the user to fit models like zero-altered Poisson models, zero-altered negative binomial models, and zero-altered gamma models to analyse zero-inflated count data and zero-inflated continuous data.

R-INLA also has facilities for multivariate priors and in combination with the multiple-family option it is possible to fit multivariate GLMMs. These models are explained in Zuur et al. (2016) although MCMC is used via JAGS to execute the models.

A useful reference is Ruiz-Cárdenas et al. (2012); they show how dynamic regression models can be fitted in R-INLA.

14.6 AR1 trends

14.6.1 AR1 trend for regularly spaced time-series data

The last technique that we discuss in this chapter is the use of an AR1 process to estimate the temporal trend. This is an alternative approach for the rw1 or rw2 processes. The reason for discussing the AR1 trend is that we will see it again when analysing spatial-temporal data in the next two chapters. We will use the sockeye salmon arrival date data set to illustrate the implementation of a model with an AR1 trend. The model specification is as follows.

$$\begin{aligned}
&MigDay_t = \text{Intercept} + \mu_t + \varepsilon_t \\
&\mu_t = \rho \times \mu_{t-1} + v_t \\
&\varepsilon_t \sim N(0, \sigma_\varepsilon^2) \quad \text{and} \quad v_t \sim N(0, \sigma_v^2)
\end{aligned} \tag{14.11}$$

The parameter ρ determines how much the trend in year t depends on the trend from the previous year. If ρ is close to 0, then there is not much temporal correlation. The closer ρ is to 1, the more similar are the trends at time t and $t - 1$, and the smoother the trend, though the v_t also plays a role in this. If we set the parameter ρ equal to 1 we get the rw1 model.

In our experience, using relative short time series of 15–30 years (which are typical for ecological field studies), it can be quite a challenge to avoid an AR1 trend that gives a perfect fit (a line that goes from point to point). We have also seen examples of Poisson GLMs in which ρ was estimated as 0, and a large σ_v (a Poisson GLM does not have the ε_t and σ_ε terms); that is the same model as a so-called GLMM with an observation-level random intercept (Elston et al. 2001).

To execute the model in Equation (14.11) in R-INLA we use the code below. We assume that the salmon data have already been imported.

```
> f1 <- MigDay ~ f(Year, model = "ar1")
> I1 <- inla(f1,
             control.compute = list(dic = TRUE),
             family = "gaussian",
             data = Salmon)
```

The fitted values of the model are presented in Figure 14.15. The shape of the curve is rather wiggly. To control the wiggliness it is an option to specify different priors for σ_v (which is called `theta1` in the AR1 model specification) or ρ (which is called `theta2` in the AR1 model specification). The confusing thing is that you cannot specify a prior for ρ directly; instead you need to specify a prior for `theta2` which is defined as $\log((1 + \rho) / (1 - \rho))$.

There is no immediate benefit of an AR1 trend over an rw1 or rw2 model. The AR1 is slightly more flexible as it has the extra parameter ρ, but the downside of an extra parameter is that you have to fight an extra battle with the priors of ρ (unless you are lucky and the model with default priors seems to work, as in Figure 14.15).

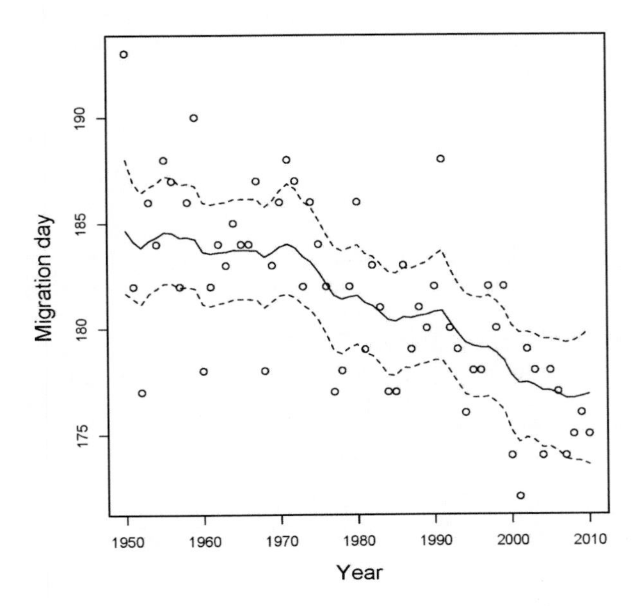

Figure 14.15. Fitted values of the model with the AR1 trend. The R code to create this figure is available on the website for this book.

14.6.2 AR1 trend for irregularly spaced time-series data

The AR1 trend is designed for the analysis of regularly spaced time series. What can we do if the data are irregularly spaced in time, if they are regularly spaced but contain many missing values, or if the AR1 process computing time is long? In this section we will explain what to do in such cases. The approach that we will follow matches that for the spatial models where we defined a mesh. In short, we will create regularly spaced time points, and define the trend on these knots. If there are also covariates then we are in for a little bit of mesh and stack headache again. The knots play the same role as the nodes of a mesh.

Defining regularly spaced knots in time is also a handy feature in Chapters 15 and 16, where we will analyse spatial-temporal data. The models that will be applied in those two chapters use a spatial random field for each year. If we have 50 years of data and the spatial random field for each year consists of a mesh with 500 nodes, then we have to estimate $50 \times 500 = 25{,}000$ spatial parameters. That requires a lot of computing time. Defining the temporal trend on alternative and fewer time points will save a lot of time!

To explain the AR1 models for irregularly spaced temporal data we use a data set from Etheridge et al. (1998) who looked at historical CO_2 records from three ice cores from Law Dome, which is a large ice dome in Antarctica. Atmospheric CO_2 was reconstructed from a large number of air bubbles sampled from the ice cores. The mixing ratio of CO_2 was determined for each air bubble and the samples were also dated. Full details on the data can be found in the Etheridge paper or at http://cdiac.ornl.gov/trends/co2/lawdome.html.

Figure 14.16 shows the plot of the time series for each of the three cores. The major growth in atmospheric CO_2 levels has been linked to the start of the industrial period.

We would like to obtain a trend for the CO_2 data. The first question is whether we should combine the data from the three cores or analyse them separately. If we combine them then we can check whether the patterns are different by fitting a model with three trends, and a model with one trend and see which one is better. If we do this then it may be better to focus only on the data sampled after 1810. But obviously, analysing the DSS core is much more interesting as the time series covers nearly a thousand years. In this section we will formulate a model on the combined data and use one trend. We invite the reader to apply the model with the three trends and compare them with DICs or WAICs.

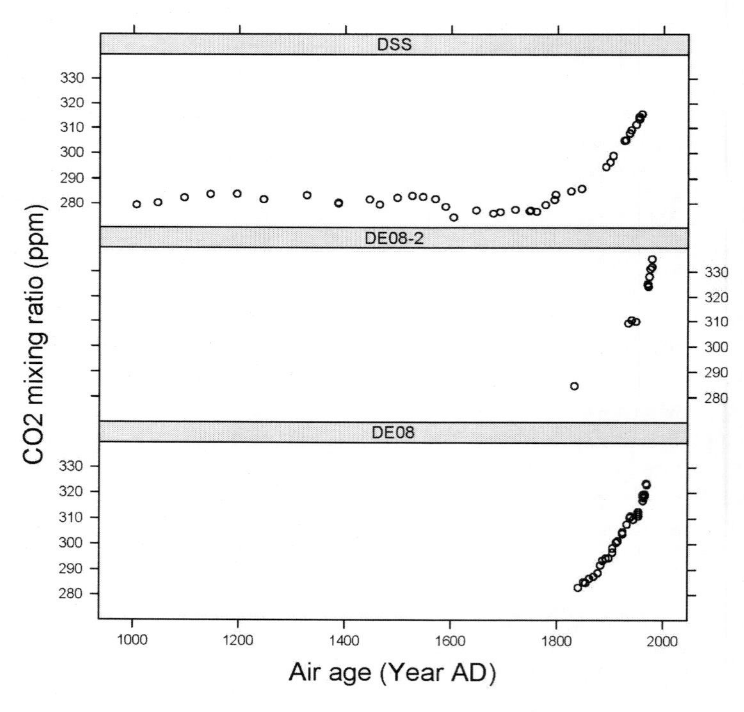

Figure 14.16. CO_2 mixing ratio (parts per million) plotted versus age (year). Each panel corresponds to a core. The R code to create this figure is available on the website for this book.

The next question is how we should implement a model with a trend as the time series are irregularly spaced. A naive approach would be to ignore the irregular nature of the data and apply the following model.

$$\begin{aligned} CO2_{ti} &= \mu_t + Core_i + \varepsilon_{ti} \\ \mu_t &= \rho \times \mu_{t-1} + v_t \\ \varepsilon_{ti} &\sim N(0, \sigma_\varepsilon^2) \quad \text{and} \quad v_t \sim N(0, \sigma_v^2) \end{aligned} \tag{14.12}$$

$CO2_{ti}$ is the CO_2 mixing ratio at time t for core i. We have three cores (hence i = 1, ..., 3) and a total of 84 observations. $Core_i$ is a categorical covariate allowing for a different mean value per core. The model assumes that all cores have the same temporal pattern, but as discussed this can be changed by extending the model so that three trends are used. One could argue that the beta distribution should be used due to the definition of a mixing ratio, but we will keep it simple and use the Gaussian distribution.

Applying the model as it is would consider the years 1006, 1046, and 1096 as three sequential years with the same time difference between them. As long as the differences are roughly comparable (40 years between the first two years and 50 years between the second two years) then there are no major problems. But there are also years separated by 60 years and others by 5 years.

One solution to avoid this is to define the trend on selected time points, also called knots; see Figure 14.17. The knot positions are regularly spaced and are the vertical dotted lines. The observed data points are represented by the '|' symbol. We choose the knot locations such that there are not too many time intervals between the knots with no observed data. Our initial aim was to have at least 30-ish knots, but that is difficult for this specific data set due to the limited observations during the first 800 years. The first knot is placed in the year 1050, and subsequent knots are 50 years apart. The last knot is in 1950.

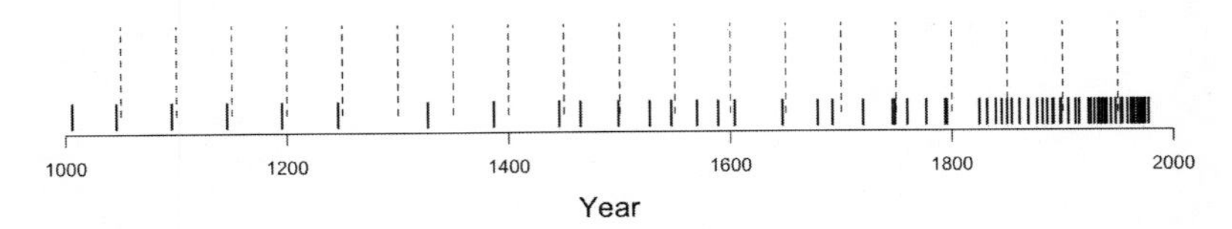

Figure 14.17. Observed time values for the samples are indicated by a '|' symbol. Knot locations are chosen at every 50 years starting at 1050 and ending in 1950. The R code to create this figure is available on the website for this book.

Let us catch up with some R code for delving more deeply into the setup of the model. We first import the data.

```
> IC <- read.csv(file = IceCoresV2.csv",
                header = TRUE)
```

For plotting purposes it is convenient to sort the data along time, but this is not required.

```
> I1 <- order(IC$MeanAirAgeYear)
> IC <- IC[I1,]
```

We also shorten the variable names.

```
> IC$Year <- IC$MeanAirAgeYear
> IC$CO2  <- IC$CO2MixingRatio
```

We can fit the model in Equation (14.12) with the following R code.

```
> f1 <- CO2 ~ -1  + Core + f(Year, model = "ar1")
> I1 <- inla(f1,
             control.compute = list(dic = TRUE),
             family = "gaussian",
             data = IC)
```

And you may want to convince yourself that when we change the Year variable to a variable with sequential numbers (1006 is 1, 1046 is 2, 1096 is 3, etc.), we get exactly the same results.

```
> IC$Year123 <- as.numeric(as.factor(IC$Year))
> f2 <- CO2 ~ -1  + Core + f(Year123,
                             model = "ar1")
> I2 <- inla(f2,
             control.compute = list(dic = TRUE),
             family = "gaussian",
             data = IC)
```

Next we create the knots.

```
> Knots <- seq(from = 1050, to = 1950, by = 50)
> Knots

1050 1100 1150 1200 1250 1300 1350 1400 1450
1500 1550 1600 1650 1700 1750 1800 1850 1900
1950
```

The next thing we do is to determine the interval into which an observation falls. We use the R function `findInterval` for this.

```
> Intervals <- findInterval(IC$Year,
                            c(1000, Knots, 2000))
> Intervals

 1  1  2  3  4  5  7  8  8  9 10 10 11 11 12 12 13
13 14 14 15 15 15 16 16 16 16 17 17 17 17 18 18 18
```

```
18 18 18 18 18 18 18 18 19 19 19 19 19 19 19 19 19
19 19 19 19 19 19 19 19 19 19 20 20 20 20 20 20 20
20 20 20 20 20 20 20 20 20 20 20 20 20 20 20 20
```

This information tells us that the first two observations (from the years 1006 and 1046) fall in the first interval (1000–1050), the third observation (from 1096) falls in the second interval (1050–1100), etc.

If we were to use the values in `Intervals` in the ar1 model (see the model specification below) then all the 23 observations between 1950 and 2000 are in the 20th interval, and would be considered as coming from the same time point by the ar1 model.

```
> f3 <- CO2 ~-1 + Core + f(Intervals, model="ar1")
```

When we applied models with a spatial correlation we defined a two-dimensional mesh. In some of the meshes the sampling locations were inside a triangle, and weighting factors w_{ik} were defined for observation i that were proportional to the distance of a sampling location to each triangle. The fitted values were then given by a projector matrix $\boldsymbol{A}$ times these waiting factors; $\boldsymbol{A} \times \boldsymbol{w}$. We will do something similar here. We make a one-dimensional mesh depending on the knot values, and these are used to calculate weighting factors.

```
> mesh1d <- inla.mesh.1d(loc = Knots)
> A4     <- inla.spde.make.A(mesh = mesh1d,
                             loc = IC$Year)
```

```
> dim(A4)
```

```
84 19
```

The projector matrix A4 has 84 rows (one for each observation in the data set) and 19 columns (corresponding to the knots). The matrix A4 contains the weighting factors that are as follows.

```
> head(A4, 10)
```

```
1.00 0.00 .    .    .    .    .    .    .    . . .
1.00 0.00 .    .    .    .    .    .    .    . . .
0.08 0.92 .    .    .    .    .    .    .    . . .
.    0.08 0.92 .    .    .    .    .    .    . . .
.    .    0.08 0.92 .    .    .    .    .    . . .
.    .    .    0.08 0.92 .    .    .    .    . . .
.    .    .    .    .    0.46 0.54 .    .    . . .
.    .    .    .    .    .    0.26 0.74 .    . . .
.    .    .    .    .    .    0.26 0.74 .    . . .
.    .    .    .    .    .    .    0.08 0.92 . . .
```

Each row in this matrix corresponds to an observation, and each column to a knot. The dots represent zeros. The weighting factors are defined as follows. Weighting factors for observations in the outer intervals are set to

1. On the first two lines we have examples of this; the weighting factor of 1 for these two years (1006 and 1046) refer to the first knot.

The third row has the values 0.08 and 0.92. These are the weighting factors for the third observation (from the year 1096). It has a weighting factor for knot 1 and also for knot 2 (because these values are in columns 1 and 2). Why the values of 0.08 and 0.92? The year 1096 falls in the interval 1050–1100, which is based on the knots 1050 and 1100. The weighting factors are (inversely) proportional to the distance between the sampling year and the knots. The year 1096 is close to the knot 1100, and is relatively far away from 1050. The 0.02 is calculated as 1 – 46/ 50, and the 0.98 comes from 1 – 4 / 50 = 0.92. This process is then repeated for every observation.

The trend is defined on the knot positions. Once the model is finished we end up with the posterior mean values of the ar1 trend on the 19 knots; see Figure 14.18. The solid line is the ar1 trend. In the spatial models we used $\boldsymbol{A} \times \boldsymbol{w}$ to get the fitted values, where $\boldsymbol{w}$ was the spatial field. For the time series we use $\boldsymbol{A} \times$ trend to get the fitted values, where the trend is the line in Figure 14.18. So, the fitted values for 1006 are given by $1 \times$ the trend value at the first knot, plus a core effect. The same holds for the fitted value for 1046. The fitted value for the third observation (from the year 1096) is given by $0.02 \times$ the trend value of the first knot plus $0.98 \times$ the trend value of the second knot plus a core effect.

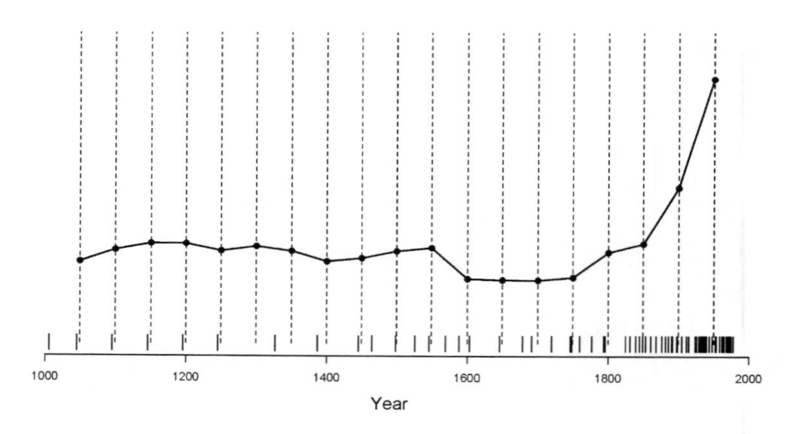

Figure 14.18. Knot positions (vertical dotted lines), observed data points (|), and posterior mean trend values at the knots (solid line). The R code to create this figure is available on the website for this book.

Because there are other terms in the model we need the `inla.stack` function to specify that core effect is linked to the observed data points and the temporal uses a projector matrix A4. The only problem is that the `inla.stack` function doesn't like factors, which is the reason that we manually convert the categorical variable Core (with three levels) into two dummy variables.

The trend at time 1 takes care of the role of the intercept, which is the reason that we dropped it from the model.

```
> f4 <- CO2 ~ -1  + Core2 + Core3 +
               f(Year, model = "ar1")
> X <- data.frame(
          Core2 = model.matrix(~Core,data=IC)[,2],
          Core3 = model.matrix(~Core,data=IC)[,3])
> StackFit <- inla.stack(
              tag = "Fit",
              data = list(CO2 = IC$CO2MixingRatio),
                A = list(1, A4),
                effects = list(
                     X = X,
                     Year = Knots))
> I4 <- inla(f4,
              control.compute = list(dic = TRUE),
              control.predictor = list(
                A = inla.stack.A(StackFit)),
                family = "gaussian",
                data = inla.stack.data(StackFit))
```

The information for the trends can be found in

```
> I4$summary.random$Year
```

The whole problem with this data set is that the majority of the observations are concentrated towards the 1800–1950 period. It makes more sense to focus the analysis only on this time period or get more cores covering data from earlier time periods. For what it is worth, the posterior mean values of the trend with 95% credible intervals are given in Figure 14.19.

The pattern in the graph matches the well-described industrial effect on CO_2 mixture in the literature. It is just a pity that no more air bubbles from earlier time periods were analysed.

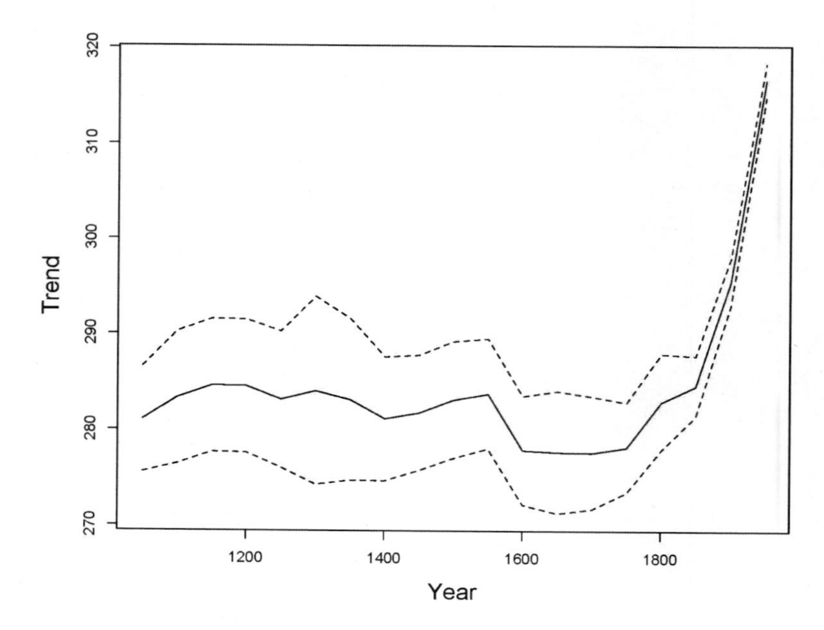

Figure 14.19. Posterior mean values and 95% credible intervals for the trend. The R code to create this figure is available on the website for this book.

15 Spatial-temporal models for orange crowned warblers count data

In Chapters 12 and 13 we explained how to fit models with a spatial correlated random effect. In Chapter 14 we illustrated the use of auto-regressive models for time series. It should not come as a surprise that in this chapter we combine the spatial and temporal elements of Chapters 12–14 and focus on the analysis of spatial-temporal data.

Prerequisite for this chapter: We assume that you are familiar with the principle of spatial-temporal correlation as explained in Chapter 6, Section 6.3. You also need to be familiar with the material discussed in Chapters 12– 14.

15.1 Introduction

The data used in this chapter are taken from Sofaer et al. (2014), who investigated the effect of competition and predation on the numbers of offspring in a population of breeding orange-crowned warblers (*Oreothlypis celata*) on Santa Catalina Island, California, US. The number of young produced in a nest was modelled as a function of population density and annual rainfall. Sampling took place from 2003 to 2009. Figure 15.1 shows the sampling locations for each of the 7 years.

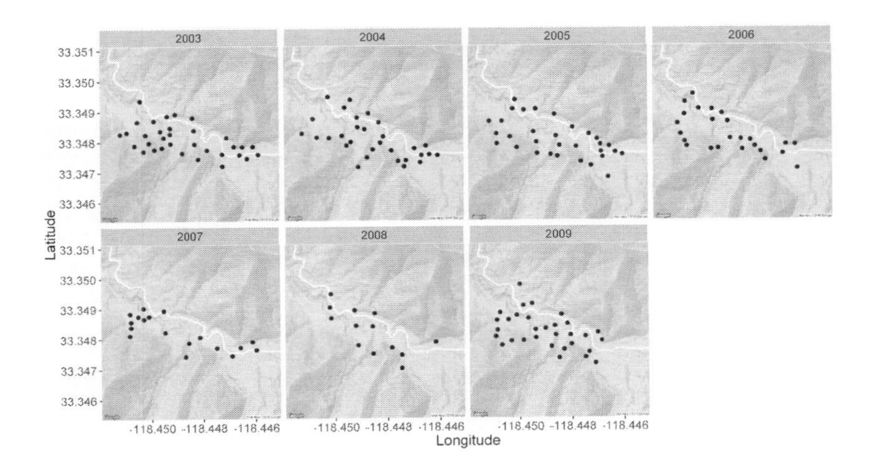

Figure 15.1. Sampling locations on Santa Catalina Island by year.

The R code to create Figure 15.1 is as follows. First we import the data with the `read.table` function. We load the required packages and use the `get_map` function from `ggmap` (Kahle and Wickham 2013) to visualise the sampling locations using Google maps. The rest is a matter of some standard `ggplot2` coding.

```
> OCW <- read.table(file = "OCWarblers.txt",
                    header = TRUE,
                    dec = ".")
> library(ggplot2)
> library(ggmap)
> library(INLA)
> source(file = "HighstatLibV10.R")
> glgmap <- get_map(
              location = c(-118.453, 33.346,
                           -118.444, 33.350),
              maptype = "terrain", zoom = 17)
> p <- ggmap(glgmap)
> p <- p + geom_point(aes(Lon, Lat), data = OCW)
> p <- p + xlab("Longitude") + ylab("Latitude")
> p <- p + facet_wrap(~ Year, nrow = 2)
> p
```

The graph indicates that the sampling locations and the number of observations differ per year. The latter can be also seen from the following `table` output.

```
> table(OCW$Year)

2003 2004 2005 2006 2007 2008 2009
  33   30   31   25   18   13   31
```

We have two covariates, breeding density and annual precipitation. These have the same value for all sampling locations within a year.

We have counts on the number of fledged chicks per nest from 7 years. In each year around 13–33 nests were sampled. There are two covariates.

15.2 Poisson GLM

To investigate whether there is a relationship between the number of fledged chicks and breeding density and annual precipitation, Zuur et al. (2016a) started with the following generalised linear model (GLM) with a Poisson distribution.

$$\begin{aligned} &FL_i \sim Poisson(\mu_i) \\ &\log(\mu_i) = \beta_1 + \beta_2 \times BreedingD_i + \beta_3 \times Rain_i \end{aligned} \quad (15.1)$$

FL_i is the number of fledged chicks for observation i, which is assumed to follow a Poisson distribution with mean μ_i. The mean is modelled as a function of breeding density and rainfall. We have 181 observations, hence $i = 1, \ldots, 181$.

The R code to execute the Poisson GLM in a frequentist setting with the `glm` function and in a Bayesian setting using JAGS is presented in Chapter 4 of Zuur et al. (2016a). In later sections of this chapter we will apply models with spatial and with spatial-temporal correlation using R-INLA. In order to compare models with and without such correlation structures, we execute the model in Equation (15.1) with R-INLA in this section.

We first standardise the two covariates to avoid potential numerical estimation problems. The function `MyStd` standardises each covariate (mean delete and divide by the standard deviation) and is taken from our support file `HighstatLib10.R`.

```
> OCW$BreedingDens.std<-MyStd(OCW$BreedingDensity)
> OCW$Rain.std         <-MyStd(OCW$Precip)
```

We execute the model in R-INLA.

```
> f1 <- NumFledged ~ 1 + BreedingD.std + Rain.std
> I1 <- inla(f1,
             family = "poisson",
             data = OCW,
             control.compute = list(dic = TRUE))
```

The first thing we need to do is model validation, for which we need Pearson residuals, and to calculate these we need the expected (i.e. fitted) values. The fitted values are in the `summary.fitted.values` object. Pearson residuals are defined as observed values minus the expected values, divided by the square root of the variance (which are the expected values again as this is a Poisson GLM).

```
> Fit1 <- I1$summary.fitted.values[1:181, "mean"]
> E1   <- (OCW$NumFledged - Fit1) / sqrt(Fit1)
```

Overdispersion — Approach 1

As part of the model validation process we need to verify whether the Poisson distribution is a valid choice. In a frequentist analysis this is done by calculating the dispersion statistic and comparing its value to one. A value large than one indicates overdispersion, and a value smaller than one means underdispersion. The dispersion statistic is calculated as the sum of squared Pearson residuals divided by sample size minus the number of parameters. This expression originates from a (frequentist) chi-square statistic, and one can ask the question whether it is valid to use the same expression in a Bayesian analysis. Although frequentist and Bayesian analysis are based on rather different philosophies, using diffuse priors in the Bayesian analysis means that the results of the two analyses will be similar. So, if the dispersion statistic obtained by a Bayesian analysis (with diffuse priors) is 1.58 (see below), then we expect that a frequentist analysis gives a similar value (and indeed it does for this data set), and both indicate overdispersion trouble.

```
> sum(E1^2) / (nrow(OCW) - 3)
```

```
1.587006
```

Overdispersion — Approach 2

When using Markov chain Monte Carlo (MCMC) techniques we can simulate numbers of fledged chicks from the model in each MCMC iteration. To assess whether the model is over- or underdispersed we can compare the variation in the MCMC simulated data with that in the observed data, and this leads to a Bayesian *p*-value. Examples of this approach can be found in Ntzoufras (2009) or Zuur et al. (2014; 2016a). As a second approach to assess over- or underdispersion for a Poisson GLM fitted with R-INLA, we can also simulate data from the model with R-INLA. To do this we need to add the option `config = TRUE` to the `control.compute` argument. We could have added this the first time we executed the model, but from a didactical point of view we prefer to do things step by step.

```
> I1 <- inla(f1,
             family = "poisson",
             data = OCW,
             control.compute = list(
                  config = TRUE,
                  dic = TRUE))
```

In Chapter 13 we used the function `inla.posterior.sample` to simulate parameters from posterior distributions. Here, we will simulate 1,000 times a set of three regression parameters (one intercept and two slopes).

```
> set.seed(12345)
> NSim <- 1000
> Sim  <- inla.posterior.sample(n = NSim,
                                result = I1)
```

We can calculate fitted values for each of these 1,000 sets of parameters. To do this we need a matrix with covariate values.

```
> Xm <- model.matrix(~BreedingD.std + Rain.std,
                     data = OCW)
```

Once we have the fitted values for each set of simulated regression parameters, we simulate the number of fledged chicks using the `rpois` function. The block of code below creates space for 1,000 simulated data sets.

```
> N     <- nrow(OCW)
> Ysim <- matrix(nrow = N, ncol = NSim)
> mu.i <- matrix(nrow = N, ncol = NSim)
> Xm    <- as.matrix(Xm)
```

We execute a loop in which we extract the three simulated regression parameters, calculate the fitted values, and simulate number of fledged chicks. Once the loop is finished, each column in Ysim contains a simulated data set of 181 observations. And we have 1,000 such simulated data sets.

```
> for (i in 1:NSim){
    Betas      <- Sim[[i]]$latent[c(182, 183, 184)]
    mu.i[,i] <- exp(Xm %*% Betas)
    Ysim[,i] <- rpois(n = nrow(OCW),
                          lambda = mu.i[,i])
  }
```

For this specific example rows 182–184 contain the simulated regression parameters, but for a data set of different size these numbers will change. Just type Sim[[1]]$latent to see which rows you need. See also Chapter 10.

We can do many things with the simulated data. For example, we can count the percentage of zeros in each simulated data set and compare that to the observed number of zeros. Or we can calculate the dispersion statistic for each simulated data set and compare that to the dispersion statistic for the observed data. Or we can make an average frequency table for the simulated data set and compare that to the frequency table of the observed data. All three approaches indicate problems for the model. Here we only show the results of the last option that we mentioned; the average frequency table for the simulated data and the frequency table for the observed data are given in Figure 15.2. It seems that the model does not generate enough zeros, it simulates too many ones, too many threes, and too many higher values.

The expression for the dispersion statistic finds its origin in frequentist analysis. There is no harm in applying the same expression in a Bayesian environment. But there is no harm either in backing up any conclusions with a simulation study. See also Zuur et al. (2016b).

Other model validation steps (e.g. plotting Pearson residuals versus fitted values, and plotting Pearson residuals versus each covariate in the model) did not indicate any problems. A sample variogram of the residuals was inconclusive. Because there is overdispersion, there is no point in presenting the posterior mean values and 95% credible intervals of the regression parameters.

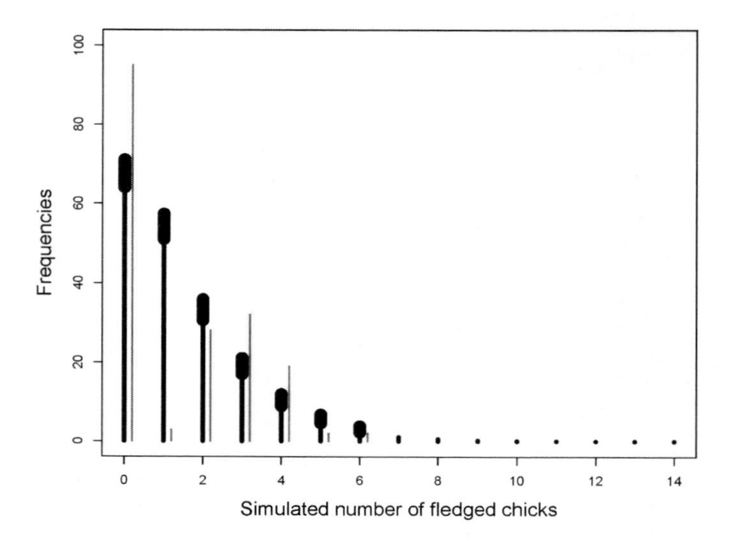

Figure 15.2. Simulated frequencies (thick lines) and observed frequencies (thinner and red lines). The large dots on top of a line for the simulated frequencies represent the standard deviation in the simulated values.

15.3 Model with spatial correlation

Results presented in the previous section indicated problems with the Poisson GLM; the model produced not enough zeros, too many ones, and the variation in the simulated data was larger than that in the observed data. A sample variogram of the Pearson residuals did not immediately set off the 'spatial correlation' alarm bells, so it may seem odd to continue the analysis with a model that contains a spatial correlation term. However, the reason to continue the analysis along this path comes partly from biological knowledge, partly from knowing the results of the models presented in Zuur et al. (2016a), and partly because we are working towards a model with spatial-temporal correlation. As to the biological motivation to add spatial correlation to the model, the distance between the sites is small and it is likely that the numbers of fledged chicks at nearby nests are similar. The reason for this may be the presence of unmeasured micro-scale covariates (e.g. some parts of the study area may provide better food or shelter from predators), competition, social status of birds, etc. In addition, we only have covariates that change over time and this means that any variation within a year contributes towards overdispersion. The spatial correlation term therefore captures real spatial dependency but also unmeasured spatial covariates.

Models with a spatial correlated random effect were discussed in Chapters 12 and 13. Recall that in such a model we add a random effect u_i to the model.

$$
\begin{aligned}
&FL_i \sim Poisson(\mu_i) \\
&\log(\mu_i) = \beta_1 + \beta_2 \times BreedingD_i + \beta_3 \times Rain_i + u_i \qquad (15.2) \\
&\boldsymbol{u} \sim GMRF(0, \boldsymbol{\Omega})
\end{aligned}
$$

The only new component is the spatial correlated random effect u_i. We will have 181 of these; $\mathbf{u} = (u_1, \ldots, u_{181})$. As explained in Chapters 12 and 13, we do not estimate the u_is directly. Instead we define a mesh, which contains L nodes. Using this mesh, small Lego pieces are calculated. These form the spatial field and we call these Lego pieces the w_ks. The Matérn correlation function is used to parameterise its covariance matrix. A projector matrix $\mathbf{A}$ links the spatial field $\mathbf{w} = (w_1, \ldots w_L)$ to the random effects $\mathbf{u} = (u_1, \ldots, u_{181})$, as sketched below.

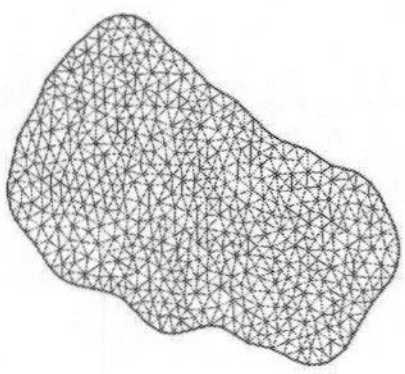

$$
\times A = \begin{pmatrix} u_1 \\ u_2 \\ \\ u_{180} \\ u_{181} \end{pmatrix}
$$

Let's do this in R. We will follow the steps enumerated in Figure 12.2. It probably does no harm to reproduce these steps again.

1. Make a mesh.
2. Define the weighting factors a_{ik} (also called the projector matrix $\mathbf{A}$).
3. Define the SPDE.
4. Define the spatial field.
5. Make a stack to tell R-INLA at which points on the mesh we sampled the response variable and the covariates.
6. Specify the model formula in terms of the response variable, covariates, and the spatial correlated term.
7. Run the spatial model in R-INLA.
8. Inspect the results.

Let us start with the mesh. Technical details for doing this were discussed in Chapter 12. We tried a variety of different meshes, and we eventually settled for the one in Figure 15.3. The selected mesh has 571 nodes. This means that we will end up with a spatial field consisting of 571 elements $\mathbf{w} = (w_1, \ldots w_{571})$.

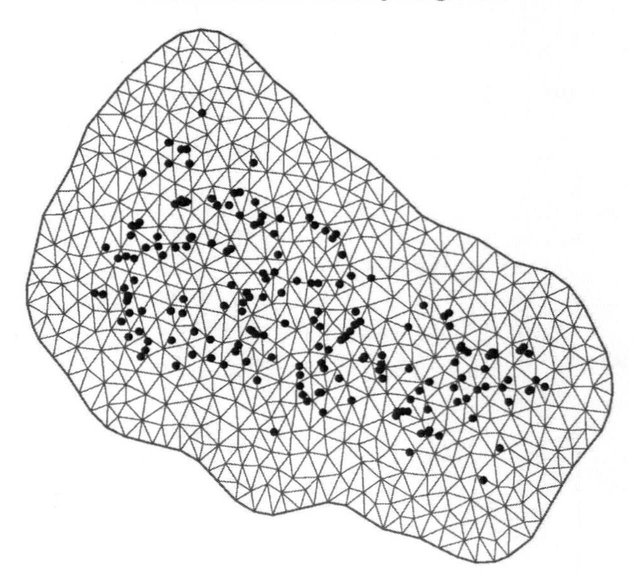

Figure 15.3. Mesh for the 181 sampling locations, which are presented as dots. The mesh has 571 nodes.

The R code to create the mesh is as follows.

```
> Loc       <- cbind(OCW$Xloc, OCW$Yloc)
> ConvHull <- inla.nonconvex.hull(Loc)
> mesh      <- inla.mesh.2d(boundary = ConvHull,
                      max.edge = c(30), cutoff   = 1)
```

You may wonder why we selected this mesh. Or formulated differently, why do we use `max.edge = c(30)` and `cutoff = 1`? We first made a histogram of the distances between all nests to get an impression of the distances between sites; see also Figure 13.6A. The smaller the `max.edge` value, the smaller the triangles that make up the mesh, the finer the mesh, and more accurate the R-INLA results. Using `max.edge = c(30)` and `cutoff = 1` means that the mesh has 571 nodes. Computing time for models using this mesh is in terms of seconds, which is good for a book chapter as some readers reproduce the results (and don't want to wait half a day for the computer to finish). But when writing a final draft of a manuscript we suggest using a finer mesh. It is also an option to investigate the effect of the mesh size of the results of the model (i.e. DIC, posterior mean values, and 95% CIs).

Once we have the mesh we define the projector matrix. This matrix links the elements of the spatial random field consisting of 571 nodes (the w_1, ..., w_{571}) to the 181 random intercepts u_i. Because this is the second model that we execute, we call the projector matrix `A2`.

```
> A2 <- inla.spde.make.A(mesh, loc = Loc)
```

The dimensions of A2 are as follows.

```
> dim(A2)
181 571
```

We suggest that you memorise the numbers 181 and 571 for the remaining part of this section.

We can rewrite Equation (15.2) using matrix notation to

$$\mathbf{FL} \sim Poisson(\boldsymbol{\mu})$$
$$\log(\boldsymbol{\mu}) = \mathbf{X} \times \boldsymbol{\beta} + \mathbf{u} \qquad (15.3)$$
$$\mathbf{u} = \mathbf{A}_2 \times \mathbf{w}$$

where $\mathbf{FL} = (FL_1, \ldots, FL_{181})$. Note the difference between the μ and u terms! The first one represents the mean and the second one represents the spatial correlated random effects. $\mathbf{X}$ is a matrix with three columns; the first column contains the value 1 (for the intercept), and the second and third columns contain the observed breeding density and rainfall values. The vector $\boldsymbol{\beta} = (\beta_1, \beta_2, \beta_3)$ contains the three regression parameters and $\mathbf{A}_2$ contains all the a_{ik} values.

In order to define the SPDE, we specify some of the parameters of the Matérn correlation function.

```
> spde <- inla.spde2.matern(mesh, alpha = 2)
```

We also define the spatial field, which we call w2.

```
> w2.index <- inla.spde.make.index('w', mesh$n)
```

Next, we define a data frame containing the covariates.

```
> N <- nrow(OCW)
> X <- data.frame(Intercept = rep(1,N),
                  BreedingD.std = OCW$BreedingD.std,
                  Rain.std = OCW$Rain.std)
```

Finally, we make a stack to match information defined on the mesh (the spatial random field) and the response variable and covariates.

```
> Stk2 <- inla.stack(
              tag = "Fit",
              data = list(y = OCW$NumFledged),
                A = list(A2, 1),
                effects = list(
                     w2.index,
                     X))
```

We now have all the ingredients to execute the model in R-INLA.

```
> f2 <- y ~ -1 + Intercept + BreedingD.std +
             Rain.std +
             f(w, model = spde)
> I2 <- inla(f2,
              family = "poisson",
              data = inla.stack.data(Stk2),
              control.compute = list(dic = TRUE),
              control.predictor = list(
                          A = inla.stack.A(Stk2)))
```

There is no point in spending lots of time focussing on the results of the model if it is not an improvement of the Poisson GLM in Equation (15.1). We therefore compare the DICs first.

```
> c(I1$dic$dic, I2$dic$dic)
```

```
541.0309 525.0761
```

The difference between the DICs from the model without and the model with the spatial correlation is about 16. The lower the DIC, the better the model. But there is no golden rule *how* much lower a DIC has to be in order for the model to be really better. A safe threshold is 10.

In this case there seems to be support for the model with the spatial correlated random effect. Hence, we continue with the validation of this model. We first obtain the fitted values, and calculate the Pearson residuals and the dispersion statistic.

```
> F2 <- I2$summary.fitted.values[1:181, "mean"]
> E2 <- (OCW$NumFledged - F2) /sqrt(F2)
> sum(E2 ^2)  / (N - 3 - 2)
```

```
1.024774
```

The dispersion statistic is close to 1; that is good. However, a dispersion statistic that is larger than 1 indicates problems, but a dispersion statistic close to 1 does not imply that there are no other problems. We may still have non-linear patterns and dependency issues. Hence, a detailed model validation is still required.

We used three regression parameters and two parameters for the spatial random field; hence the '–3' and '–2' when calculating the dispersion statistic. The '–2' for the spatial random field is a rather quick and dirty approach. There is an ongoing debate in the mixed-effects literature how to quantify degrees of freedom of random effects. The same problem applies to spatial correlated random effects!

The second approach to verify overdispersion (or underdispersion) is to carry out a simulation study similar to the one in Section 15.2. The function `inla.posterior.sample` will not only simulate regression parameters but also spatial random fields. Hence, if we simulate 1,000 sets of regression parameters, we also get 1,000 spatial random fields, which are added to the simulated mean values. Results are not presented here, but these are comparable to the ones presented in Figure 15.2. Hence adding the spatial random field to the Poisson GLM does give a lower DIC, but we still have problems (the model produces not enough zeros, too many ones, and larger variation than observed).

Instead of presenting the posterior mean values and 95% credible intervals of the regression parameters here, we present them later in this chapter, together with those obtained by the other models. We will do the same with the hyperparameters.

The model also provides the posterior mean values of the spatial random field. These are the w_1, ..., w_{571} values. Using standard tools for presenting three-dimensional data (e.g. the `levelplot` from the `lattice` package) we can plot them; see Figure 15.4. The contour plot is obtained by interpolating. It is also possible (and recommended) to make a contour plot of the posterior standard deviations of the spatial random field.

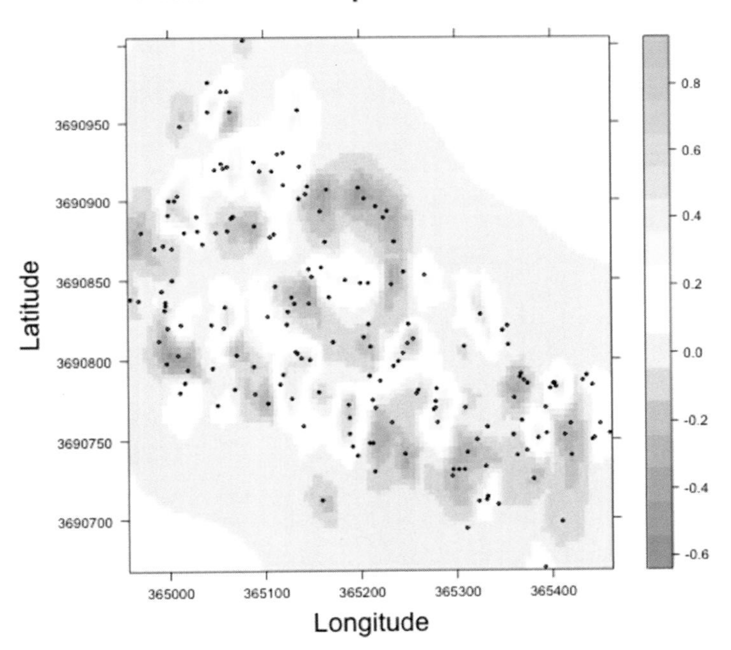

Figure 15.4. Posterior mean values of the spatial random field. These are the w_ks interpolated. There are small hotspots of higher values for the spatial random field.

The spatial random field in Figure 15.4 shows hotspots of positive values of around 0.8. This value does not sound large, but keep in mind that we use a log link, which means that we can write the model as

$$\begin{aligned}\mu_i &= e^{\beta_1+\beta_2\times BreedingD_i+\beta_3\times Rain_i+u_i} \\ &= e^{u_i}\times e^{\beta_1+\beta_2\times BreedingD_i+\beta_3\times Rain_i}\end{aligned} \qquad (15.4)$$

A value of u_i = 0.8 at a nest in one of these hotspots has an expected number of fledged chicks that are around exp(0.8) ≈ 2 larger than at a nest not in such a hotspot.

The sample variogram of the Pearson residuals (`E2`) does not give any indication to suspect any remaining spatial correlation.

15.4 Spatial-temporal correlation: AR1

15.4.1 Why do it?

You may ask why this chapter continues with a section on spatial-temporal correlation because the results in the previous section did not show any remaining spatial correlation in the Pearson residuals. The answer is again (partly) driven by biology. The spatial random field in Figure 15.4 represents real dependency but also unmeasured covariates. Why would the effect of these two components be constant over time? Perhaps the spatial random field is changing over time! Another reason for digging deeper into the dependency structure is that the sample variogram was applied on all 181 Pearson residuals. But there is a blocking structure in these residuals; they come from 7 years. If the data contained more spatial observations per year, then ideally we should make sample variograms for the residuals of each year. That would give us seven variograms.

In this section we will present tools that allow the spatial random field to change over time. We will present three scenarios. In the first and second scenarios we will put some structure on how the spatial random field changes over time. In the third approach there is no structure on how it changes over time.

The models that we will use can cope with different numbers of locations sampled in each year. It is even possible to have a complete mismatch of sampling locations between the years. For the orange-crowned warbler data, sampling locations do indeed differ between the years.

15.4.2 Explanation of the model

We introduce a model in which the spatial random field is allowed to change over time. There are various ways to implement a spatial-temporal correlation structure into our model and in this subsection we will use the residual autoregressive correlation of order 1 (AR1) for the temporal part of the model. We already discussed the underlying mathematical expressions for AR1 correlation in Chapter 6, Section 6.3, and we also presented a small simulation study. Before reading on we suggest that you get familiar with the material in Section 6.3.

The model with the spatial-temporal AR1 correlation is specified as follows.

$$\begin{aligned} &FL_{tj} \sim Poisson(\mu_{tj}) \\ &\log(\mu_{tj}) = \beta_1 + \beta_2 \times BreedingD_{tj} + \beta_3 \times Rain_{tj} + v_{tj} \\ &v_{tj} = \phi \times v_{t-1,j} + u_{tj} \end{aligned} \tag{15.5}$$

FL_{tj} is the jth observation in year t, where t = 1, ...7 (we have 7 years of data), and j = 1, ..., n_t. The n_ts are the number of observations in year t and were presented in the `table` output in Section 15.1. The subscript t = 1 corresponds to the year 2003.

The new part in the model is the $v_{tj} = \phi \times v_{t-1,j} + u_{tj}$. We first give a non-technical explanation. The map below shows a weather forecast for Europe. Suppose that it is for today. On television the forecast is sometimes shown as a movie. Multiple maps are presented in sequence; one for today, one for tomorrow morning, one for tomorrow afternoon, etc. The second author of this book lives in Spain. If it is hot today then it will most likely be hot tomorrow. The first author lives in Scotland; if it rains today, it snows tomorrow, and it will be hot the day after.

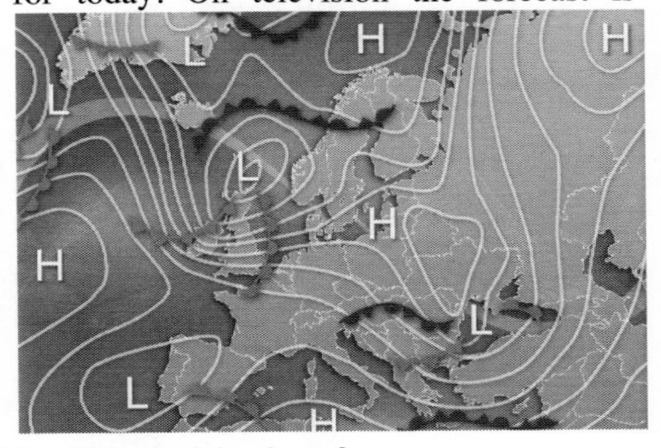

To model the weather for tomorrow we need to figure out how much it depends on today's weather. This is the auto-regressive part, which is quantified via the parameter ϕ. If it is close to 1 then it is a slow-moving weather system; tomorrow depends very much on today. If it is close to 0 then the weather system does not have an auto-regressive memory.

The weather story explains the first two components in the expression $v_{tj} = \phi \times v_{t-1,j} + u_{tj}$. The u_{tj} is a spatial random field that is not correlated over time. It is a pure spatial process. Each day it causes local variations in the weather. It causes cold weather in the north of Europe and warm weather in the south. But it changes randomly from day to day. It is as if someone were sitting on the clouds in Scotland affecting only the Scottish weather and someone else is sitting on the South-European clouds affecting the weather here. But whoever it is, every day is a different story.

To summarise this story: The slow-moving weather maps are the v_{tj}s, the ϕ

determines how much tomorrow depends on today, and the local man in the clouds with his hammer causes spatial variation and is the u_{tj}.

15.4.4 Simulating a spatial-temporal AR random field

In case you are still confused about how exactly the spatial-temporal correlation structure works, we simulate some data in the subsection. We only focus on the AR1 part of Equation (15.5), which is this part.

$$v_{tj} = \phi \times v_{t-1,j} + u_{tj}$$

In Section (15.7) we will extend the simulation study and also add covariate effects and simulate numbers of fledged chicks. But in this subsection we only focus on the mechanics of the spatial-temporal AR1 correlation.

Because we want to produce contour plots of the simulation results, we need to have the spatial coordinates on a grid. We therefore create a 25 by 25 grid with longitude and latitude values. This is only for plotting and illustration purposes.

```
> X1 <- seq(364959, 365460, length = 25)
> X2 <- seq(3690669, 3691003, length = 25)
> xy <- expand.grid(X1, X2)
```

We now have coordinates that can be used by the `contourplot` function from the `lattice` package. Using the new coordinates we create a mesh. The `max.edge` and `cutoff` values are identical to the mesh of the orange-crowned warblers, but that is just for convenience.

```
> Loc1 <- cbind(xy[,"Var1"], xy[,"Var2"])
> ConvHull1 <- inla.nonconvex.hull(Loc1)
> meshN      <- inla.mesh.2d(boundary = ConvHull1,
                             max.edge = c(30),
                             cutoff   = 1)
```

We now start with the core part of the R code for the simulation (everything up to now was for `contourplot` convenience). The online SPDE manual, which is available from the INLA website, comes with an R support file. We need to source this file. We will simulate data for nine time units (e.g. days or years). This is the `k` in the code below. The SPDE tutorial file that we just sourced has a function `rspde`; for a given set of locations, mesh, and Matérn correlation parameters it simulates a spatial correlated random field. These are the u_{tj}s. Because we are using a set of 625 spatial coordinates, this random field has 625 values. We have chosen the Matérn correlation parameters such that we have a strong spatial correlation. So, we have now simulated one spatial random field.

```
> source("spde-tutorial-functions.R")
> set.seed(1)
> k <- 9
```

```
> u.t <- rspde(coords = Loc1,
               kappa = 0.34 * 0.5,
               variance = 0.001,
               n = 1,
               mesh = meshN,
               return.attributes = TRUE)
```

Next we choose the value for ϕ. In the first instance we use $\phi = 0.9$, which means that we will end up with nine highly temporal correlated random fields.

```
> phi <- 0.9
```

We are now ready to execute a loop that calculates the v_{tj}s for nine different time points. First we create the space to store the v_{tj} values.

```
> v <- matrix(nrow = 625, ncol = 9)
> v[,1] <- u.t
```

With AR1 models there is always some trouble as to what to do at time $t = 1$; we set the v_{1j} values to the `u.t` that we just simulated. We then execute a loop. In each iteration t of the loop we use the `rspde` function to simulate a spatial random field; these are the u_{tj}s and consist of 625 values. And these u_{tj}s are strongly spatially correlated. In each iteration we have a different simulated spatial field. The Matérn correlation parameters can be changed to increase the magnitude of the u_{tj} values and also the extent of the spatial correlation. The v_{tj}s are then calculated using the AR1 equation.

```
> for (t in 2:9){
     u.t <- rspde(coords = Loc1,
                  kappa = Kappa * 0.5,
                  n = 1 ,
                  variance = 0.001,
                  mesh = mesh,
                  return.attributes = TRUE)
   v[,t] <- phi * v[,t-1] + u.t  #*sqrt(1-phi^2)
  }
```

Each column in `v` contains a temporal correlated random field.

The spatial random field consisting of the u_{tj}s is *simulated.* The temporal correlated random field consisting of the v_{tj}s is *calculated* using the AR1 expression. Note the difference between 'simulated' and 'calculated'.

If we put the nine temporal correlated random fields in one long vector and copy the spatial coordinates and time unit next to it, then we can use the contourplot function from the `lattice` package to produce Figure 15.5. See the R code below.

```
> MyData <- data.frame(
                 x    = rep(Loc1[,1], k),
                 y    = rep(Loc1[,2], k),
                 z    = as.vector(v),
                 Year = factor(rep(1:k, each = 625)))
> contourplot(z ~ x + y | Year,
                 data = MyData,
                 cuts = 5, pretty = TRUE,
                 region = TRUE, labels = FALSE)
```

There are nine panels in Figure 15.5. Note that sequential panels are rather similar! This is because we used ϕ = 0.9, as if the weather is changing rather slowly from day to day.

Next, we investigate the situation when the temporal correlation is small. We reran the code with

```
> phi <- 0.1
```

The resulting graph is presented in Figure 15.6. Note that sequential panels show rather different patterns; there is hardly any temporal correlation. There is strong spatial correlation in each panel though. That is due to the u_{ij}s.

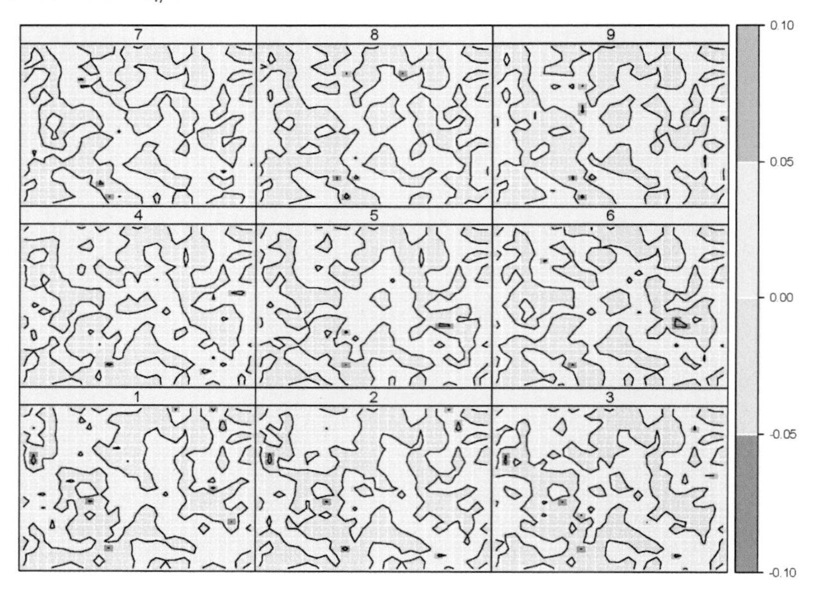

Figure 15.5. Temporal spatial random field with ϕ = 0.9.

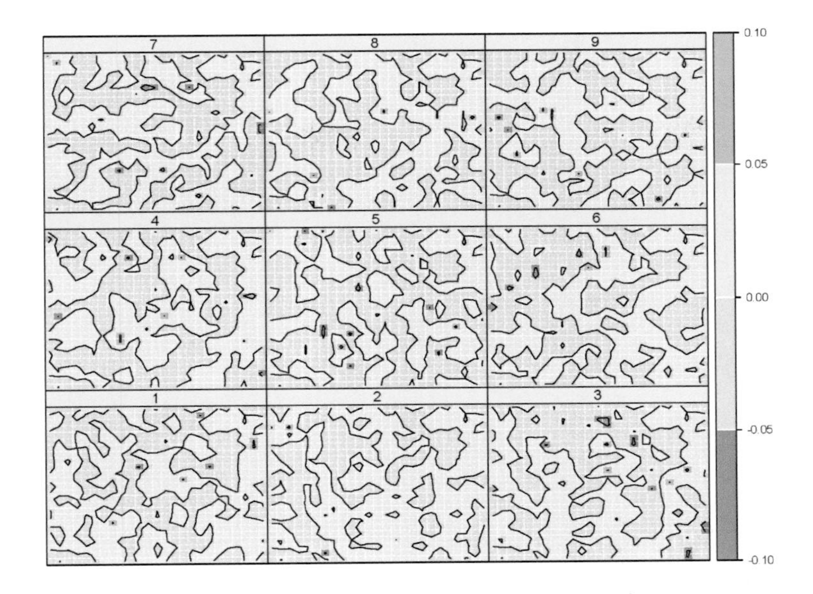

Figure 15.6. Temporal spatial random field with $\phi = 0.1$.

For plotting purposes we simulated data on a 25 by 25 grid. We originally tried to do it for the 181 sampling locations for the orange-crowned warbler data, but this gave us all kinds of contour plotting trouble. R-INLA applies the AR1 process on w_ks of the mesh. This also means that there are no problems if certain nests are not observed in certain years.

It can be quite useful to change some of the Matérn correlation parameters and settings in our simulation study and see what happens. A common question we face is what happens if ϕ is negative; use $\phi = -0.9$ and see what happens in the contour plots.

You may have noticed that we hashed out the sqrt(1-phi^2) in the AR1 code. You need to add this if you want to apply R-INLA on the simulated data and compare the original and estimated parameters. The reason for using sqrt(1-phi^2) is explained in the helpfile of the AR1 model in R-INLA.

When we run R-INLA we don't choose parameters and simulate random effects like we did above. Instead R-INLA will estimate the parameters as described in Chapter 7.

15.4.5 Implementation of AR1 model in R-INLA

Let's fit the model in Equation (15.5) in R. The code to run the model requires small modifications from the code for the spatial Poisson model that was presented in the previous section.

We first make a variable `Group3` with the values 1 1 1 … 2 2 2 … 3 3 3 … 4 4 4, etc. The value of 1 is for observations from 2003, the 2s are for

2004, and the 7s are for the 2009 data. In this case the data are sorted by year, which explains why we have sequential numbers. But there is no need to work with sorted data.

```
> Group3  <- OCW$Year - 2002
> NGroups <- length(unique(Group3))
```

We have seven groups (NGroups). Now comes the crucial part. When making a projector matrix, which we will call A3, we need to tell R-INLA which observations belong to a specific year. This is done with the group argument in the function inla.spde.make.A.

```
> A3 <- inla.spde.make.A(mesh,
                      group = Group3,
                      loc = Loc)
```

The structure of this matrix is rather confusing and in the next subsection we will discuss it in more detail.

The function inla.spde.make.index defines the spatial random field, which we will call w3.index. We need to specify the number of groups.

```
> w3.index <- inla.spde.make.index(
                        name = 'w',
                        n.spde = mesh$n,
                        n.group = NGroups)
```

The w3.index$w.group object tells the inla function which rows in the data belong to the same year, and there is no need to sort the data in blocks of the same year.

Next we define the stack; it links the spatial field and covariates to the observed data.

```
> Stk3 <- inla.stack(
            tag = "Fit",
            data = list(y = OCW$NumFledged),
             A = list(A3, 1),
             effects = list(
                  w3.index,
                  X))
```

And we define the formula. The new parts are the group argument and the control.group argument with the ar1 argument. It tells R-INLA that we have a whole group of observations that follow the temporal AR1 correlation structure.

```
> f3 <- y ~ -1 + Intercept + BreedingD.std +
           Rain.std +
           f(w,
             model = spde,
             group = w.group,
```

```
                control.group = list(model = 'ar1'))
```

We are now ready to run the model using nearly identical code as in the previous subsection.

```
> I3 <- inla(f3,
            family = "poisson",
            data = inla.stack.data(Stk3),
            control.compute = list(dic = TRUE),
            control.predictor = list(
                A = inla.stack.A(Stk3)))
```

The DICs indicate that the model with the spatial-temporal correlation is slightly better than the model with the spatial correlation. The difference is less than 10, so the gain of adopting a considerably more complicated model is small!

```
> c(I1$dic$dic, I2$dic$dic, I3$dic$dic )

541.0309 525.3635 519.3477
```

Pearson residuals are obtained in exactly the same way as before. As to the dispersion statistic, calculating the number of parameters consumed by the spatial-temporal correlated random effects is open for debate. There is the parameter ϕ, and two Matérn correlation parameters for the spatial random field for the u_{ij}s. It is probably easier to simulate data from the model (using exactly the same code as before) and compare this with the observed data.

Figure 15.7 shows the temporal correlated spatial random fields for each year. These are the w_ks that are used to calculate the v_{ij}s. Compare it with the movie of the weather forecast for the next couple of days.

The grey-scale version of the graph in the printed book is not visually appealing. The online R code can be used to produce a nice colour version.

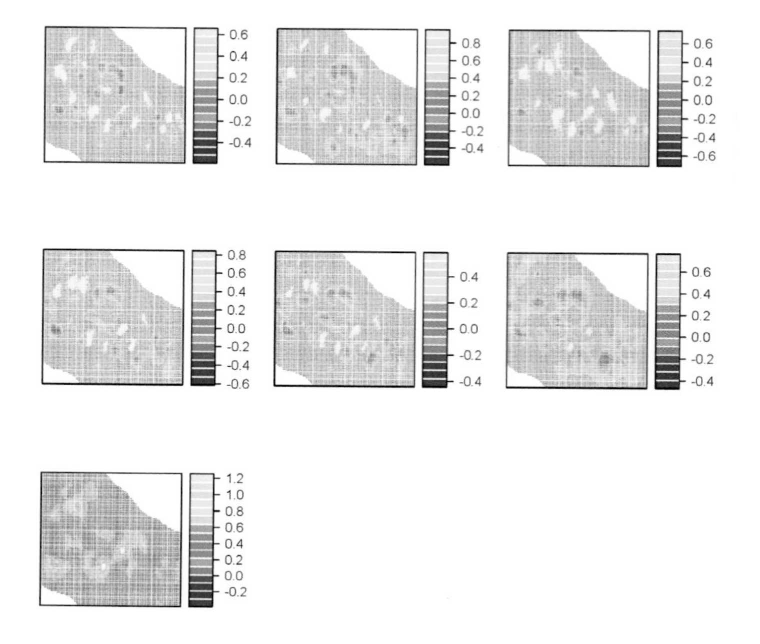

Figure 15.7. Temporal correlated spatial random fields for each year. Top left: 2003. Top right: 2005. Middle left: 2006. Bottom left: 2009.

15.4.6 More detailed information on the code

In this subsection we provide a more detailed explanation of the R code that we used in the previous subsection. This section is only relevant if you want to better understand the setup of the projector matrix, and if you want to create Figure 15.7. Upon first reading you may decide to skip this subsection.

Let us inspect the dimension of the projector matrix A3.

```
> dim(A3)
```

```
181 3997
```

The A3 has 181 rows (just like the projector matrix A2 from the previous subsection and this is the observed number of data points) and 3997 columns. The value of 3997 is equal to 7×571, where the 571 is the number of nodes in the mesh. The projector matrix A3 contains a projector matrix for the data of each year and it can be written as

$$A3 = \begin{bmatrix} A3_1 & A3_2 & \cdots & A3_7 \end{bmatrix}$$

By taking the first 571 columns of A3

```
> A3_1 <- A3[, 1:571]
```

we obtain the projector matrix with 181 rows and 571 columns. There are 33 rows in A3_1 that have non-zero values. These 33 rows correspond to the 2003 data. The row sums for these 33 rows are all equal to 1 (it is a projector matrix after all). The other 148 rows in A3_1 are filled with zeros (those sites were not sampled in 2003). Similar rules apply for columns 572 to 1142, which correspond to the 2004 data.

```
> A3_2 <- A3[,572:1142]
```

Now there are 30 rows with non-zero values (the 2004 data), and 151 rows with only zeros (the non-2004 data). It is good to know that at the time of the writing of this book a package is in preparation that will relieve you of the projector matrix misery.

Just as in Chapters 12 and 13 we can use the projector matrix and the spatial random field(s) to calculate the temporal correlated random effects v_{tj} at the sampling locations.

```
> A3m   <- as.matrix(A3)
> w.pm <- I3$summary.random$w$mean
> v     <- A3m %*% w.pm
> length(v)
```

```
181
```

The v contains the 181 v_{tj}s. The first 33 of these are for 2003, the second 30 for 2004, etc. We have plotted them in Figure 15.8.

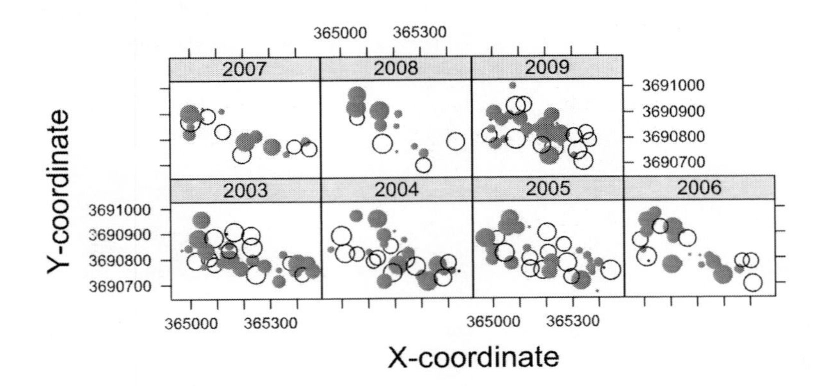

Figure 15.8. Spatial random effects v_{tj}. Open circles are used for negative random effects and filled circles for positive random effects. The size of a dot is proportional to the posterior mean value of v_{tj}.

The w3.index$w contains 3997 values; the numbers 1 to 571 are repeated seven times. So, the first block of 571 numbers are for 2003, the second block of 571 numbers is for 2004, etc. We can access the posterior mean values of the spatial-temporal random field as follows (see also above).

```
> w.pm <- I3$summary.random$w$mean
> length(w.pm)
```

```
3997
```

R-INLA gives a spatial random field consisting of 3997 values, but in fact these correspond to seven blocks of 571 values. The `w.pm` contains these seven random fields in stacked format. Just as in Chapters 12 and 13 we can plot these.

```
> wproj <- inla.mesh.projector(mesh,
                  xlim = range(Loc[,1]),
                  ylim = range(Loc[,2]))
> w7 <- list()
> for (i in 1:7){
    w7[[i]] <- inla.mesh.project(
                      wproj,
                      w.pm[w3.index$w.group==i])
  }
```

The `w3.index$w.group` is a long vector with the values 1 1 1 ... 2 2 2 ... 3 3 3 ... etc., and it identifies to which year a row in `w.pm` belongs. The `w7` contains all seven interpolated spatial random fields. We plotted these in Figure 15.7. In most analyses the prime interest is on the fixed parameters (breeding density and rainfall), but the patterns in Figure 15.7 are worthwhile to include in a paper. It is your task as an ecologist to come up with a nice explanation of the observed patterns.

We can also obtain the posterior mean value of ϕ via

```
> phi <- I3$summary.hy[3,"mean"]
> phi
```

```
0.6526637
```

The 95% credible interval for ϕ is quite large in this example, namely from –0.12 to 0.97. The mode is 0.92; hence it is worthwhile plotting the density curve. See Figure 15.9. The fact that the density curve is so wide is quite worrying. We suspect that the small number of years may account for this. In Section 15.7 we will outline a simulation study that can be used to assess the effect of the number of years on the quality of the estimates of ϕ.

Figure 15.9 was created using the following R code.

```
> plot(I3$marginals.hyper[[3]],
       xlab = expression(phi),
       ylab = 'Density', type='l')
> abline(v = rho, col = 2)
```

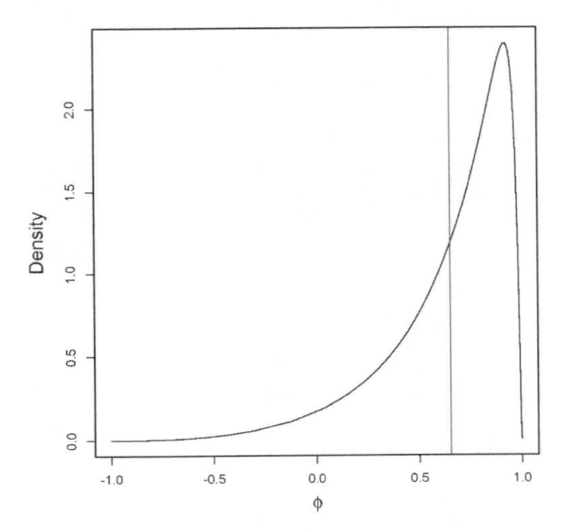

Figure 15.9. Posterior density function for ϕ. The vertical line represents the posterior mean.

15.5 Spatial-temporal correlation: Exchangeable

Another spatial-temporal correlation model that we will consider is the so-called exchangeable correlation. This model allows for spatial correlation and also temporal correlation, but the latter is not necessarily between consecutive years. Exchangeable correlation was presented in Equation (3.8); the correlation between any two temporal random effects v_{ti} an v_{si} is equal to ϕ. This correlation structure only makes sense for shorter time series.

The implementation of this model is as follows. First we define the equation. It is nearly identical to that of the AR1 mode, except that 'AR1' is replaced by 'exchangeable'. The projector matrix, stack, and `inla` command are identical as before.

```
> f4 <- y ~ -1 + Intercept + BreedingD.std +
            Rain.std +
            f(w,
              model = spde,
              group = w.group,
              control.group = list(
                 model = 'exchangeable'))
> I4 <- inla(f4,
             family = "poisson",
             data = inla.stack.data(Stk3),
             control.compute = list(dic = TRUE),
             control.predictor = list(
               A = inla.stack.A(Stk3)))
```

```
> c(I1$dic$dic, I2$dic$dic, I3$dic$dic,
    I4$dic$dic)
```

```
541.0309 525.3637 519.3477 521.7565
```

The DIC indicates that the model with the AR1 correlation is still the better one. Accessing the results and plotting the seven spatial random fields are similar to the previous subsection.

15.6 Spatial-temporal correlation: Replicated

The last model with spatial-temporal correlation that we consider in this chapter allows the spatial random field to be replicated in each year. This means that the spatial random field in each year is allowed to be different, but all spatial random fields are based on the same hyperparameters. This is essentially the residual AR1 correlation structure with $\phi = 0$. The underlying model is given by

$$\begin{aligned} &FL_{tj} \sim Poisson(\mu_{tj}) \\ &\log(\mu_{tj}) = \beta_1 + \beta_2 \times BreedingD_{tj} + \beta_3 \times Rain_{tj} + v_{tj} \end{aligned} \quad (15.6)$$

The v_{tj} term is the spatial correlated random effect. We will have seven of them, and each can be different. To gain insight into how this model works we suggest you carry out the simulation study presented in Subsection 15.4.4 and set $\phi = 0$.

The R-INLA implementation of this model is as follows. Instead of the `group` option we now use the `repl` option, which stands for replicate.

```
> Repl5 <- OCW$Year - 2002
> NRepl <- length(unique(Repl5))
> A5 <- inla.spde.make.A(mesh,
                         repl = Repl5,
                         loc = Loc)
> w5.index <- inla.spde.make.index(name = 'w',
                                   n.spde = mesh$n,
                                   n.repl = NRepl)
> Stk5 <- inla.stack(
             tag = "Fit",
             data = list(y = OCW$NumFledged),
               A = list(A5, 1),
               effects = list(
                    w5.index,
                    X))
> f5  <- y ~ -1 + BreedingD.std + Rain.std +
             f(w,
               model = spde,
               replicate = w.repl)
> I5 <- inla(f5,
             family = "poisson",
```

```
            data = inla.stack.data(Stk5),
            control.compute = list(
               dic = TRUE),
               control.predictor = list(
                   A = inla.stack.A(Stk5)))
```

Just as in the previous subsection we can plot seven random fields. Instead of spatial fields that slowly change from year to year as in Figure 15.7, we now have seven spatial fields that can change quite drastically.

We compare all models fitted so far in this section.

```
> c(I1$dic$dic, I2$dic$dic, I3$dic$dic,
    I4$dic$dic, I5$dic$dic)

541.0309 525.3637 519.3477 521.7565 525.2472
```

The best model among the models that we have fitted in this section is the spatial-temporal model with AR1 correlation.

15.7 Simulation study

The model with the AR1 correlation was deemed the best model, but the temporal component of the spatial-temporal correlated random effect (ϕ) is only based on 7 years. This is rather small for our liking. How do you know whether the small number of years is indeed a problem for the model? The odd pattern and wide range of the density curve of ϕ is a first clue. Warning messages during the estimation process may be another clue (though we did not get these).

In this section we sketch the layout of a simulation study that one could use to study the effect of sample size (or better: number of years) on modeling results. The core idea of this simulation study was taken from Section 10.1 in the online R-INLA tutorial on SPDE models (Krainski et al. 2016).

Suppose that instead of 7 years we have 25 years of data on the number of fledged chicks. Let K be the number of years.

```
> K <- 25
```

We will use the results from the model presented in Subsection 15.4.5 (the AR1 model) for the simulation study. This means that we need to extract its estimated regression parameters, the auto-regressive parameter, and the parameters for the Matérn correlation function.

```
> beta <- I3$summary.fixed[,"mean"]
> phi  <- I3$summary.hy[3,"mode"]
> SpFi.w <- inla.spde2.result(inla = I3,
                              name = "w",
                              spde = spde,
                              do.transfer = TRUE)
> Kappa <- inla.emarginal(function(x) x,
                SpFi.w$marginals.kappa[[1]])
```

As mentioned earlier in this chapter, the online SPDE tutorial (Krainski et al. 2016), which is available on the INLA website, comes with a support file that we source.

```
> source("spde-tutorial-functions.R")
```

This support file has a function `rspde`, which simulates a spatial random field for a given mesh, sampling locations, and Matérn correlation function.

```
> set.seed(12345)
> u.t <- rspde(coords = Loc,
               kappa = Kappa,
               n = K,
               mesh = mesh,
               return.attributes = TRUE)
```

We now have $K = 25$ random fields. These are spatially correlated and not temporal. They are the u_{tj}s in Equation (15.5). We can now use the following construction to calculate the temporal correlated spatial fields.

```
> v <- u.t #needed for year 1
> for (t in 2:K){
    v[,t] <- phi * v[,t-1]+u.t[,t] * sqrt(1-phi^2)
  }
```

These are the v_{tj}s, and we could plot them like we did in Figure 15.8. Now that we have the temporal correlated spatial fields we can add covariates and simulate the number of fledged chicks. Covariates can be simulated in a variety of ways. Here we keep it simple and draw K values from a uniform distribution (recall that breeding density and rainfall only had one value per year). For your own data you may want to adopt a better approach for this step. For example, you could investigate the effect of drier years or higher breeding density effects.

```
> BD <- runif(K,
              min = min(OCW$BreedingD.std),
              max = max(OCW$BreedingD.std))
> RF <- runif(K,
              min = min(OCW$Rain.std),
              max = max(OCW$Rain.std))
```

The next step is to calculate the mean value μ_{ti} and simulate data from a Poisson distribution with mean μ_{ti}. In this process we use the regression parameters from Section 15.3.

```
> mu <- matrix(nrow = 181, ncol = K)
> Y  <- matrix(nrow = 181, ncol = K)
> for (t in 1:K){
    mu[,t] <- exp(beta[1] + beta[2] * BD[t] +
                  beta[3] * RF[t] + v[,t])
```

```
    Y[,t] <- rpois(181, lambda = mu[,t])}
```

We now combine all the simulated data.

```
> MyData <- data.frame(Y    = as.vector(Y),
                       RF   = rep(RF, each = 181),
                       BD   = rep(BD, each = 181),
                       Xloc = rep(Loc[,1], K),
                       Yloc = rep(Loc[,2], K),
                       Year = rep(1:K, each = 181)
                       )
```

The data frame `MyData` contains a response variable (`Y`), two covariates (`RF` and `BD`), spatial coordinates (`Xloc` and `Yloc`), and the `Year`. It has 181 observations per year, which obviously is unrealistic. From the simulated data set we take a subset so that we have approximately 30 observations per year.

```
> N       <- nrow(Loc)
> Isel    <- sample(1:(N*K), K * 30)
> MyData2 <- MyData[Isel,]
```

The data set in `MyData2` is comparable to what was sampled by Sofaer et al. (2014), except it is for 25 years. The model in Equation (15.5) can now be applied on the data in `MyData2` using the same R code as in Subsection 15.4.5.

Depending on the focus of the simulation study we can change certain aspects of the data. For example, we noticed that the density curve for ϕ was rather large (Figure 15.9). This raises the question how the number of years affects the quality of the posterior distribution of ϕ. To answer this question we could drop x numbers of years from `MyData2`, reapply the model, and repeat this process a large number of times (e.g. 100) for different values of x. This should give an impression of how the number of years affects the posterior distribution of ϕ.

We could also investigate how the number of sampled nests in each year affects the width of the 95% credible intervals of the regression parameters. Maybe sampling 30 nests per year is not enough, or maybe this number can be reduced.

In this type of simulation studies there are no golden rules for comparing results and judging good from bad. It is just a matter of common sense, but it can be very insightful. It may even be argued that one should do it before conducting fieldwork.

15.8 Discussion

In this chapter we considered various models that contain a random intercept that captures the spatial or spatial-temporal correlation. The following five models were considered and executed in R-INLA.

1. There is no spatial or spatial-temporal correlation (this is an ordinary Poisson GLM).
2. There is spatial correlation but it is constant over time (this is the spatial Poisson model with the SPDE).
3. There is spatial correlation and it is correlated between consecutive years (this is the spatial Poisson model with the temporal AR1).
4. There is spatial correlation; it is correlated in time, but not necessarily between *consecutive* years (this is the exchangeable correlation).
5. There is spatial correlation, and it changes randomly between the years (this is the spatial Poisson model with the temporal AR1, but with $= 0$).

Posterior mean values and 95% credible intervals obtained by all techniques discussed in this chapter are presented in Figure 15.10. We also rerun the AR1 model with $\phi = 0$. It gives the same results as the model with the replicate option. For this specific data set, there are hardly any differences between the results obtained by the six models. That is probably due to the small number of years and the weak temporal correlation.

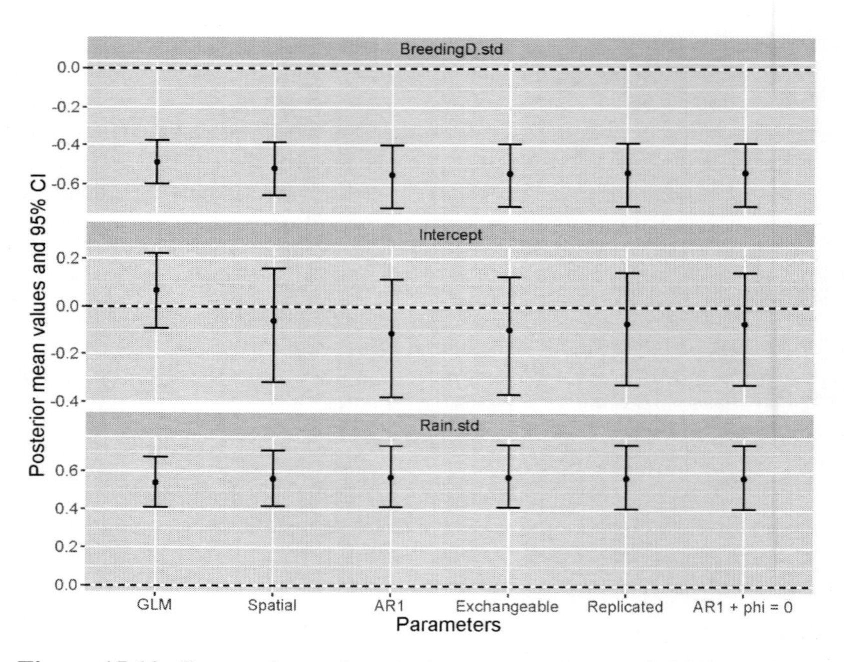

Figure 15.10. Comparison of posterior mean values and 95% credible intervals obtained by all techniques discussed in this chapter.

16 Spatial-temporal Bernoulli models for coral disease data

In this chapter we use data from Muller and van Woesik (2014) who used R-INLA to model the absence and presence of white-pox disease on *Acropora palmata* coral colonies in the Caribbean. We will use their data to illustrate the use of Bernoulli models with spatial and spatial-temporal correlation. We will minimise the technical explanation as this can be found in Chapter 15.

Prerequisite for this chapter: We assume that you are familiar with the material explained in Chapter 15. You also need to be familiar with Bernoulli GLMs; see Chapter 10.

16.1 Introduction

Infectious diseases are a major health threat for coral reefs. To protect healthy coral it is important to know whether a disease is contagious or whether it is due to environmental stress. White-pox disease, which may be caused by an infectious agent, has caused considerable coral mortality on Caribbean reefs.

Muller and van Woesik (2014) studied the absence and presence of white-pox disease on 68 *A. palmata* colonies in the US Virgin Islands. Their model contains a series of covariates that might affect the disease and also a spatial-temporal component that may indicate whether the disease is contagious (i.e. whether it was transmitted to nearby neighbours).

Between February 2003 and December 2009, 68 reef colonies in Haulover Bay on the northeast side of St. John, US Virgin Islands (Figure 16.1) were sampled on a monthly basis. The paper uses the following covariates: Northing, Easting, distance to nearest neighbour (DIST), previous incidence of disease (PIC), distance to previous infected colony (DFPIC), temperature (TEMP), solar insulation (IRR), and colony size (SIZE).

Data exploration indicates that there are no covariates with extremely small or large values, and there is no collinearity except for easting and northing. Muller and van Woesik (2014) did include these two variables as covariates in the fixed part of the model. This opens a more general discussion as to whether we should fit a model in which spatial coordinates are used as covariates and also use a spatial random field in the model. It has been suggested on the INLA discussion board to include spatial coordinates as covariates to capture the large-scale patterns and let the spatial correlation term capture any small-scale spatial patterns; apparently, this is common practice. We feel slightly uneasy about this because including the spatial random field and the spatial coordinates as covariates in the same model may add an extra level of collinearity. We will not use the two spatial coordinates as covariates, but you are free to change the model and include them. If you do this, we suggest using the cross product of longitude and latitude as the study area follows a diagonal line. Rotating the coordinates so that the rotated coordinates are either horizontal or vertical is another option.

The data are imported with the `read.csv` function. There is some R coding involved to get the years and month of sampling, but that is not shown here.

```
> CR <- read.csv(file = "CoralDisease.csv",
                 header = TRUE)
```

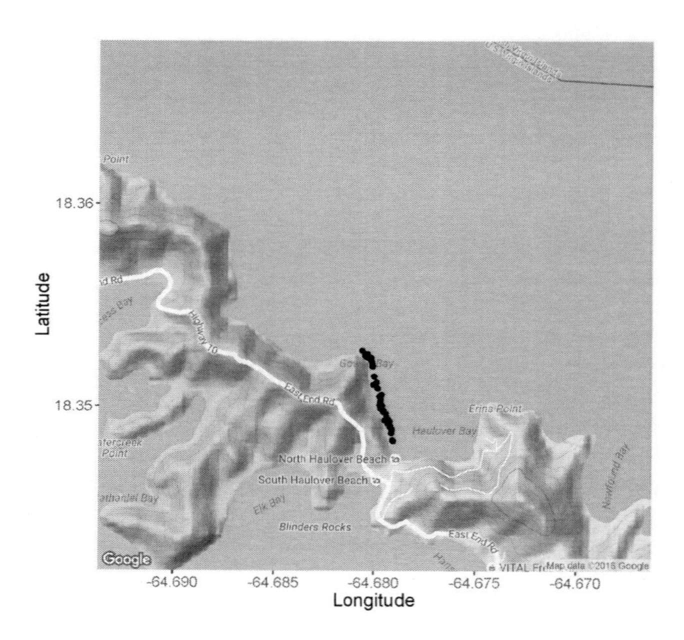

Figure 16.1. Position of 69 sampling locations.

16.2 Bernoulli model in R-INLA

Whenever we teach statistics courses on Bernoulli and binomial GLMs there is always some confusion. We therefore clarify the difference between a Bernoulli model and a binomial model before presenting the model for the coral data. If we toss a coin N times, and each time we record whether we have a head (which we label as success), and if each toss is independent, and if each toss has the same probability of a head, then we can use a binomial distribution for the number of heads (Y) out of N tosses. We can write this as $Y \sim B(\pi, N)$. If we only toss the coin once, then the number of trials is 1 and we use a Bernoulli distribution, which can be written as $Y \sim B(\pi)$. We then take multiple samples and model π_i as a function of covariates, which makes it a Bernoulli GLM.

The `glm` and `inla` functions automatically recognise that a Bernoulli distribution should be used if the response variable is a vector with zeros and ones. Executing a GLM for binomial data with the `glm` function is discussed, for example, in Zuur et al. (2013). In R-INLA a binomial GLM can be executed by specifying the number of successes as the response variable and defining `Ntrials` as the number of trials.

The response variable for the coral data is absence or presence of a disease, and this means that we need a Bernoulli distribution. We now apply a Bernoulli GLM on the coral data, and for the moment we will ignore the spatial and spatial-temporal correlation. The model is given by

$$\begin{aligned}
& WP_{ts} \sim B(\pi_{ts}) \\
& E(WP_{ts}) = \pi_{ts} \\
& \text{logit}(\pi_{ts}) = \beta_1 + DIST_{ts} \times \beta_2 + PIC_{ts} \times \beta_3 + DFPIC_{ts} \times \beta_4 + \\
& \qquad TEMP_{ts} \times \beta_5 + IRR_{ts} \times \beta_6 + SIZE_{ts} \times \beta_7
\end{aligned} \tag{16.1}$$

WP_{ts} is the absence (coded with a 0) or presence (coded with a 1) of the white-pox disease in colony s at time t, where s = 1, …, 68 and t = 1, .., 76. The 76 comes from 12 months of sampling in 7 years, but there were a few months in which no sampling took place, as can be seen from the table below.

```
> table(CR$MonthN, CR$Year)

    2003 2004 2005 2006 2007 2008 2009
 1     0   68   68   68   68    0   68
 2     0   68   68   68   68   68   68
 3    68   68   68   68   68   68   68
 4    68   68   68   68   68   68   68
 5    68   68   68   68   68   68   68
 6     0   68   68   68   68   68   68
 7    68   68   68   68    0   68   68
 8    68   68   68   68    0   68    0
 9    68   68   68   68   68   68   68
 10   68   68   68    0   68   68   68
 11   68   68   68   68   68   68   68
 12   68   68   68   68   68   68   68
```

We use standardised covariates in the model.

```
> CR$DIST.std   <- MyStd(CR$DIST)
> CR$PIC.std    <- MyStd(CR$PIC)
> CR$DFPIC.std  <- MyStd(CR$DFPIC)
> CR$TEMP.std   <- MyStd(CR$TEMP)
> CR$IRR.std    <- MyStd(CR$IRR)
```

```
> CR$SIZE.std  <- MyStd(CR$SIZE)
```

The model in Equation (16.1) can be executed with the `glm` function using

```
> M1 <- glm(WP ~ DIST.std + PIC.std + DFPIC.std +
                 TEMP.std + IRR.std + SIZE.std,
            family = binomial,
            data = CR)
```

or we can use R-INLA.

```
> library(INLA)
> f1 <- WP ~ 1 + DIST.std + PIC.std + DFPIC.std +
             TEMP.std + IRR.std + SIZE.std
> I1 <- inla(f1,
             Ntrials = 1,        #default value
             family = "binomial",
             data = CR,
             control.compute = list(dic = TRUE))
```

We typed `Ntrials = 1` in the R-INLA command below, but there is no need to do this for binary data as this is the default value. We do not want to digress too much into plain Bernoulli GLMs (we already did this in Chapter 10), and we therefore continue the analysis by adding a spatial correlation term to the model. If it turns out that the model in Equation (16.1) is the optimal model, then we will apply a model validation and model interpretation later in this chapter. Another reason for extending the model with a spatial term is that due to the binary nature of the response variables, it is rather difficult to inspect Pearson residuals of the Bernoulli GLM and assess them for spatial or spatial-temporal correlation.

16.3 Spatial correlated Bernoulli model

We now extend the model with a spatial correlated random effect. This model is given by

$$\begin{aligned}
&WP_{ts} \sim B(\pi_{ts}) \\
&E(D_{ts}) = \pi_{ts} \\
&\text{logit}(\pi_{ts}) = \beta_1 + DIST_{ts} \times \beta_2 + PIC_{ts} \times \beta_3 + DFPIC_{ts} \times \beta_4 + \\
&\qquad TEMP_{ts} \times \beta_5 + IRR_{ts} \times \beta_6 + SIZE_{ts} \times \beta_7 + u_{tj} \\
&u_{ts} \sim GMRF(0, \Sigma)
\end{aligned} \tag{16.2}$$

This is nearly the same expression as in Equation (16.1) except for the spatial correlated random intercept u_{ts}. We are now back to Section 13.5; see, for example, Equation (13.2). The fact that we now use two subscripts instead of one subscript is irrelevant with respect to the spatial correlation term, and so is the use of the Bernoulli distribution and the logistic link function. The whole approach of getting the spatial random field in a Bernoulli GLM is identical to getting it for a Gaussian model as in Chapter 12. We can follow the flowchart in Figure 12.2 to implement the SPDE approach to estimate a spatial random field $w_1, \ldots, w_L$ at the L nodes of a mesh. And once we have these L w_ks we can calculate the spatial correlated random effects u_{ts} via a projector matrix.

In terms of R coding, we need to define a mesh, the projector matrix, the spde, the spatial field, the stack, a formula, and then we can execute the model in R-INLA and inspect the results. These are steps 1– 8 in Section 12.5.

We start with the mesh. For this step it is handy to know distances between the sampling locations; see Figure 16.2. The sites are all within 500 m, and 20% of the sites are separated by less than 50 m.

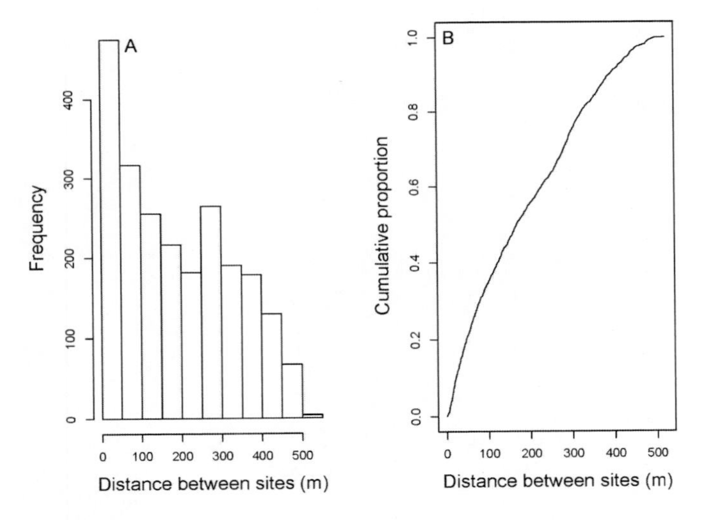

Figure 16.2. A: Histogram of distances between sites. B: Cumulative distribution of distances between sites.

The R code used by Muller and van Woesik (2014) is available online, which means that we can reproduce their mesh; see Figure 16.3. It has 504 nodes. This mesh looks slightly odd for our liking. It was obtained with the following code.

```
> mesh <- inla.mesh.create.helper(
                points = cbind(CR$Northing,
                               CR$Easting),
                offset = c(10, 140),
                max.edge = c(50, 1000),
                min.angle = c(26, 21),
                cutoff = 0,
                plot.delay = NULL)
```

The `min.angle` specifies the minimum angle in a triangle (inside and on the outer part of the mesh).

We initially went for a much finer mesh with 1000–1500 nodes. Computing time for the spatial model was still in terms of seconds, but the spatial-temporal model that will be applied in the next section, which uses the same mesh, took hours. This is because such a model uses 76 months × L nodes, which is a very large number of w_ks to estimate! If we use a mesh with 1,000 nodes, then a model with a spatial correlation needs to estimate 1,000 w_ks, but the model with a spatial-temporal correlation uses 76,000 w_ks! If you select a fine mesh then we hope you have a fast computer. We will adopt the mesh in Figure 16.3.

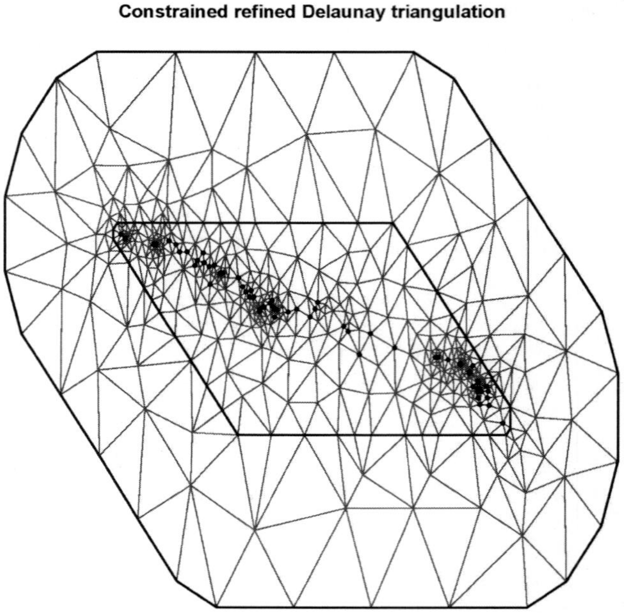

Figure 16.3. Mesh used by Muller and van Woesik (2014).

Next we make the projector matrix A2. We have 5,168 observations and the mesh has 504 nodes. Recall from Chapter 12 that the projector matrix A2 links the w_1, …, w_{504} values to the 5,168 spatial random effects u_{tj}.

```
> A2 <- inla.spde.make.A(mesh, loc = Loc)
> dim(A2)

5168  504
```

The SPDE needs to be defined and also the spatial random field (the $w_1,\ldots,w_{504}$).

```
> spde <- inla.spde2.matern(mesh, alpha = 2)
> w.index <- inla.spde.make.index(
                   name    = 'w',
                   n.spde  = spde$n.spde,
                   n.group = 1,
                   n.repl  = 1)
```

We make a date frame containing all the covariates and the intercept.

```
> N <- nrow(CR)
> X <- data.frame(Intercept = rep(1, N),
                  DIST.std  = CR$DIST.std,
                  PIC.std   = CR$PIC.std,
                  TEMP.std  = CR$TEMP.std,
                  DFPIC.std = CR$DFPIC.std,
                  IRR.std   = CR$IRR.std,
                  SIZE.std  = CR$SIZE.std)
```

The stack links the spatial random field on the mesh and the covariates at the sampling locations to the response variable WP.

```
> Stk2 <- inla.stack(
             tag  = "Fit",
             data = list(y = CR$WP),
               A  = list(A2, 1),
               effects = list(
                    w = w.index,
                    X = as.data.frame(X)))
```

We define the formula, which is nearly the same as for the model in Section 16.2, except for the last term, the spatial random field.

```
> f2 <- y ~ -1 + Intercept + DIST.std + PIC.std +
            DFPIC.std + TEMP.std + IRR.std +
            SIZE.std + f(w, model = spde)
```

And we finally execute the model in R-INLA.

```
> I2 <- inla(f2,
             family = "binomial",
             data = inla.stack.data(Stk2),
             control.compute = list(dic = TRUE),
             control.predictor = list(
                          A = inla.stack.A(Stk2)))
```

The DIC indicates that the model with the spatial random field is better, indicating that there is spatial correlation.

```
> c(I1$dic$dic, I2$dic$dic)

3093.927 2823.660
```

Before spending time on model interpretation we investigate whether the spatial random field is constant over time or whether it changes over time.

16.4 Spatial-temporal correlated Bernoulli model

We now extend the model by allowing for a spatial-temporal correlated random effect; see Equation (16.3).

$$\begin{aligned} & WP_{ts} \sim B(\pi_{ts}) \\ & E(WP_{ts}) = \pi_{ts} \\ & \text{logit}(\pi_{ts}) = \beta_1 + DIST_{ts} \times \beta_2 + PIC_{ts} \times \beta_3 + DFPIC_{ts} \times \beta_4 + \\ & \qquad TEMP_{ts} \times \beta_5 + IRR_{ts} \times \beta_6 + SIZE_{ts} \times \beta_7 + v_{ts} \\ & v_{ts} = \phi \times v_{t-1,s} + u_{ts} \end{aligned} \tag{16.3}$$

We focus on the explanation of the spatial-temporal random effect. For each month t we end up with 68 random effects $\mathbf{v_t} = (v_{t1}, \ldots, v_{t,68})$. These 68 random effects are equal to ϕ times $\mathbf{v_{t-1}}$ (the 68 random effects from month $t - 1$) plus a purely spatially correlated term $\mathbf{u_t} = (u_{t1}, \ldots, u_{t,68})$. If ϕ is close to 1 then there is a strong temporal correlation. It is an option (though not our favourite) to include a temporal smoother in the fixed part of the model to capture the global temporal patterns and let the spatial-temporal correlated random field capture the deviations from it.

The implementation of this model in R-INLA is as follows. We have 76 sampling months and for the moment we will assume that these are regularly spaced with no gaps. We need to specify a vector with values 1 1 1 . . . 2 2 ... 3 3 3 ... 76 76 76, though not necessarily in this order.

We can execute the model in Equation (16.3) with the following code. We first define the vector MonthSeq, which defines the sampling month. We can't use Time as it has gaps and doesn't start on month 1. The number of groups is 76, and the projector matrix `A3` is created with the `inla.spde.make.A` function. Although not relevant for running the code and getting the output, recall from Chapter 15 that `A3` essentially consists of 76 mini projector matrices. They are all next to each other.

```
> CR$MonthSeq <- as.numeric(as.factor(CR$Time))
> NMonths <- length(unique(CR$MonthSeq))
> A3 <- inla.spde.make.A(mesh,
                          loc = Loc,
                          group = CR$MonthSeq,
                          n.group = NMonths)
```

We create the spatial random field. Again, it is not immediately relevant to know but the `w.st$group` variable defines the 76 spatial random fields.

```
> w.st <- inla.spde.make.index(
                    name    = 'w',
                    n.spde  = spde$n.spde,
                    n.group = NMonth)
```

We define the stack matrix to link the spatial random field to the observed data.

```
> Stk3 <- inla.stack(
              tag  = "Fit",
              data = list(y = CR$WP),
              A    = list(A3, 1),
              effects = list(
                     w.st,
                     X = as.data.frame(X)))
```

The model formula contains the group argument.

```
> f3 <- y ~ -1 + Intercept + DIST.std +
            PIC.std + DFPIC.std +
            TEMP.std + IRR.std + SIZE.std +
            f(w,
              model = spde,
              group = w.group,
              control.group = list(model='ar1'))
```

Now we are ready to run the model. Little did we realise the first time that we ran the code that computing time would be hours. That is because 76 spatial random fields $\mathbf{w}_1$ to $\mathbf{w}_{76}$ are estimated, and each spatial random field contains 504 nodes. This means that for month 1 we end up with $\mathbf{w}_1 = (w_{1,1}, \ldots, w_{1,504})$, for month 2 we have $\mathbf{w}_2 = (w_{2,1}, \ldots, w_{2,504})$, and for month 76 we have $\mathbf{w}_2 = (w_{76,1}, \ldots, w_{76,504})$. Combined this means that $76 \times 504 = 38{,}304$ parameters are estimated. That is a lot!

```
> I3 <- inla(f3,
           control.inla = list(
                  reordering = "metis"),
           family = "binomial",
           data=inla.stack.data(Stk3),
           control.compute = list(dic = TRUE),
           control.predictor = list(
                  A = inla.stack.A(Stk3))
           )
```

The first thing we do is to compare the DIC of the three models and see which one is the best. It shows that there is strong support for the model with the spatial-temporal correlation.

```
> c(I1$dic$dic, I2$dic$dic, I3$dic$dic)

3093.927 2823.661 2684.870
```

The next thing we are interested in is to see how the parameters for the fixed terms differ from the three models. We have plotted them next to each other in Figure 16.4. The widths of the 95% credible intervals for the model with the spatial-temporal correlation tend to be wider, and this is what we expected. It also means that basing the conclusions and providing advice on the results of the GLM or the model with the spatial correlation is likely to result in errors. For example, the temperature effect in the spatial-temporal model is not important whereas in the other two models it is (incorrectly).

All models show an important size effect and a previous incidence of disease effect.

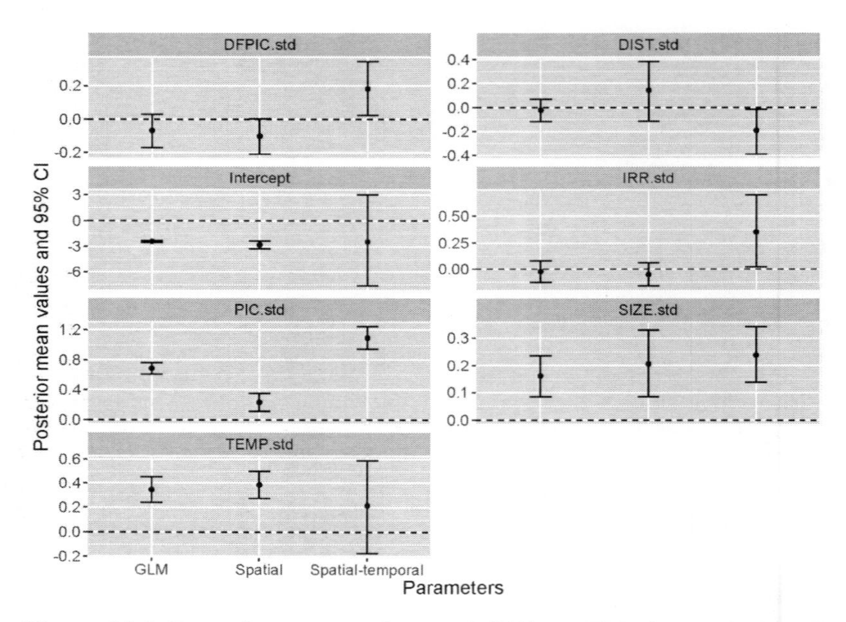

Figure 16.4. Posterior mean values and 95% credible intervals for all regression parameters obtained by the model without correlation, with spatial correlation and with spatial-temporal correlation.

We show the full numerical output of the model with the spatial-temporal correlation. Time used is in seconds. You can shorten this by using a mesh with fewer nodes (or get a faster computer).

```
Time used:
 Pre-processing    Running inla Post-processing      Total
          4.199      16944.0013           3.351  16951.552

Fixed effects:
             mean    sd  0.025q    0.5q   0.975q   mode   kld
Intercept -2.434  3.096  -7.655  -2.496   3.007 -2.576  0.012
DIST.std  -0.190  0.095  -0.390  -0.185  -0.017 -0.175  0.000
PIC.std    1.095  0.075   0.945   1.096   1.240  1.096  0.000
DFPIC.std  0.181  0.081   0.023   0.180   0.345  0.179  0.000
```

```
TEMP.std    0.214  0.192  -0.171   0.216    0.589  0.220  0.000
IRR.std     0.351  0.171   0.024   0.348    0.699  0.341  0.000
SIZE.std    0.238  0.051   0.137   0.238    0.339  0.239  0.000

Random effects:
Name       Model
 w    SPDE2 model

Model hyperparameters:
                  mean    sd     0.025q    0.5q     0.975q   mode
Theta1 for w     3.230 0.340     2.585   3.220     3.922  3.188
Theta2 for w    -5.436 0.221    -5.887  -5.429    -5.019 -5.409
GroupRho for w   0.982 0.022     0.922   0.989     0.999  0.999

Expected number of effective parameters(std dev):
127.00(16.67)
Number of equivalent replicates: 40.69

Deviance Information Criterion (DIC) ...: 2684.87
Effective number of parameters .........: 125.33

Marginal log-Likelihood:  -1450.03
Posterior marginals for linear predictor and fitted values
computed
```

Note that the posterior mean value of ϕ is close to 1, indicating strong temporal correlation. We now face the daunting task of visualising 76 spatial random fields. One option is to make a jpg file for the spatial random field for each year and put the 76 jpg files in a movie. Obviously, this is impractical for a book, but it would be useful on a website. Somehow we managed to get all 76 random fields into one graph; see Figure 16.5. If you see this graph in colour then there is a clear correlation over time. Large negative values (purple in case you have the colour version of the graph) of the spatial field correspond to low probability of presence of white-pox disease, whereas positive values (blue) corresponds to large probability of presence of white-pox disease. The graph can be improved by adding the coastline and clipping out the part of the spatial random field that covers land.

To get an idea how well (or badly) the model is performing, we suggest carrying out a simulation study like we did in Chapter 13. For this you need to add the following code to the `inla` function

```
control.compute = list(config = TRUE)
```

and rerun the code again. Then you can create 1000 simulated sets for each regression parameter and also for the spatial random field with the following code.

```
> Sim <- inla.posterior.sample(n = 1, result = I3)
```

You can now compare the observed numbers of zeros and ones with the simulated zeros and ones, and get some sense of the variation in this. Alternative tools for a sensitivity analysis are described in Roos and Held (2011).

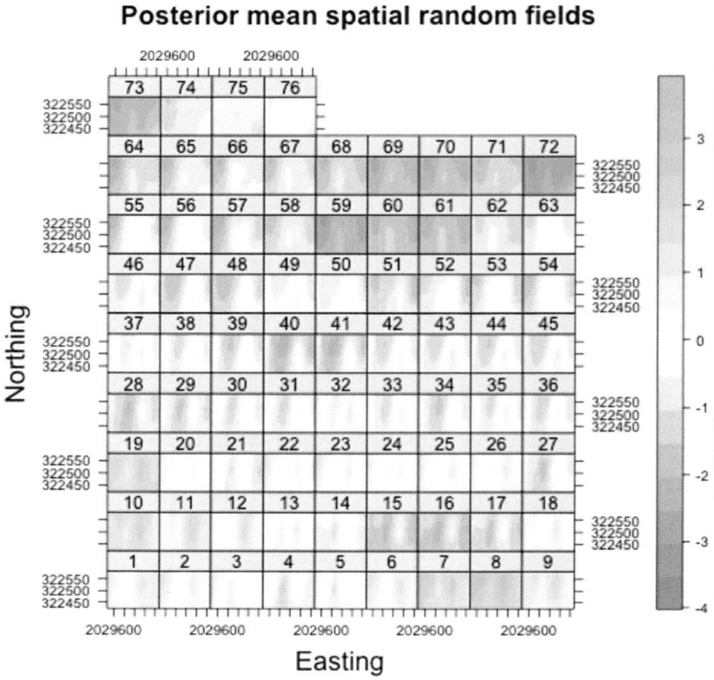

Posterior mean spatial random fields

Figure 16.5. Spatial random fields for sampling months 1 to 76. The R code to create this graph is on the website for this book.

References

Akaike H (1973). Information theory as an extension of the maximum likelihood principle. Pp. 267–281 in Petrov BN, Csaki F (eds). *Second International Symposium on Information Theory*. Budapest: Akadémiai Kiadó.

Bakka H, Vanhatalo J, Illian J, Simpson D, Rue H (2016). Accounting for physical barriers in species distribution modeling with non-stationary spatial random effects. arXiv:1608.03787.

Barbraud C, Weimerskirch H (2006). Antarctic birds breed later in response to climate change. *Proceedings of the National Academy of Sciences USA* 103: 6048–6051.

Bates D, Maechler M, Bolker B, Walker S (2015). Fitting linear mixed-effects models using lme4. *Journal of Statistical Software* 67: 1–48. doi:10.18637/jss.v067.i01.

Berridge DM, Crouchley R (2011). *Multivariate Generalized Linear Mixed Models Using R*. Boca Raton, FL: CRC Press.

Bivand R, Keitt T, Rowlingson B (2016). `rgdal`: Bindings for the Geospatial Data Abstraction Library. R package version 1-1-10. http://CRAN.R-project.org/package=rgdal.

Blangiardo M, Cameletti M (2015). *Spatial and Spatio-temporal Bayesian Models with R – INLA*. Chichester: John Wiley & Sons.

Bolker BM (2008). *Ecological Models and Data in R*. Princeton, NJ: Princeton University Press.

Bouriach FS, Samraoui F, Souilah R, Houma I, Razkallah I, Alfarhan AH, Samraoui B (2015). Does core-periphery gradient determine breeding performance in a breeding colony of White Storks *Ciconia ciconia*? *Acta Ornithologica* 50: 149–156.

Burnham KP, Anderson DR (2002). *Model Selection and Multimodel Inference,* 2nd ed. Berlin: Springer.

Carroll R, Lawson AB, Faes C, Kirby RS, Aregay M, Watjou K (2015). Comparing INLA and OpenBUGS for hierarchical Poisson modeling in disease mapping. *Spatial and Spatio-temporal Epidemiology* 14–15: 45–54.

Chatfield C (2003). *The Analysis of Time Series: An Introduction*, 6th ed. London: Chapman & Hall.

Cressie N (1993) *Statistics for Spatial Data,* rev. ed. Wiley: New York.

Crozier LG, Scheuerell MD, Zabel RW (2011). Using time series analysis to characterize evolutionary and plastic responses to environmental change: A case study of a shift toward earlier migration date in sockeye salmon. *The American Naturalist* 178: 755–773.

Cruikshanks R, Laursiden R, Harrison A, Hartl MGH, Kelly-Quinn M, Giller PS, O'Halloran J (2006). Evaluation of the use of the Sodium Dominance Index as a potential measure of acid sensitivity (2000-LS-

3.2.1-M2). 26 pp. Synthesis Report, Environmental Protection Agency, Dublin, Ireland.

Dale MRT, Fortin MJ (2014). *Spatial Analysis. A Guide for Ecologists*, 2nd ed. Cambridge, UK: Cambridge University Press.

Diggle PJ, Ribeiro Jr, PJ (2007). *Model-based Geostatistics*. New York: Springer.

Durbin J, Koopman SJ (2012). *Time Series Analysis by State Space Methods*, 2nd ed. Oxford: Oxford University Press.

Fox J (2016). *Applied Regression Analysis and Generalized Linear Models*, 3rd ed. London: Sage Publications.

Freeberg TM, Lucas JR (2009). Pseudoreplication is (still) a problem. *Journal of Comparative Psychology* 123: 450–451

Fukuda Y, Manolis C, Saalfeld K, Zuur AF. (2015) Dead or alive? Factors affecting the survival of victims during attacks by saltwater crocodiles (*Crocodylus porosus*) in Australia. *PLoS One* 10: e0126778.

Fuller RA, Bearhop S, Metcalfe NB, Piersma T (2013). The effect of group size on vigilance in Ruddy Turnstones *Arenaria interpres* varies with foraging habitat. *Ibis* 155: 246–257.

Geisser S (1980). Discussion of 'Sampling and Bayes' inference in scientific modelling and robustness' (by G. E. P. Box). *Journal of the Royal Statistical Society: Series A* 143: 416–417.

Gelman A, Carlin JB, Stern HS, Dunson DB, Vehtari A, Rubin DB (2013). *Bayesian Data Analysis,* 3rd ed. Boca Raton, FL: Chapman & Hall.

Gelman A, Lee D, Guo J (2015) Stan: A probabilistic programming language for Bayesian inference and optimization. *Journal of Educational and Behavioral Statistics* 40: 530–543.

Genz A, Bretz F, Miwa T, Mi X, Leisch F, Scheipl F, Hothorn T (2016). mvtnorm: Multivariate Normal and t Distributions. R package version 1.0-5. URL http://CRAN.R-project.org/package=mvtnorm.

Gneiting T, Raftery AE (2007). Strictly proper scoring rules, prediction, and estimation. *Journal of the American Statistical Association* 102: 359–378.

Haining R (2003). *Spatial Data Analysis: Theory and Practice*. Cambridge, UK: Cambridge University Press.

Halsey LG, Curran-Everett D, Vowler SL, Drummond GB (2015). The fickle *P* value generates irreproducible results. *Nature Methods* 12: 179–185.

Hanke AR, Lauretta M, Andrushchenko I (2015). A preliminary western Atlantic bluefin tuna index of abundance based on Canadian and USA rod and reel fisheries data: 1984–2014. *Collective Volume of Scientific Papers ICCAT* 72: 1763–1781.

Harvey AC (1989). *Forecasting, Structural Time Series Models and the Kalman Filter*. Cambridge, UK: Cambridge University Press.

Hilbe JM (2014). *Modelling Count Data*. Cambridge, UK: Cambridge University Press.

Hopkins WD, Reamer L, Mareno MC, Schapiro SJ (2013). Genetic basis in motor skill and hand preference for tool use in chimpanzees (*Pan troglodytes*). *Proceedings of the Royal Society B* 282: 20141223.

Hurlbert SH (1984). Pseudoreplication and the design of ecological field experiments. *Ecological Monographs* 54: 187–211.

Hurlbert SH (2004). On misinterpretations of pseudoreplication and related matters: A reply to Oksanen. *Oikos* 104: 591–597. doi:10.1111/ j.0030-1299.2004.12752.x.

Irl SDH, Harter DEV, Steinbauer MJ, Puyol DG, Fernandez-Palacios JM, Jentsch A, Beierkuhnlein C (2015). Climate vs. topography – Spatial patterns of plant species diversity and endemism on a high-elevation island. *Journal of Ecology* 103: 1621–1633.

Johnson JB, Omland KS (2004). Model selection in ecology and evolution. *Trends in Ecology and Evolution* 19: 101–108.

Jolliffe IT (2002). *Principal Component Analysis*. New York: Springer.

Kahle DF, Wickham H (2013). `ggmap`: A package for spatial visualization with Google Maps and OpenStreetMap. R package version 2.3. http://CRAN.R-project.org/package=ggmap.

Krainski ET, Lindgren F, Simpson D, Rue H (2016). The R-INLA tutorial on SPDE models. www.r-inla.org/examples/tutorials/spde-tutorial.

Ieno EN, Zuur AF (2015). *Beginner's Guide to Data Exploration and Visualisation with R*. Newburgh, UK: Highland Statistics.

Kéry M, Royle JA (2016). *Applied Hierarchical Modeling in Ecology: Analysis of Distribution, Abundance and Species Richness in R and BUGS*, Vol. I: *Prelude and Static Models*. London: Academic Press.

Lajeunesse MJ, Fox GA (2015). Statistical approaches to the problem of phylogenetically correlated data. Chapter 11 in Fox GA, Negrete-Yankelevich S, Sosa VJ (eds). *Ecological Statistics: Contemporary Theory and Application*. Oxford: Oxford University Press.

Lawson AB (2013). *Bayesian Disease Mapping: Hierarchical Modeling in Spatial Epidemiology*, 2nd ed. London: Chapmann & Hall.

Ligas A (2008). Population dynamics of *Procambarus clarkia* (Girard, 1852) (Decapoda, Astacidea, Cambaridae) from southern Tuscany (Italy). *Crustaceana*, 81: 601–609.

Lindgren F, Rue H, Lindstrom J (2011). An explicit link between Gausian fields and Gaussian Markov random fields: The stochastic partial differential equation approach. *Journal of the Royal Statistical Society B*, 73: 423–498.

Lindgren F, Rue H (2015). Bayesian spatial modelling with R-INLA. *Journal of Statistical Software* 63: 1–25.

Lunn D, Thomas A, Best N, Spiegelhalter D (2000). WinBUGS—A Bayesian modeling framework: Concepts, structure, and extensibility. *Statistics and Computing* 10: 325–337.

Lunn D, Spiegelhalter D, Thomas A, Best N (2009). The BUGS project: Evolution, critique and future directions. *Statistics in Medicine* 28: 3049–3082.

Maes D, Jacobs I, Segers N, Vanreusel W, Van Daele T, Laurijssens G, Van Dyck H (2014). A resource-based conservation approach for an endangered ecotone species: The Ilex Hairstreak (*Satyrium ilicis*) in Flanders (north Belgium). *Journal of Insect Conservation* 18: 939–950.

Millar RB, Anderson MJ (2004). Remedies for pseudoreplication. *Fisheries Research* 70: 397–407.

Millar RB (2009). Comparison of hierarchical Bayesian models for overdispersed count data using DIC and Bayes' factors. *Biometrics* 65: 962–969.

Montgomery DC, Peck EA (1992). *Introduction to Linear Regression Analysis*. New York: Wiley.

Muller EM, van Woesik R (2014). Genetic susceptibility, colony size, and water temperature drive white-pox disease on the coral *Acropora palmate*. *PlosOne* 9: e110759.

Mundry R, Nunn Cl (2009). Stepwise model fitting and statistical inference: Turning noise into signal pollution. *The American Naturalist* 173: 119–123.

Murtaugh PA (2009). Performance of several variable-selection methods applied to real ecological data. *Ecology Letters* 12: 1061–1068.

Ntzoufras I (2009). *Bayesian Modeling Using WinBUGS*. Hoboken, NJ: Wiley.

Nychka D, Furrer R, Paige J, Sain S (2015). "fields: Tools for spatial data." doi: 10.5065/D6W957CT. URL: http://doi.org/10.5065/D6W957CT. R package version 8.4-1, URL: www.image.ucar.edu/fields.

Oksanen L (2001). Logic of experiments in ecology: Is pseudoreplication a pseudoissue? *Oikos* 94: 27–38.

Pebesma EJ (2004). Multivariable geostatistics in S: The `gstat` package. *Computers & Geosciences* 30: 683–691.

Pebesma EJ, Bivand RS (2005). Classes and methods for spatial data in R. *R News* 5 (2), http://cran.r-project.org/doc/Rnews/.

Petty SK, Zuckerberg B, Pauli JN (2015). Winter conditions and land cover structure the subnivium: A seasonal refuge beneath the snow. *PlosOne*. DOI:10.1371/journal.pone.0127613

Pinheiro J, Bates D (2000). *Mixed Effects Models in S and S-Plus*. New York: Springer.

Pinheiro J, Bates D, DebRoy S, Sarkar D and R Core Team (2016). nlme: Linear and Nonlinear Mixed Effects Models. R package version 3.1-128.

Plummer M (2003). JAGS: A program for analysis of Bayesian graphical models using Gibbs sampling. Proceedings of the 3rd International Workshop on Distributed Statistical Computing (DSC 2003), March 20–22, Vienna.

Quinn GP, Keough MJ (2002). *Experimental Design and Data Analysis for Biologists*. Cambridge, UK: Cambridge University Press.

R Core Team (2016). R: A language and environment for statistical computing. R Foundation for Statistical Computing, Vienna, Austria. URL https://www.R-project.org/.

Reed JM, Elphick CS, Ieno EN, Zuur AF (2011). Long-term population trends of endangered Hawaiian waterbirds. *Population Ecology* 53: 473–481.

Ribeiro PJ Jr, Diggle PJ (2016). geoR: Analysis of Geostatistical Data. R package version 1.7-5.2. https://CRAN.R-project.org/package=geoR.

Roos M, Held L (2011). Sensitivity analysis in Bayesian generalized linear mixed models for binary data. *Bayesian Analysis* 6: 259–278.

Roulin A, Bersier LF (2007). Nestling barn owls beg more intensely in the presence of their mother than their father. *Animal Behaviour* 74: 1099–1106.

Rue H, Martino S, Chopin N (2009). Approximate Bayesian inference for latent Gaussian models using integrated nested Laplace approximations. *Journal of the Royal Statistical Society B* 71: 319–392.

Sarkar D (2008). *Lattice: Multivariate Data Visualization with R*. New York: Springer.

Schabenberger O, Pierce FJ (2002). *Contemporary Statistical Models for the Plant and Soil Sciences*. Boca Raton, FL: CRC Press.

Shumway RH, Stoffer DS (2017). *Time Series Analysis and Its Applications*, 4th ed. New York: Springer Verlag.

Sick C, Carter AJ, Marshall HH, Knapp LA, Dabelsteen T, Cowlishaw G (2014). Evidence for varying social strategies across the day in chacma baboons. *Biology Letters* 10: 0140249.

Simpson DP, Rue H, Martins TG, Riebler A, Sorbye SH (2014). Penalising model component complexity: A principled, practical approach to constructing priors. arXiv preprint arXiv:1403.4630.

Sofaer HR, Sillett TS, Langin KM, Morrison SA, Ghalambor CK (2014). Partitioning the sources of demographic variation reveals density-dependent nest predation in an island bird population. *Ecology and Evolution* 4: 2738–2748

Spiegelhalter DJ, Best NG, Carlin BP, Van der Linde A. (2002). Bayesian measures of model complexity and fit (with discussion) *Journal of the Royal Statistical Society Series B* 64: 583–639.

Steidl RJ, Griffin CR, Niles LJ (1991). Contaminant levels of osprey eggs and prey reflect regional differences in reproductive success. *The Journal of Wildlife Management* 55: 601–608.

Sturrock AM, Hunter E, Milton JA, EIMF, Johnson RC, Waring CP, Trueman CN (2015). Quantifying physiological influences on otolith microchemistry. *Methods in Ecology and Evolution* 6: 806–816.

Timi JT, Lanfranchi AL, Etchegoin JA, Cremonte F (2008). Parasites of the Brazilian sandperch *Pinguipes brasilianus* Cuvier: A tool for stock discrimination in the Argentine Sea. *Journal of Fish Biology* 72: 1332–1342.

Venables WN, Ripley BD (2002). *Modern Applied Statistics with S*, 4th ed. New York: Springer.

Watanabe S (2010). Asymptotic equivalence of Bayes cross validation and widely applicable information criterion in singular learning theory. *Journal of Machine Learning Research* 11: 3571–3594.

Whittingham MJ, Stephens PA, Bradbury RB, Freckleton RP (2006). Why do we still use stepwise modelling in ecology and behaviour? *Journal of Animal Ecology* 75: 1182–1189.

Wickham H (2007). Reshaping data with the reshape package. *Journal of Statistical Software*, 21: 1–20.

Wickham H (2009). *`ggplot2`: Elegant Graphics for Data Analysis*. New York: Springer.

Wickham H, Francois R (2016). `dplyr`: A Grammar of Data Manipulation. R package version 0.5.0. https://CRAN.R-project.org/package=dplyr.

Zuur AF, Fryer RJ, Jolliffe IT, Dekker R, Beukema JJ (2003a). Estimating common trends in multivariate time series using dynamic factor analysis. *Environmetrics* 14: 665–685.

Zuur AF, Tuck ID, Bailey N (2003b). Dynamic factor analysis to estimate common trends in fisheries time series. *Canadian Journal of Fisheries and Aquatis Science* 60: 542–552.

Zuur AF, Pierce GJ (2004). Common trends in Northeast Atlantic squid time series. *Journal of Sea Research* 52: 57–72.

Zuur AF, Ieno EN, Smith GM (2007). *Analysing Ecological Data*. New York: Springer.

Zuur AF, Ieno EN, Walker N, Saveliev AA, Smith GM (2009a). *Mixed Effects Models and Extensions in Ecology*. New York: Springer.

Zuur AF, Ieno EN, Meesters EHWG (2009b). *A Beginner's Guide to R*. New York: Springer.

Zuur AF, Ieno EN, Elphick CS (2010). A protocol for data exploration to avoid common statistical problems. *Methods in Ecology and Evolution* 1: 3–14.

Zuur AF, Saveliev AA, Ieno EN (2012). *Zero Inflated Models and Generalized Linear Mixed Models with R*. Newburgh, UK: Highland Statistics.

Zuur AF, JM Hilbe, EN Ieno (2013). *Beginner's Guide to GLM and GLMM with R*. Newburgh, UK: Highland Statistics.

Zuur AF, Saveliev AA, Ieno EN (2014). *Beginner's Guide to Generalized Additive Mixed Models with R*. Newburgh, UK: Highland Statistics.

Zuur AF, Ieno EN (2016a). *Beginner's Guide to Zero-Inflated Models with R*. Newburgh, UK: Highland Statistics.

Zuur AF, Ieno EN (2016b). A protocol for conducting and presenting results of regression-type analyses. *Methods in Ecology and Evolution* 7: 636–664.

Index

W

Z

Books by Highland Statistics

Available exclusively from: www.highstat.com

Zero Inflated Models and Generalized Linear Mixed Models with R. Zuur AF, Saveliev AA, Ieno EN. (2012)

Chapter 1 provides an introduction to Bayesian statistics and Markov chain Monte Carlo simulation. In Chapter 2 we analyse nested zero-inflated data of sibling negotiation in barn owl chicks. We explain application of a Poisson GLMM for one-way nested data and discuss the observation-level random intercept to allow for overdispersion. We show that the data are zero inflated and introduce zero-inflated GLMM.

Data of sandeel otolith presence in seal scat is analysed in Chapter 3. We present a flowchart of steps to selecting the appropriate technique: Poisson GLM, negative binomial GLM, Poisson or negative binomial GAM, or GLMs with zero-inflated distribution.

Chapter 4 is relevant for readers interested in the analysis of zero-inflated two-way nested data. This chapter takes us to marmot colonies: multiple colonies with multiple animals sampled repeatedly over time.

Chapters 5–7 address GLMs with spatial correlation. Chapter 5 presents an analysis of common murre density data and introduces hurdle models using GAM. Random effects are used to model spatial correlation. In Chapter 6 we analyse zero-inflated skate abundance recorded at approximately 250 sites along the coastal and continental shelf waters of Argentina. Chapter 7 involves spatial correlation in parrotfish abundance data collected around islands, increasing the complexity of the analysis. GLMs with residual conditional auto-regressive correlation structures are used. In Chapter 8 we apply zero-inflated models to click beetle dispersal data.

Chapter 9, analysing a time series of zero-inflated whale strandings, is relevant for readers interested in GAM, zero inflation, and temporal auto-correlation. Chapter 10 demonstrates that an excessive number of zeros does not necessarily mean zero inflation. We also discuss whether the application of mixture models requires that the data include false zeros, and whether the algorithm can indicate which zeros are false.

A Beginner's Guide to Generalized Additive Models with R. Zuur AF. (2012)

This is the first book in Highland Statistics' Beginner's Guide series.

A Beginner's Guide to Generalized Additive Models with R is, as the title implies, a practical handbook for the non-statistician. The author's philosophy is that the shortest path to comprehension of a statistical technique without delving into extensive mathematical detail is through programming its basic principles in, for example, R.

This book is not a series of 'cookbook exercises'. The author uses data from biological studies to go beyond theory and immerse the reader in real-world analysis with its inherent messiness and challenges. The book begins with a review of multiple linear regression using research on human cranium size and ambient light levels and continues with an introduction to additive models based on deep sea fishery data. Research on pelagic bioluminescent organisms demonstrates simple linear regression techniques to program a smoother. In Chapter 4 the deep sea fishery study is revisited for a discussion of generalised additive models. The remaining chapters present detailed case studies illustrating the application of Gaussian, Poisson, negative binomial, zero-inflated Poisson, and binomial generalised additive models using seabird, squid, and fish parasite studies.

A Beginner's Guide to GLM and GLMM with R. Zuur AF, Hilbe JM, Ieno EN. (2013)

This is the second book in Highland Statistics' Beginner's Guide series.

This book presents generalised linear models (GLM) and generalised linear mixed models (GLMM) based on both frequency-based and Bayesian concepts. Using ecological data from real-world studies, the text introduces the reader to the basics of GLM and mixed-effects models, with demonstrations of binomial, gamma, Poisson, negative binomial regression, and beta and beta-binomial GLMs and GLMMs.

A Beginner's Guide to Generalised Additive Mixed Models with R. Zuur AF, Saveliev AA and Ieno EN. (2014)

This is the third book in Highland Statistics' Beginner's Guide series, following the well-received *A Beginner's Guide to Generalized Additive Models with R* and *A Beginner's Guide to GLM and GLMM with R*.

In this book we take the reader on an exciting voyage into the world of generalised additive mixed models (GAMM). Keywords are GAM, mgcv, gamm4, random effects, Poisson and negative binomial GAMM, gamma GAMM, binomial GAMM, NB-P models, GAMMs with generalised extreme value distributions, overdispersion, underdispersion, two-dimensional smoothers, zero-inflated GAMMs, spatial correlation, INLA, Markov chain Monte Carlo techniques, JAGS, and two-way nested GAMMs. The book includes three chapters on the analysis of zero-inflated data.

Throughout the book frequentist approaches (gam, gamm, gamm4, lme4) are compared with Bayesian techniques (MCMC in JAGS, and INLA). Data sets on squid, polar bears, coral reefs, ruddy turnstones, parasites in anchovy, Common Guillemots, harbour porpoises, forestry, brood parasitism, maximum cod length, and Common Scoters are used in case studies. The R code to construct, fit, interpret, and comparatively evaluate models is provided at every stage (either in the book or on the website for the book).

A Beginner's Guide to Data Exploration and Visualisation with R. Ieno EN and Zuur AF. (2015)

This is the fourth book in Highland Statistics' Beginner's Guide series.

While teaching statistics to ecologists and reviewing manuscripts, the authors of this book have noticed common statistical problems that could have been avoided. That was one of the reasons why we wrote a paper in 2010 that presented a protocol for data exploration; see Zuur et al. (2010). This paper is the most downloaded paper in the history of the journal in which it was published, indicating the need and niche for a more detailed text on data exploration and visualisation. In this book we follow the data exploration protocol paper and discuss the important steps in separate chapters.

Detecting outliers is fundamental and should be the first thing one looks at (Chapter 2). Homogeneity (Chapter 3) is an important assumption for regression type techniques. Heterogeneity might produce estimated parameters with standard errors that are too small. Traditional solutions include data transformation, but we try to avoid this for as long as possible.

Normality is discussed in detail in most statistics textbooks and undergraduate statistics courses. In Chapter 3, we emphasise that when applying regression models you do *not* test normality of the raw data but rather their residuals.

The inspection of data to determine whether variables are associated constitutes a fundamental part of any data exploration process (Chapter 4). An important step is identifying linear or nonlinear relationships, or even more complex relationships that could, in principle, define what type of mathematical model needs to be formulated.

In ecological studies one tends to have a large number of explanatory variables, and evaluating collinearity is a major challenge. In Chapter 5 we explain how to detect collinear variables.

In Chapter 6 we use a detailed case study to show how the `ggplot2` package (Wickham 2009) in R can be used for data exploration and visualisations of the results for a linear regression model as well as a linear mixed-effects model.

In Chapter 7 we show how you can end up with the wrong ecological conclusions if good data exploration is not applied. In Chapter 8 we reproduce some graphs (using the `lattice` and `ggplot2` packages) that we have used in scientific publications.

A Beginner's Guide to Zero Inflated Models with R. Zuur AF and Ieno EN. (2016)

This is the fifth book in Highland Statistics' Beginner's Guide series.

In Chapters 2 and 3 we provide brief explanations of the Poisson, negative binomial, Bernoulli, binomial and gamma distributions, and revise the Poisson generalised linear model (GLM) and the Bernoulli GLM, followed by a gentle introduction to zero-inflated Poisson models.

Chapters 4 and 5 contain detailed case studies using count data of orange-crowned warblers and sharks. In Chapter 6 we use zero-altered Poisson models to deal with the excessive number of zeros in count data. In Chapter 7 we analyse continuous data with a large number of zeros. Biomass of Chinese tallow trees is analysed with zero-altered gamma models.

In Chapter 8, which begins the second part of the book, we explain how to deal with dependency. Mixed models are introduced, using beaver and monkey datasets. In Chapter 9 we encounter a rather complicated data set in terms of dependency. Reproductive indices are sampled from plants, but the seeds come from the same source and are planted in the same bed in the same garden. We apply models that take care of the excessive number of zeros in count data, crossed random effects, and nested random effects.

Up to this point we have done everything in a frequentist setting, but at this stage of the book you will see that we are reaching the limit of what we can achieve with the frequentist software. For this reason we switch to Bayesian techniques in the third part of the book. Chapter 10 contains an excellent beginner's guide to Bayesian statistics and Markov chain Monte Carlo techniques. In Chapter 11 we show how to implement the Poisson, negative binomial, and ZIP models in MCMC. We do the same for mixed models in Chapter 12.

In Chapter 13 we discuss a method, called the 'zero trick', that allows you to fit nearly every distribution in JAGS.

A major stumbling block in Bayesian analysis is model selection. Chapter 14 provides an easy-to-understand overview of various Bayesian model selection tools. Chapter 15 contains an example of Bayesian model selection using butterfly data.

In Chapter 16 we discuss methods for the analysis of proportional data with a large number of zeros. We use a zero-altered beta model with nested random effects. Finally, in Chapters 17 and 18 we discuss various topics, including multivariate GLMMs and generalised Poisson models. We also discuss zero-inflated binomial models.